ISBN 978-1-332-15530-9
PIBN 10292086

PURDUE
LIBRARY
LIBRARY
SCIENCE
TECHNOLOGY
AGRICULTURE
PURDUE

PURDUE
LIBRARY
LIBRARY
SCIENCE
TECHNOLOGY
AGRICULTURE
PURDUE

PURDUE
LIBRARY
LIBRARY
SCIENCE
TECHNOLOGY
AGRICULTURE
PURDUE

PURDUE
LIBRARY
LIBRARY
SCIENCE
TECHNOLOGY
AGRICULTURE
PURDUE

WITHDRAWN

PURDUE
LIBRARY
LIBRARY
SCIENCE
TECHNOLOGY
AGRICULTURE
PURDUE

PURDUE
LIBRARY
LIBRARY
SCIENCE
TECHNOLOGY
AGRICULTURE
PURDUE

PURDUE
LIBRARY
LIBRARY
SCIENCE
TECHNOLOGY
AGRICULTURE
PURDUE

PURDUE
LIBRARY
LIBRARY
SCIENCE
TECHNOLOGY
AGRICULTURE
PURDUE

PURDUE
LIBRARY
LIBRARY
SCIENCE
TECHNOLOGY
AGRICULTURE
PURDUE

PURDUE
LIBRARY
LIBRARY
SCIENCE
TECHNOLOGY
AGRICULTURE
PURDUE

PURDUE
LIBRARY
LIBRARY
SCIENCE
TECHNOLOGY
AGRICULTURE
PURDUE

PURDUE
LIBRARY
LIBRARY
SCIENCE
TECHNOLOGY
AGRICULTURE
PURDUE

AZAMA

A Record of Mountaineering in the Pacific Northwest

NESIKA KLATAWA SAHALE

Published by THE MAZAMAS
213 NORTHWESTERN BANK BUILDING
PORTLAND, OREGON, U. S. A.

Fifty Cents

IN THE MT. ADAMS COUNTRY.
BEAUTIFUL STREAMS—haunt of gamiest trout—chanting their way through noble forests, interesting geological formations, lava caves, Huckleberry Mountain, where friendly Indians gather for their potlatches, etc., and many other desirable features for outings and camps.
Mt. Adams (12,307 feet high), third of the greatest peaks of the Northwest, with one of the greatest glacial fields for alpine sport, adventure.
All this within 2¼ hours by rail from Portland on
The North Bank Road
Request your ticket to the North Pacific Coast be routed Spokane, Portland & Seattle Ry., between Spokane and Portland, and Clatsop Beach on Pacific Ocean
PORTLAND TICKET OFFICE—5th and Stark Streets

MAZAMA

A RECORD OF MOUNTAINEERING IN THE PACIFIC NORTHWEST

Publication Committee

ALFRED F. PARKER   MARY C. HENTHORNE, *Chairman*   BEULAH F. MILLER

VOLUME V   PORTLAND, OREGON, DECEMBER, 1916   NUMBER 1

Contents

Something calls and whispers along the city street
Through shrill cries of children and soft stir of feet
And makes my blood to quicken and makes my flesh to pine.
The mountains are calling. The wind wakes the pine.

—Georgiana Goddard King.

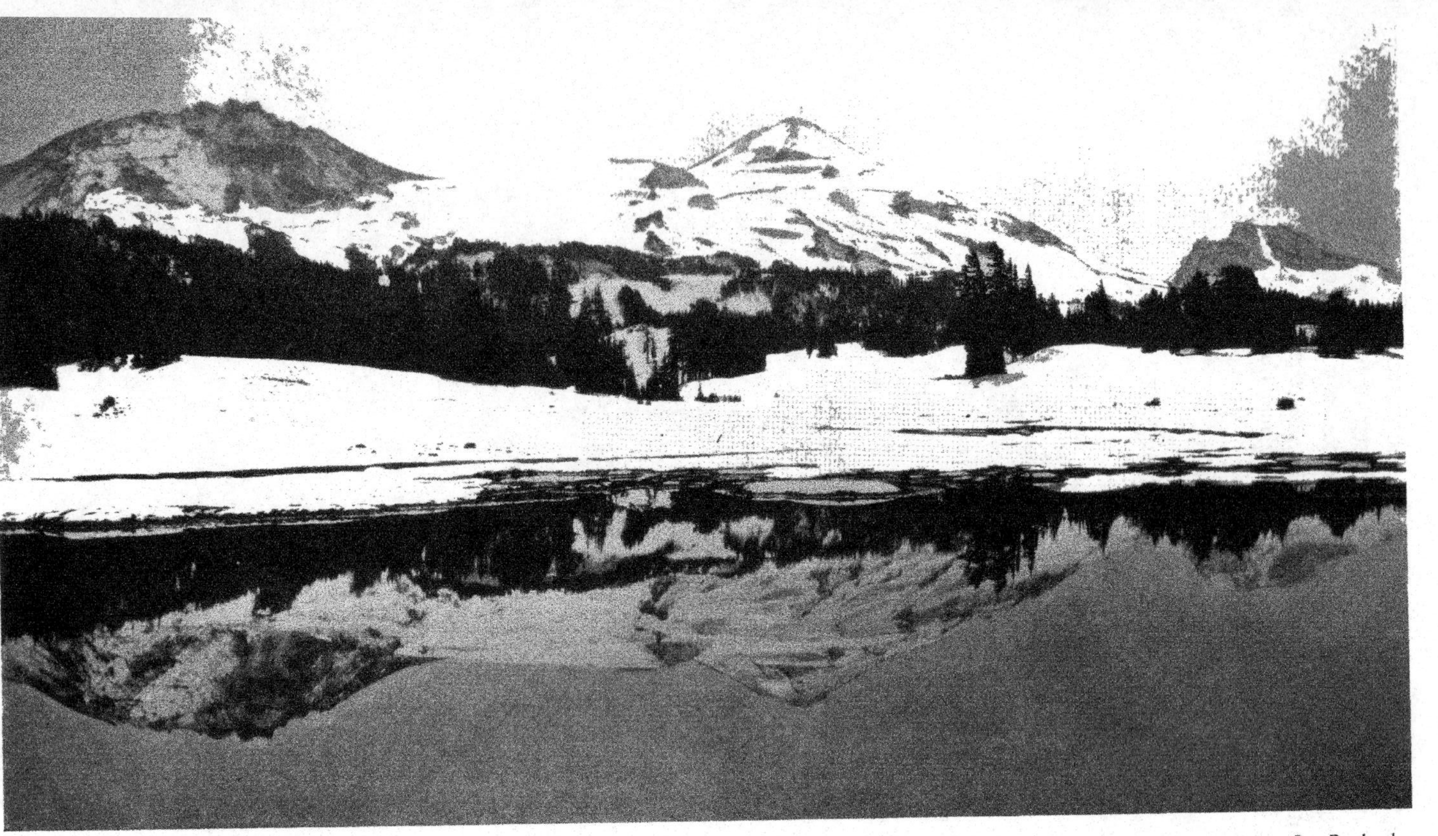

Three Sisters, ooking east om Mazama camp, afte the snowstorm

MAZAMA

VOLUME V — DECEMBER, 1916 — NUMBER 1

The Three Sisters Outing, 1916

By MARY C. HENTHORNE

Only those who have answered the persistent invitation of the mountains can appreciate fully the joy of preparation for camp and trail. There is a fine flavor of anticipation even in the preliminary overhauling of equipment. Everyone knows how futile is the effort to cram the first selection of clothing and miscellaneous articles into one small dunnage bag. In addition to those things really necessary for a two weeks' existence remote from civilized centers, each person is certain to include his own particular and peculiar contrivances for camp comfort. As the initial weighing usually shows about twenty pounds overweight, one cherished article after another is removed until the scales show no excess. Joyously the bag is tied and sent off, not to be seen again until the first night in camp—possibly not then, if a belated pack-train forces a few unfortunates to spend the night huddled around the campfire.

The Three Sisters country was entirely new territory to the majority of those on this year's outing. Twice before, in 1903 and in 1910, this section was the objective of the summer's trip. Except for short accounts written by Mazamas, very little has been published about this part of the state until the Oregon Bureau of Mines and Geology devoted part of its May, 1916, issue to a description of the scenery and the most important geological features of the region.

The Three Sisters, a trinity of snow-capped mountains, form part of the Cascade range, lying between Crook and Lane counties about one hundred miles south of Mt. Hood, on the line separating the Deschutes and Cascade National Forests. All are about the same height, the Middle Sister being 10,038 feet high, the North Sister 10,067, and the South Sister 10,352. Seven peaks constitute the entire group—the North, Middle and South Sisters, Bachelor Butte, the Husband and the Wife, and Broken Top. Several lakes in the vicinity provide good fishing.

After a night spent luxuriously in the Pullman speeding from Portland to Eugene, over one hundred Mazamas and prospective Mazamas left the college town early on the morning of August the sixth. Through the courtesy of the Eugene Commercial Club, automobiles were provided for the twenty-five mile ride to the Poujade fish hatchery, where an appetizing breakfast was served in a beautiful grove.

The famous McKenzie river is first seen about twelve miles east of Eugene—a lovely, brawling stream that falls noisily down its rocky bed in numberless feathery cascades. The road follows beside it for over fifty miles to McKenzie Bridge, winding through a thick forest of tall fir, spruce and cedar trees. In many places there is a heavy undergrowth of ferns, Oregon grape and salàl.

The flat country of the lower valley disappears imperceptibly, the hills becoming higher and more rugged. Cedar and spruce mingle with the hemlock, pine and Douglas fir. A little later these are replaced by thickets of mountain balm, or greasewood, alder, and vine maple, with occasional madrona, dogwood, and white-bark pine. Tantalizing views of the snow caps are obtained at intervals, but it is not until after crossing Lost creek that there is an unobstructed sight of the Middle and North Sisters from a long stretch of open road. A brief glimpse of the South Sister may be caught from Alder Spring near the foot of Deer Butte.

Luncheon was eaten at the bridge, and muscles were stretched after the long confinement in the large automobile trucks which had superseded the lighter machines on leaving the hatchery. In the afternoon it became increasingly difficult for the trucks to travel, as the grade is much greater in the twenty miles between McKenzie Bridge and Frog Camp in Lake valley. First one and then another was disabled, so that finally but a single truck rolled triumphantly across the flat meadow to Frog Camp, leaving a comet-like trail of pedestrians in its wake.

The start on the final five mile walk to permanent camp was made as the sun was setting in a glow of amethyst, rose and gold. With alpenstocks clicking an irregular march, the party proceeded across the grassy park, through shadowy pines and hemlocks. The trail was well-defined until the White Branch lava flow was reached. Here it became difficult to follow the crooked path in the waning light.

This lava bed is part of a large area over which, during comparatively recent times, molten lava from the numerous vents spread like a blanket. From a geologic standpoint, this region is newer than any other in the Cascades.

After crossing White Branch, which issues from the snout of Collier glacier, the trail to camp ascended a ridge covered eight to ten feet deep with snow. This covering was heavier last winter than at any time since the advent of white settlers in the Willamette valley. Meadows which in other years have been brilliant flower gardens were bleak and barren snow wastes. The last mile lay through a forest dim as a cathedral in the twilight. At intervals candles placed in the snow lighted weary stragglers upward to the waiting campfire.

Camp was made, at an altitude of 6,400 feet, in an alpine meadow situated on a bench of obsidian on the west side of the Middle Sister. There are several of these interesting terraces, one above the other, each composed of dark, shiny lava, that is, in fact, natural glass. The three peaks towered above the plateau—"Three mountain tops, three silent pinnacles of ancient snow," "with the primal unrest locked away in their breasts." It was difficult for the advance party to find enough places free from snow in which to pitch the tents. These showed white in relief against their background of fir and hemlock, each on its small island in the surrounding sea of snow. Water was obtained by digging holes in the snow where the lakes should have been. This provided enough for the necessary camp purposes until, toward the end of the outing, the sun melted the snow sufficiently to form several miniature lakes which then afforded an ample supply for both culinary and bathing purposes.

The first day was spent in camp-making, each group of five or six persons vying with another in making the most attractive home. Luxurious beds of springy, aromatic boughs were painstakingly built close to the dressing tents. During the morning the weather was very warm, but in the afternoon a heavy fog closed in. As this turned to rain during the night, the novices in camping had a real initiation.

Tryout trips began the next day. A small group of hardy climbers ascended the Middle Sister. A large party, numbering fifty-two, made an all day trip to Red Lava Crags under the leadership of Mr. W. C. Yoran. This bold, rocky spur stands near the tongue of Collier glacier, with its crenellated summits sharply outlined against the sky. On the trip Mr. Yoran found a curious lava formation resembling a squash, which he had seen there several years ago. It has since been placed in the University of Oregon and labeled "Yoran's squash."

Trips to the Husband proved popular. This is a rough, jagged-topped mountain lying several miles south of camp. From the summit the view of the Middle and South Sisters is especially fine. A long snow field furnished splendid glissading as the descent was made. Knapsack trips were made by small parties to Broken Top, and also fishing trips to several of the lakes beyond the South Sister.

Enthusiastic reports from those who journeyed to the "Flower valley" led many to this Mecca for botanists. Everyone started with the intention of reaching Lost Creek falls, but "the way was long and weary," so only the sturdiest plodders reached the goal and very few penetrated beyond. Moreover, as the way to the valley is nearly all down grade, the return necessitates several miles of steady climbing when muscles are already tired by the day's work. Yet it is worth the effort. As we read Mrs. Attilla Norman's account of her trips there, we feel again the great charm of the beautiful, flower-strewn meadow, and we find in Dr. J. Duncan Spaeth's poem the very embodiment of its spirit.

A day was spent in visiting the Cinder Cone, a beautifully colored pyramid of cinders and ashes on which practically no vegetation has gained a foothold. The line of the crater is broken on the west side in such a way that the interior reminds one of a vast amphitheatre with the stage placed in the opening. To reach the cone the party crossed the Collier glacier where the ice-stream is deflected from its straight course by the huge cinder pile. Under ordinary conditions much of the surface of the ice flow is badly broken by crevasses. This year comparatively few had opened on the lower glacier. However, enough were seen to interest those who had never before been on a real live glacier.

On August the tenth six men and seven women, under the guidance of Mr. Harley H. Prouty, made an early start from camp for the North Sister, the sternest, most forbidding-looking of the trio. The peak is a rough, ridged framework of lava, crowned by jagged pinnacles. The route of the mountaineers led them across the néve of Collier glacier to the saddle between the Middle and the North Sisters. After crossing the saddle the way continues up a knife-edged ridge on the south shoulder, where there is barely room enough to climb. The slope on either side is steep, with a sheer drop of thousands of feet in several places. The ropes were used before reaching the base of the highest pinnacle. A small glacier with a slant of sixty-five or seventy degrees hangs almost in mid-air just at the base of the rock spire. Mr. Prouty and Mr. E. F. Peterson cut steps across its lower face and then up to the rocks at the top. From here the two made their way slowly to the topmost point. They were followed by Mr. A. S. Peterson and Mr. Thomas R. Jones. Few firm rocks were available for foot and handholds, and some of these were wet and icy. As the rock is disintegrating very rapidly, the climb is becoming increasingly difficult and will probably be impossible to make in a few years, if the process of weathering continues at the present rapid rate. Mr. Prouty found it much more dangerous than it was on his two former ascents. In 1910 he made the first recorded ascent of the pinnacle. There are twelve names in the record book, Mr. Prouty's occurring three times.

The rest of the climbers sat quietly on the narrow ledge at the base of Prouty Pinnacle for nearly three hours. Their perch was precarious, but they felt refreshed for the descent, which had to be made as slowly and carefully as the ascent.

Later Messrs. R. L. Glisan, John A. Lee, C. H. Sholes and Guy

Photograph by Winter Photo Co.

North and Middle Sisters from a high ridge south of camp.

Photograph by Winter Photo Co.

The Husband, looking south from point near Mazama camp

Thatcher ascended Glisan Pinnacle from the north side. They found the climb equally hazardous from there. As they did not reach the summit until five o'clock, no attempt was made to climb the highest pinnacle.

With Mr. Roy W. Ayer as chief guide, on August the twelfth at 6:45 a. m., fifty-two people left Camp Riley on the official trip to the Middle Sister. The climbers, divided into numbered companies, marched rapidly in single file to timberline, then more slowly and with frequent pauses—up, up, across the snow fields, past the Folding Rock, a curious formation showing the strata, to the final steep ridge leading to the summit. As the party climbed carefully up over the loose lava blocks, a thunderstorm was gathering over the shoulder of the South Sister. Billowy clouds began to roll across the space intervening between the peaks, and distant thunder sounded menacing warnings. Just after luncheon and registration were completed at midday, these masses of blue-black mist closed in around the top of the Middle Sister, obscuring the country below and enfolding everything in a heavy blanket of fog. Electricity crackled and snapped from finger tips, alpenstocks and flying locks of hair. It seemed to pulsate through the air with recurrent force. Vivid lightning flashes were succeeded by peals of thunder that reverberated from peak to peak. The din and roar of the storm were almost dismaying to the climbers, who precipitately began a retreat. Great hailstones pelted unmercifully upon their shoulders and a beating rain, descending in wind-driven sheets, drenched them as they crossed the snow fields. Every opportunity to glissade was welcomed, even by those who had looked askance at this most exhilarating sport. One long-legged individual made a record trip, reaching camp in less than an hour after leaving the summit. Chef Weston, with his hot oyster soup, administered most adequate first aid treatment to the dripping, shivering crowd. A half dozen of the men remained on the top a little longer than the rest of the party to watch the awe-inspiring electrical display. Evidences of the disastrous work of the lightning in the valley were seen the next day. Large trees had been struck and shattered, and the debris covered the ground for many yards in all directions.

Other parties, led by Mr. W. C. Yoran and Mr. L. E. Anderson, were favored with fine weather for their ascents. A glorious panorama of snow-capped peaks stretching north and south as far as the eye can reach may be seen on a clear day. Shasta, Thielsen, Diamond Peak, then Mt. Washington, Three-fingered Jack, Jefferson, Hood, St. Helens, and—yes—the tiniest tip of Rainier, rise majestically from the far-reaching blue ridges. To the west the country rolls in undulating waves to the fertile green valleys that lie between the Cascades and the Coast range. To the east may be seen the broad stretches of irrigated wheat land lying serenely in the sunshine, the flat plains unflecked by cloud shadows. The joy of attainment becomes a very tangible thing when the reward of the climb is such a colorful, living picture outspread in all directions.

The official climb of the South Sister was not so eventful. On the afternoon of August the fifteenth, thirty-five people with packs on backs started on a seven mile walk to the bivouac camp, which was located well up on the slope of the mountain at timberline. The knapsack trip was another experience new and strange to the uninitiated. No soft beds of boughs to sleep on after the campfire—just the bare ground. Though the air was chilly, few stayed awake long enough to realize the wonder of the night. The wind crept rustling through the trees, which were silhouetted against the sky by the fitful light flashes from the dying fire. Little sharp noises out of the darkness served to accent the great stillness that pervaded the forest.

The ascent was commenced at 6:20 the next morning. Thirty fell in line at the leader's call, the other five having decided to return to the main camp. The greater part of the climb was made on Lost Creek glacier. It may also be made almost entirely on the rocks. The summit is a well-defined crater about a third of a mile in diameter. It presents a level surface of snow that forms a glittering setting for a tiny, gem-like lake, wonderful in its deep sapphire blue coloring.

The view was disappointing, for the fog lay thick and heavy below. There were occasional tantalizing glimpses of snow peaks and green valleys through rifts in the clouds, but not enough to make a long stay on the summit desirable. Others who made the ascent on clear days saw the same wonderful picture that is visible from the Middle Sister, with a fine view of both the Middle and North sisters in addition. On the descent several beautifully colored crevasses were investigated. During the tramp to camp the fog settled lower. The line of people became a shadowy procession of spectral shapes moving in the mist. Camp was a welcome sight after the eerie journey in the gloom.

A small party of especially hardy climbers left camp at 3:00 a. m. the day of the ascent, made the summit by a different route, and returned to camp an hour in advance of the main party.

For the first time in Mazama history there was no evening campfire. The fog still hung dismally among the trees, making warm beds seem the most comfortable places for the tired mountaineers. There was even a little rain to add variety to the fog.

No one thought of the possibility of snow. Yet, when heads

Ascent of Middle Sister. Upper—Side view of "Folding Rock." (*Photograph by Charles J. Merten.*) Middle—End view of same. (*Photograph by John R. Leach.*) Lower—Crossing Renfrew glacier. (*Photograph by R. L. Glisan.*)

Upper—Broken Top and The Bachelor, looking south from summit of Middle Sister Middle—Crater on summit of South Sister, looking southeast (*Photographs by A L Roberts*) Lower—South Sister enveloped in clouds (*Photograph by W C Yoran*)

emerged from sleeping bags next morning, what a sight greeted the eyes of the summer campers! Winter had arrived out of season and had covered the ground with a deep blanket of snow. The trees "were ridged inch deep with pearl" and the skies were gray with the swirling flakes. Those who had slept in the open were dug out by the moving-picture men, who were astir early in their search for laughable situations. Only the very wise, provident in fine weather, had dry wood hidden away to start fires. Tables and benches were covered with eight or nine inches of snow, so breakfast was eaten standing.

The morning was devoted to various snow sports. Commencement exercises were held in the afternoon. Addresses were made and the class song sung with the snow dripping from the trees down the collars of the unwary. There were twenty-three in the graduating class who became eligible to membership in the Mazamas. Summit badges were given to all who were entitled to them.

On Friday morning twenty-seven members, fearing a long storm, left camp, but those hardy spirits who remained indulged in unlimited glissading and waged mimic battles in snow forts, to the great delight of the moving picture operators. The sun came out, the snow melted, and the flowers reappeared during the last two days in camp.

Nearly everyone made a last visit to "The end of the world" to see the sunset. This name had been given to the cliff which marked the termination of the obsidian ridge on which we camped.

> "'T was a glorious scene—the mountain height
> Aflame with sunset's colored light.
>
> Even the black pines, grim and old,
> Transfigured stood with crowns of gold.
>
> There on a hoary crag we stood
> When the tide of glory was at its flood,

looking away across the dusky valley to where the sun was going down in a sea of gorgeous color that softened gradually to shades of palest blue and gray.

In the long winter evenings many will recall the campfire sessions of Camp Riley. There were splendid talks on the geology, the botany and the zoölogy of the region by those who could speak with authority on the subjects. Original songs, impromptu plays, and readings from favorite poets made the evenings pass swiftly.

Sometimes, in the midst of the jollity, there came a silence as each watched for the rising of the moon. First there shone a dusky light, softening the stark outlines of the great peaks, that stand "like huge waves petrified against the sky." Then a silvery radiance filled the whole arch, as the brilliant disk rose over the crest and swept across the sky. The campfire seemed but a tiny prick of red in the white light, but it was warm, and in its pleasant heat the fun was soon resumed.

Camp was broken on the morning of August the twentieth. The Mazamas and their friends returned to sea-level tanned and hardy, re-vitalized in mind and body, and feeling that the outing was well worth while, in spite of rain and hail and snow.

△△△

Glaciers of the Three Sisters

By IRA A. WILLIAMS
OREGON BUREAU OF MINES AND GEOLOGY

Thirty years ago in his report to the Director of the U. S. Geological Survey, Captain C. E. Dutton expressed the opinion that, "There are few localities equal in geologic interest to the neighborhood of the Three Sisters." To those of us who have had the pleasure of visiting this region since that time, the overflowing truth of that opinion has certainly never been questioned. There we see displayed in all clearness not only the present characteristics of the three great peaks themselves, but also the story of their life, their birth, growth and beginning decay, eloquently laid bare, only waiting our attention to be read.

Nor does it require the critical, penetrating eye of the scientist to decipher the story. Its facts are so persistently thrust beneath one's very eyes that it remains only for "him who walks" to connect them into sentences, and the paragraphs and chapters, that spell out a record tinged with the romantic, 'tis true, but replete with tragedy.

We camp securely in this day at the feet and in the shadow of these towering snow-striped peaks. Mazamas make intimate friends of them. As between friends when intimacy develops, they respond with confidences that the eye and the ear attuned to the "various language" that nature speaks can appreciate. As with intimate friends, we cultivate their acquaintance and are permitted to learn of events, and of the exigencies and crises in their lives, that are not entrusted to the onlooker and the passer-by.

The Three Sisters peaks of the central Oregon Cascades, and the immediately surrounding country, is a region in which an unusual number of recent geologic events have taken place. There are two main reasons why the nature of these events may be so readily seen. In the first place, so short a time has passed since many of them occurred that the evidence of what has happened is not yet obscured by the effects of the weather, by soil accumulation or dense forest growth. Rock surfaces are bare and clean and possess their original characteristics. Lava flows are still but little touched by erosion, and the glaciers are as busy as when they were much more extensive. Then, again, the differing nature of the various events happens to be such that the results of one have not covered up beyond recognition those of another.

What is the language in which the story of the region is made clear? Every stream in its canyon speaks out in terms unmistakable; every flow of lava with its twisted, ropy and broken surface tells its tale; each smooth and glistening ice-scored ledge, each frigid clinging glacier, the shapes of the mountains themselves with their radiating ridges, intervening snowfields, and cruelly riven sides appeal in terms not to be misread. The language, thus, while one of tongues not few, nevertheless calls out to us in universal tones that all may understand. Transformed to words of human speech, what then of romance or of tragedy is revealed? The temptation to undertake the writing of this story in all its parts is one strong to resist, so interwoven and closely related are its various incidents, and so all-absorbing is the plot and nature's setting of it. For the present moment, however, but one feature of this grand drama must claim attention, one scene only in a single act on which the curtain is not yet drawn. It is a scene of ice and snow.

And we get our first clew from the diminutive glaciers that are precariously gnawing away at the flanks of each of the Three Sisters. Diminutive they are, for the reason that, whichever one we examine, abundant evidence is found that they are each of them only the wasting remnants of ice masses, once, and not long ago, of much greater extent. Precarious is their position, for so despoiled have they become by the growing warmth of sun since the continuous winter of glacial times held sway, that some of them are now relatively, in truth, mere grasping and shriveled icy shreds of their former selves.

We need look for no hidden signs of former extensive glaciation. By no direction of approach to the Sisters can we escape their glaring testimony. Twenty miles west of the Cascade summit along the McKenzie road we begin to see the marks of glacier work. Lost Creek canyon throughout practically its entire length from where it heads against the slopes of South Sister to its union with the McKenzie, a distance of twenty miles or thereabouts, is deeply glacier cut and its U-shaped cross section is not to be mistaken. Can we conceive of the day when this great rock-walled trough was filled to the brim with frigid blue ice, was indeed even a mere corrugation in the uneven old lava surface over which the ice spread, inundating all but the highest elevations for a great many square miles along the west slope of the Cascade range? And when we now perilously search our way down its crumbling cliffs for a thousand and more of feet into its depths to where in August the brilliant flowers of springtime that adorn acres of its level floor lure us on,—can we realize that these perfect little meadows with their sod of green through which scattered boulders peep, and across which sparkling rivulets wind a graceful way,—is it possible that they, too, are to be accounted for by the former presence and work of the same chilling stream of ice? There is no question that this is so.

Shortly above Alder spring on the McKenzie road we pass bare, hummocky rock surfaces which bear the indelible imprint of glacial ice. They are scored, plowed and rounded as though some gigantic rasp of uneven grain had irresistibly borne down upon them. In its passage across the summit, bare or sparsely wooded glaciated hills and knobs are seen on every hand along the McKenzie road but a few miles to the north of the Sisters group of peaks. Similarly to the south for many miles along the crest of the Cascades are uncounted lakes, in size from miles in diameter to but a few feet across, that occupy rock-bound sags or shallow pits in the hard rock. Some of them have outlets, many not; and all are eloquently reminiscent of a time not very long ago when the whole summit of the range was buried beneath an immense roof of ice and snow.

In our examination of the Three Sisters and their environs we find that very important events have transpired since the time of widespread glaciation. Great masses of liquid lavas have issued at the foot of, and in some cases upon, their slopes. So fresh are some of these flows that it is rather difficult to believe eruption is no longer taking place. The lavas have come out and spread over large areas of the glaciated country. One may be almost certain in some places that the lavas appeared while frigid conditions were still present. Can we picture the spectacular display that must have accompanied the issuance of the glowing hot lavas as they melted their way up through and flowed out upon the surface of the accumulated arctic snows of we know not how many winters? Scarcely. Yet in so many places may we see the ice-scored rock surfaces passing directly under the borders of the new lava flows that from any and every reasoning standpoint we are unable to scout the probability that glaciation and volcanism were vigorously contesting processes here in the not distant past. The net result of their contentions to date is expressed in the character of the region at the present time. Have either of these two differing forces of nature so gained the ascendency as to henceforth discourage opposition from the other; has the battle been fought to a draw; or is it as if the declaration of a truce has temporarily allayed the conflict?

So far as man may judge, the last is the most probable status of the situation. But while peace between these two may be the order of the present day, just as in warfare among men, if not the victors, others promptly enter and begin to clear away the wreckage and debris, to remodel and to reconstruct; so that in course of time but scattered sign remains perhaps to tell the story of the past. The active erosion of the streams, the cutting action of what is left of the glaciers, and the crumbling effect of the weather, are the agents of reconstruction that are slowly, 'tis true, but so surely revolutionizing the surface features of the Sisters region that, unless they are deterred, even the mountains themselves are in the end doomed to obliteration. Such is the outcome in the measured terms of earth history, though no one of us need have serious concern that these magnificent peaks may be lost to us; for while years and hundreds of years are our units, their passage is but a mere tick of the clock of epochal time by which the crucial periods of earth events are measured.

With the superficial satisfaction of distant inspection or philosophizing, we shall not, however, be content. A real speaking acquaintance with the Sisters, all three, must be gained if to us is to be yielded up many of the intimacies of their life careers. As we approach them one may seem more or less communicative than another as to the past, but when we reflect that the facts we learn depend less upon their inclination to impart than our own ability to comprehend, it is plain after all that it is our own keenness of sense that is to determine the pleasure and satisfaction we enjoy in our association with these, our friends, the Three Sisters.

All are easy of approach, if too great intimacy at the start is not attempted. Records at the summit of South Sister show that it has been climbed from the east, south and west sides without difficulty. Upon its slopes are five living glaciers. All of them may be seen to be much shrunken from their former size and extent when examined at close range. Their lower borders are frequently rimmed with sharp ridges of loose rock detritus, that rise from a few to a hundred feet higher than the present surface of the ice. Their extremities are as a rule so entirely obscured by the accumulated rock materials which they themselves have brought down, that their limits can rarely be definitely made out. They are, in other words, blocking their own courses and, as it were, burying themselves beneath a load of their own hauling. A great deal of water is of course produced by melting of the ice in summer, and that which issues from beneath the glaciers goes away charged and milky with the finely pulverized rock powder that the moving ice has etched from the sides and bottom of its channel. But these waters have little or no power to carry away much of the vast quantity of coarse material which rides down frozen in or upon the top of the glacier, and it therefore heaps up in ridges and embankments where it is dropped as the ice melts. At a little distance these moraines, for such they are called, are a prominent feature.

One may with slight difficulty walk clear round South Sister in a day's trip, in which each of its five glaciers may be crossed at altitudes between 8,000 and 9,000 feet. In places the going is across the glaring snowfields, elsewhere a meandering path upon bare ice amongst the crevasses where calks and alpenstock, if not indispensable, are certainly reassuring safeguards to secure progress. Again it is a scramble over long rock slides or the scaling of the cliffs of crumbling lava that frequently rise from the glacier's edge and separate one from another.

Against the northwest slope of South Sister and in plain view in the photograph (opposite page 18) Lost Creek glacier clings. It has diminished greatly in size in recent times, and although it displays many crevasses in the ordinary season, in 1916 little else could be seen than a wide expanse of boulder-strewn snow. At its head it is cruelly eating into the mountain by a process of undercutting known as "plucking" where by freezing fast to the rocks the ice of the glacier literally plucks out great masses as gravity draws it down the steep slope of the mountainside. The distal or lower end of Lost Creek glacier is so thoroughly obscured by rock detritus that the exact location of the front of the ice is not to be seen. The snowfield of this glacier connects with the great mass of snow which fills the crater of South Sister.

Nestling in a shallow rock-walled niche of its own on the north slope is a mass of snow and ice whose glacial character is not very evident except in seasons when melting has been unusually active so as to expose the ice near its lower end. At such times a series of crevasses is in view, and while the area covered by both snowfield and the ice tongue itself is small, the common characteristics of the alpine glacier are present. The most typical and unmistakable confirmation of our observations as to its glacial nature is the presence about and below its lower border of well-defined embankments of morainal materials. Below it, too, are one or more small lakes into which the waters from its melting seep their way.

High up on the northeast slope of South Sister is the most extensive snowfield on the mountain. It is more than a mile across and from it as a feeder four small tongues of glacial ice creep well down the moun-

Summit of South Sister looking across Lost Creek glacier.

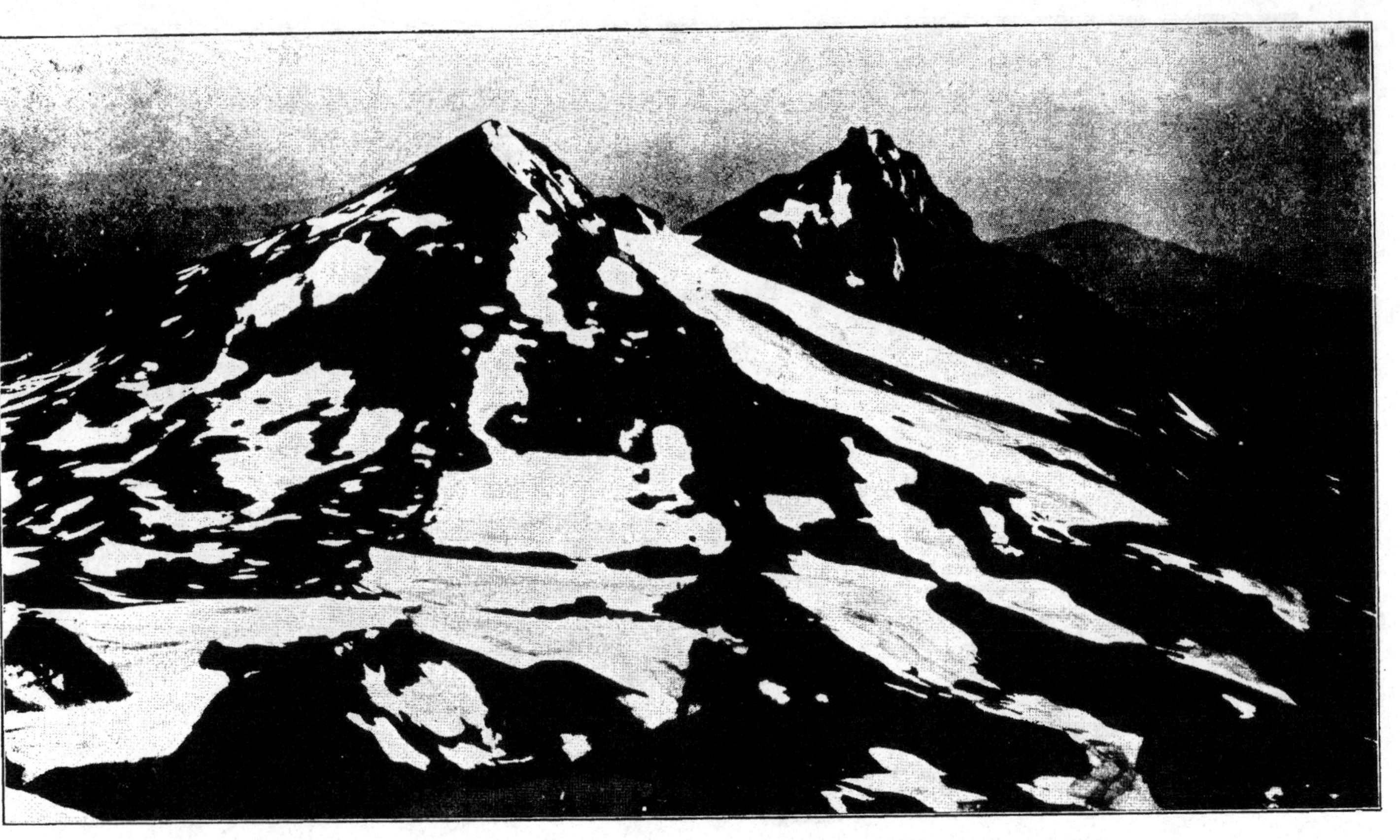

Middle and North Sisters from summit of South Sister, showing Hayden and Diller glaciers on the right.

tain side. At the present time these individual ice lobes are separated only by what appear to be elongated morainal ridges, that is, heaps of broken rock, gravel pebbles, boulders of all sizes, and sand, that the ice itself has dug out, carried down and deposited. We are left no other inference here than that the position of what we see today as four separated small ice streams was formerly occupied by one large mass of moving glacial ice that emanated from a snow-filled amphitheatre of probably considerably greater dimensions. In the course of its shrinkage this once large glacier has been forced to seek its way out through a series of channels rather than a single one, on account of the obstructions which it, itself, has dropped in its old course.

Could one visualize the present condition of this great body of ice and snow that is slowly smothering itself with its own burden, no better picture of its position and outlines could be found perhaps than to conceive of the snowfield area as the palm of a huge hand from which the ice streams push out as the fingers. Between each two fingers is a morainal ridge, often rising a hundred feet higher than the fingers themselves. Against the uphill ends of these ridges, where the fingers attach to the hand, as it were, the ice mass splits and, as if crowded far beyond its plastic limit, here is usually a group of radiating wide open crevasses running, not across, but up and down the slope of the mountain and of the glacier. Each finger of ice reaches down to nearly 7,500 feet until dwindled by melting and obscured by its own load; its waters, usually surcharged with fine sediment, accumulating in a series of little lakes, either directly or by seepage, drain away into some of the smaller head-streams of Squaw creek and thence to the Deschutes.

The size of this glacial field on the northeast slope of South Sister, and the distinctive and typical features of the ice streams leading from it, are such as should give it a recognized standing among the glaciers of, at least, the Oregon Cascades. In my study of the region the past summer, the desirability of dignifying it with a name came to me very strongly. Little then did I suspect that by now there might be occasion to commemorate the passing of any one of the congenial group gathered in the 1916 Mazama camp. Were Mr. Prouty here today I know that his modesty would urge against such recognition. What to us were his superior attainments seemed to him mere incidents in his everyday life. In acknowledgment of those attainments, however, which were his only because he possessed a poise and judgment that never failed him in the most crucial of moments, and as a testimony to his character and his knowledge of mountaineering and of the mountains of the Pacific coast ever placed, in his most kindly manner, at the service of whomsoever he might assist, the writer of this paper proposes to christen this largest of the glaciers on South Sister, Prouty glacier. A humble testimonial this, perhaps, yet one by which, it is hoped, Mazamas and others of Mr. Prouty's acquaintance may in the future be frequently reminded of him whose memory will no doubt be perpetuated in other more conspicuous manner, yet by no monument more fitting or substantial.

Against the east of south slope of South Sister in a niche which it has excavated for itself is a small glacier that, in a similar way, terminates in two narrow extending fingers that reach very little below 9,000 feet. Between the two fingers is a high morainal ridge against which the ice mass appears to split into the two parts. This entire body of ice and snow has an exceedingly steep slope, in general too steep to traverse with perfect security, and does not occupy, all told, more than a few acres. At the tip, or snout, of the easterly branch of the two ice streams is a sheer front of clean ice forty to fifty feet in height which shows many of the characteristics of the true alpine glacier. It is jointed, broken by crevasses and exhibits the horizontal bandings that mark the cumulative snowfalls of successive seasons. Below this ice front the solid lava slope is so steep that morainal material cannot remain, but when released from its icy bond rolls, or is promptly moved by the series of copious glacial streams of water far down towards the base of the mountain. Blocks of ice at times part from the parent mass to be similarly precipitated headlong down for a thousand feet or more to more stable positions and to where once, without question, the glacier itself extended and dumped its load.

Upon the southwest slope of South Sister is yet another small glacier. Its surface is usually pretty thoroughly snow-covered, the crevassed blue ice showing toward its lower end only in late summer when melting has exposed it. About the upper rim of its cirque is another remarkable example of undercutting or plucking of rock masses by freezing and the downward gravitational movement. A very good idea of the character of the volcanic materials of which the top part of the mountain is composed may be obtained by a study of these great overhanging cliffs that are developed to a greater or less extent round the head of all the glacial fields on South Sister.

We may again say, then, that this peak has five glaciers of a size worthy of recognition. Each of them is but what is left of ice streams once more extensive. We may properly regard what we see today as the dwindled remnants of a series of feeders that contributed, from this elevated peak, to the more widespread glacial fields that buried the whole Cascade summit for a hundred miles or more. It seems highly probable that South Sister was an actively erupting volcano during, at least, the period of greatest glaciation; otherwise we would expect to find it more deeply cut into and its cone shape more seriously marred than it is at the present time.

Middle Sister is the source of four active glaciers. Of these, two are on the east side, one upon the north of west, and the fourth and largest passes down to the northwest, its lower portion being along the west base of the North Sister. Each of these glaciers has been mentioned in the past by various writers, so no detailed separate descriptions will be given. In the photograph (page 19) the position of the two glaciers on the east slope of Middle Sister may be seen. The one to the south is Diller glacier while Hayden glacier passes to the north near the foot of North Sister. The two spring from the same gathering ground above, but separate against a great jutting crag to become thence individual streams of moving ice. Both are excellent examples of the alpine glacier, exhibiting well in their various parts all stages of consolidation from granular snow, half-ice, half-snow or névé, to the solid ice of blue or blue-green cast. Fissuring is a common feature and in places near the extremity of Hayden glacier particularly, in a season of ordinary melting and flowage, wonderful development of seracs, a pinnacled maze of crevassed and broken ice, may usually be seen. Deserted and ancient moraines lead from their present termini far down the lower mountain slopes and into the bordering forests.

Renfrew glacier hangs against the north of west side of Middle Sister. It is, in part, across its snowfields that two Mazama official ascents of the mountain have been made, in 1910 and 1916. The Renfrew presents a most striking display of both lateral and terminal moraines around its borders, and at its lower edge appears to be split into two or more separate tongues of ice. It is ordinarily so obscured by its mantle of snow that, aside from the broad snowfield, few distinctive features are in evidence. It is worthy of mention that at the south side and near its head the rim of its amphitheatre is made of a much more recent lava than is the bulk of the mountain. A great quantity of viscous lava has oozed from a subsidiary vent at a little over 9,000 feet at the west side of the peak, and the present surface of the main rock ridge down that slope for 1,500 to 2,000 feet is due to the fresh veneer of what appears to be a porphyritic andesite of cellular texture extruded in this final eruptive paroxysm of Middle Sister. This ridge which in places has been narrowed by glacier cutting to a vertically walled causeway of but comfortable width, offers a most attractive course of ascent for those who prefer solid rock to a climb across the snow.

Collier glacier originates on the west of north slope of Middle Sister, flows along the west foot of North Sister, and terminates rather more than three miles from the upper rim of its cirque, giving rise to White Branch, a stream of fair size and one of the headwaters of Lost creek which flows into the McKenzie river. In length and volume the Collier is as large as any other, if not the largest, glacier in the Oregon Cascades. Nor is there one that exhibits more typically the many interesting features of a stream of flowing ice. Great moraines, single, double, even triple crested, fringe its lower borders, rising in places more than 100 feet above its surface. This of course indicates a great shrinkage in its bulk within comparatively recent times. In August, 1916, practically its entire surface was covered with snow. In 1915 and in 1910 to the writer's knowledge, practically the lower mile of its length was an expanse of boulder-strewn firm ice, pinnacled and crevassed in part, elsewhere coursed with uncountable hurrying rivulets of snow water. Its front was then one high wall of solid ice, down the face of which when the sun shone bright, glinted and gleamed and flashed innumerable rills, brooks, cascades of purest quill, as if in uncontrolled haste to join the mud-reeking waters of White Branch, which was milk white indeed with its charge of rock "flour" from the ponderously grinding glacier mill, as it issued with sullen gurgle from its somber cavern at the glacier's base. In the view we can see far up its icy surface, which is an estimated full mile or more in width, a cross-break which on approach proves to be an open crevasse of the type known as bergschrund. The bergschrund differs from the ordinary crevasse or yawning fissure in that the wall on the lower side of the break has settled down, sometimes slightly, sometimes many feet, so as to leave a bare upstanding wall of ice of corresponding height on the up-hill side of the opening.

Although North Sister seems from all appearances to have received far more harsh treatment at the hands of the erosive agencies, it does not have upon its slopes today a glacier to compare with those on the other two mountains. The upper portion of North Sister is a jagged rock ridge with exceedingly steep faces on all sides. The precipitancy of its slopes very likely accounts in part for the lack of glaciers of any size, since there is little space sufficiently flat for snow to accumulate. There is evidence, however, of the existence of a once full-fledged glacier on its northeast side where a succession of well-defined moraines may be seen extending down beyond the range of vision and into the forest. Within a sharp deep cleft in the mountain, which was no doubt once perennially filled with snow, there is now a small body of glacial ice. From a little distance the writer could discern some blue ice cut by crevasses at its lower end. The surrounding walls of this former cirque are so precipitous and crumbly that both snow and ice are abundantly over-strewn with rock debris. In fact the place itself fairly resounds with the crash of falling boulders. Each in its descent starts a myriad of others, all joining in a grand competitive relay of midair leaps from cliff to crag, from crag to snow, and then a racing ricochet across snow and ice far down, often to the limit of one's vision, to be finally lost within the depths of yawning crevasses, or mingle with others of its kind and become a part of some morainal heap at the glacier's border. Where is there more exhilarating sport than to watch boulders take their bounding course, after having been unfastened from their resting place by our puny human efforts? But what is this compared to the thrill of delight and pure satisfaction that is ours when we catch glimpses, as here, of nature's own invisible hand at work!

In a similar cleft on the south of east slope of North Sister there is what appears, when viewed from the country to the east, a second glacial remnant. The deserted moraines appear to be there, and although the writer has not had opportunity to make a close examination of this portion of the mountain, its general appearance is such as to strongly suggest that here too is the site of a former glacier of some magnitude that contributed its mite to the summit ice-cap, and that was an important factor in reducing North Sister to its present jagged, eaten-away, and relatively almost skeletal condition of decay.

A recounting of the glaciers of the Three Sisters thus totals eleven. All are within a north-south limit of not over seven miles, and fifteen square miles will amply enclose the entire group, as they stand, with space to spare. No other area in the United States so small, and withal so accessible, surrounds so many glaciers as the Three Sisters region affords. Besides these, it is the overwhelming presence of the great peaks, the widespread new lava flows with their tale of smoldering fires now burned low, far-away rock-cut canyons, and the enchantment of lake, and of flower-besprinkled meadow where winding rivulets purl a restless way—these it is that call us back again and again to this one of Oregon's beauty spots.

The thronging mountains, crowding all the scene,
Are like the long swell of an angry sea,
Tremendous surging tumult that has been
Smitten to awful silence suddenly.
—*Celia Thaxter.*

A Geologist's Thoughts on Returning From the Mazama Outing of 1916

By Warren D. Smith

If the reader were to come into a theatre just after the actors had finished speaking their parts in a great tragedy, but before the curtain had fallen, he would know how I felt when I got a close view of the Three Sisters for the first time. I had long seen them from a distance and they were cold, distant—they were more— they were remote. It has been my privilege to see volcanoes in action in Italy, in the Far East, and therefore I viewed the majestic Sisters with something akin to respect, almost with awe, wondering how long they had slept, how long they would be sleeping.

The Middle Sister looked comely and placid and so I chose to place my offering upon her chaste white brow. The South Sister lived too far out in the country and the North Sister reminded me of Medusa of old with her wild snaky locks—death lurked there in each of those jagged points. Even as I looked at the Sister of my choice a disquieting thought came to me—what if the majestic, but sleeping princess of the forest should wake from her hundred year's slumber! Many have wooed her and placed their offering upon her head, but might not one of us be the prince whose kiss would wake her? God forbid that she should awaken again in our time or at least while we are so close by! What an appalling spectacle those angry Sisters would make no one can adequately imagine who has not seen one of those furnaces in action. I once saw three Tagalog women haranguing one another on the street corner and the eruption of words (no dictograph could keep up with the steady flow) was volcanic, epithets were hurled forth, and streams of vituperation poured out and overwhelmed all within hearing. Whenever I think of the Three Sisters and try to picture their turbulent past I am reminded of those three little brown sisters in far off Luzon.

The dormant (I am not so sure that they are extinct) volcanoes of the Cascades are a part of that great ring of vents about the Pacific ocean known to geologists as the "Circle of Fire." We cannot exactly say how long since they were active (we do know that it has been only a few hundred years) nor how long before they may erupt again. To the south of us we see Lassen Peak in this same range, enjoying the reputation of being able to "come back"—to the north, in Alaska, are also signs of rejuvenescence. The writer knows full well the fate that overtakes all prophets, particularly those of the Cassandra class, and yet he ventures the prediction that "the worst is yet to come." When the next furnace will be "blown in" he does not care to attempt to guess. Supposing this be true, it need not be an unmixed evil. Volcanoes are spectacular; they sometimes cause no little inconvenience, as for instance in the "late unpleasantness" at Pompeii—no doubt it upset quite a number of little dinner parties and spoiled the rugs. I have a picture in my collection by *Le Roux* depicting what might have been a very common incident during that eruption. It shows milady with her maidservants around her on an eminence overlooking the stricken city. She is swooning into the arms of one of them, while others are hurrying up the slope, their arms laden with robes and trinkets. They really look quite flustered, and yet volcanoes are mere incidents in geological history and in the last analysis they have been beneficial rather than otherwise, for they doubtless afford a relief of heat and pressure in old Mother Earth which if pent up too long might cause more serious damage.

Some suggestions as to recent history of the Three Sisters region may not be without interest. As these can be conjectural only, in the absence of detailed study, it is hoped the reader will not be too critical. From the apparent general positions and topographical relations of the various peaks in the cluster known as the "Sisters" and from the rate of erosion which has taken place, it would seem that the North Sister is but a remnant of a large rim, only a small arc of which remains, of a much larger volcano than any now in the region. The South and Middle Sisters then must be considered as later and subsidiary cones. The Middle Sister shows no crater now, only a sharp peak, but the South Sister has still a well-defined crater. These, then, bear perhaps the same relationship to the north member of the group that the present active cone of Vesuvius does to the old mountain of Monte Somma. A more appropriate name for the North Sister would, in my opinion, be the "Mother."

Furthermore, the nature of much of the eruptive material in this region is such as to lead one to conclude that while there has been outpouring of molten rock at times, much of the ejecta has been blown violently from the different vents. At least in their latter days these Sisters were turbulent. It may not be improbable that the oldest Sister (North) fell into such a rage that she had a "brain storm" and blew off her own head.

As for the Husband, who certainly looks "henpecked" to say the least, we venture the opinion that he too is fairly old. There was once a well-defined crater in this gentleman's august top which has been sadly eaten into on one side. This was probably done by a glacier. It is hard to tell from a distance whether the now prominent amphitheatre on the east is due more to crater than to cirque.

This leads us to another point, namely the role played by these cirques in the resulting shapes of the peaks. Originally most of these mountains were cones and now many of them, particularly the older ones north of the Sisters, are pyramids. In Europe they would be called "horns." Now the horn is the result not of a building up, but of a tearing down process and in this the glacial cirque has had the dominant part. If there be several glaciers emanating from a mountain (and this is often the case) there will be as many cirques or amphitheatres at their heads, and if these cirques be considered as approximating the form of so many truncated and inverted cones and gradually approaching one another through erosion so as ultimately to intersect them, we may look for interesting results. Going back to your Conic sections you will recall that such intersections will produce parabolas and so in Alpine Europe where there is a perfect forest of peaks and glaciers, the parabola-shaped skyline between peaks is common. These are known as *cols* and they leave almost invariably sharp pyramidal masses between them. Such cols, as far as I know, are of limited occurrence in the Cascades, but the older peaks in most cases are pyramids rather than cones. Mt. Jefferson, which has been called the "Matterhorn of America," is a good example.

And while we are on this subject, I wish to call attention to the very fine example of a particular feature of glaciated country shown on the north side of the South Sister and known as the *bergschrund* (literally mountain gap). It is the boundary crevasse between the nevé and the glacier. Mr. Williams calls attention to this feature in his bulletin but does not use this term. This is such a marked feature of glaciated countries that it should be perfectly familiar to all Mazamas.

So much for the superstructure—what about the foundation and lower floors of this great geological edifice? The foundation of the Cascades few persons have seen. It is visible at only a few points where it outcrops deep down in the lowest-cut canyons. It is represented by several outcrops of granite or granodiorite which are offshoots, in the writer's opinion, of the great Sierra granite batholith of California. Upon this we find in places a basal conglomerate and wherever this conglomerate is found there once the sea beat in long surges, but upon what a lonely shore!

Above these come piles of sediments, sandstones, shales and more conglomerates, and all these make up the first story of the structure. On top of these and in such great heavy masses as to cause the whole substructure of sediments to sag visibly under the enormous load, which has never been calculated but which certainly amounts to billions of tons, lie sheet upon sheet of basic lavas, andesite and basalt. Interbedded with these are some hundreds of feet of volcanic ash or tuff and even coarser ejecta called agglomerate. This pile is like unto a great layer cake. Into this series of layers, numbering more than a score in places, streams large and small have made great gashes and these gashes have subsequently carried tongues of living ice which softened and rounded their contours and now we see them as beautiful valleys. We have then, as has been pointed out by many others, a dissected plateau rather than a series of ranges.

And last and uppermost we have the rock pyramids and cones of the Sisters, Jefferson, Hood, and others, rising above the noble forests like the towers and minarets above a Moslem city. Ah, did you not feel as the departing sun lit up those great rock piles with the matchless alpine glow, that indeed it was the hour for prayer and meditation!

I have said little so far about the interesting phases of past and present glaciation in this beautiful region and I do not purpose to do so in this article. Personally I do not care for glaciers; they are cold, forbidding things and their work is usually destructive, unrelenting and remorseless.

Since my return from that all too short week spent with my happy Mazama friends, I have thought how fortunate we are to have this wonderful geological laboratory right at the doors of our educational institutions. Never have I seen in so small a compass, and so accessible, such an array of things of major interest to the geologist. Were I a botanist I would doubtless be equally satisfied. Those upland meadows and pastures make it unnecessary for us to go to Europe for the Alps—they are themselves true alps—high mountain pastures. It has also occurred to me that the Three Sisters region might very appropriately be set aside as a state park.

By way of conclusion, my Mazama friends, I shall tell you what was the best part of all that experience to one member of your party, and I dare say to each one of you as well, and that was the feeling of having overcome something really formidable. Some there are I know who are either genuinely unable to do this sort of thing or are too lazy and are content to take their exercise passively at the expense of gasoline instead of sweat, who say in that tired way, "Oh! What's the use?" We are glad there are some things in life upon which a cash value has never yet been placed, which cannot be bought, sold, given away or stolen. There are times when it is good for the soul of man, and his body too, to have to wrestle with something and to conquer it, if for nothing else than just to down it. Every one of those last four thousand feet, which I put behind me and beneath me meant a greater mastery of and respect for myself and it is because of this enriching experience, with its reward at the top, that I have a little tinge of pity for those lowland peoples who have never stood on the pinnacles of the world and looked away beyond the dwellings of men—who have never grappled with Nature in her wildest moods. You who have stood amidst the crackle of electricity on the lonely mountain top, you who have looked into the bowels of the ice-stream, you who have gazed into the yawning, hissing throat of a volcano, you who have listened to the silence of the night in the fir forest far above the snowline, you have indeed had "a sense of deeper things."

> "And I have felt a presence that disturbs me with the joy
> Of elevated thoughts; a sense sublime of something far more deeply interfused
> Whose dwelling is the light of setting suns
> And the round ocean and the living air,
> And the blue sky and in the mind of man:
> A motion and a spirit that impels all thinking things,
> All objects of all thoughts,
> And rolls through all things."

And so as we go in life from gaping craters of sorrow, through dim and misty jungles of misunderstanding and doubt, out over sunlit stretches surrounded by friends and up, ever up, to the Sinai of some great inspiration, may we follow the lead of the Mazamas, and ring out the challenging call, "Is everybody happy?"

My eyes dim for the skyline
Where the purple peaks aspire,
And the forges of the sunset
Flare up in golden fire.

There crests look down unheeding
And see the great winds blow,
Tossing the huddled tree-tops
In gorges far below;

Where cloud-mists from the warm earth
Roll up about their knees
And hang their filmy tatters
Like prayers upon the trees.

I cry for night-blue shadows
On plain and hill and dome,—
The spell of old enchantments,
The sorcery of home.

—*Bliss Carman.*

Photograph by Winter Photo Co

North and Middle Sisters, showing Co ier glacier, from Cinder Cone.

North Sister and The Husband, showing Lost Creek falls.

Lost Creek Valley

By HENDERSON DAINGERFIELD NORMAN

We were not pioneers in Lost Creek valley. Of distance, elevation, and the like we have little data; but we saw the valley in fog and sunshine and this record tells of the things that are clearest in our memory.

To find the valley from Camp Riley, we set out in the general direction of Husband mountain. Straight over the ridge we went and across a long snow field. Here and there a gap in the mountains gave us a framed picture of billowy blue hills to the west, while always the Sisters stood eastward, white and glistening. We could tell which members of the party were true children of the Pacific coast and which were Easterners at heart by their choice of views,—the shining masses of eternal snow, or the waves of the dear, blue mountains.

How far it is from Camp Riley to the mountain side that slopes steeply down to Lost creek, I have no means of knowing and no more accurate gauge than the remark, hissed fiercely into my ear by a Mazama, returning fagged and footsore, "It's a right far piece to your old flower patch."

It is always "a right far piece" to the Regions of the Blessed.

The flower show of Lost Creek valley begins with the cliff that overhangs the trail. That cliff, when we first saw it in the August sunshine, was glowing pink with the most beautiful of the penstemons, rightly called "pride-of-the-mountains." Along the trail were collinsias, honey-sweet, from the tiny varieties to tall "Chinese-houses, and calochortus lilies, and purple beard-tongue, and by the edge of the little snow-stream that emptied into Lost creek were monkey flowers, large and small, pale yellow, deeper yellow with conspicuously dotted lip, and bright magenta.

Then came the first near view of the valley's flower carpet. It is possible that the people who went on the Mazama trip to this valley are the first who ever saw it in bloom, for usually, as soon as any growth begins, the sheep are turned in. This year, the government had protected it by a special order, for the sake of science and the Mazamas, and besides the snow fell so heavily and lay so late the sheepmen had of necessity taken their flocks elsewhere.

With the memory of that first vision of Lost creek, comes the baffling certainty that no one who has never seen a flower carpet can possibly believe you mean exactly what you say when you use that term. The thing is so incredibly beautiful even while you look at it. Behind, above, around, and even along the very edges of the creek itself, encroaching upon the very carpet, are miles and miles of snow. A few hundred feet higher up, the white pine is fighting for life against the bare face of a rock, and here is color and perfume and lavishness of bloom such as you have never seen before.

The grass has the vivid green of grass that has grown quickly in damp places, and yet at times that green is wholly lost and white and blue and yellow supervene.

The ground along the trail begins to be boggy and all around are white marsh marigolds. Below are buttercups, and lupine and larkspur. Closer to the water's edge are saxifrage and little, delicate lilies, stenanthella, brown and rosy pink. Now the ridge is all above us and our trail is made by following Lost creek. It runs full and crystal clear for a mile, with hardly any perceptible fall. "Go where you like," says the leader, "but keep ahead and stay on this side of the creek." It sounds easy, but Lost creek will hardly let you know which is "this side." It makes the most amazing loops and turns, and moreover the most glowing castilleja blooms always on the other side. This "Indian paint brush" of the Cascades we found in many varieties and in color from palest yellow to orange, from pale pink to flaming crimson. Along the creek, it was intermingled with the great white saxifrage, very stately, and along the water's edge was a fine fringe of tiny many-colored gentians.

But the most entrancing flower of the upper waters of Lost creek is the glacier lily. This erythronium differs from the avalanche lily both in size and color. We found myriads of its blossoms, slender, golden and dancing, growing not only through melting snow, but actually pushing their delicate scapes through ice—a very miracle of strength made perfect in weakness. The castillejas showed no less of hardihood. We measured one that had pushed through six inches of snow before it found daylight.

In these smooth upper waters of the creek, where it looped itself around little flower-crowned islands and molded mimic straits and promontories, our party found an irresistible invitation to bathe. Our forces separated: the men went up the creek; the ladies, down. Bathing suits were hurriedly improvised,—and if necessity is the mother, she must also be, *ex officio*, the chaperon, of invention.

If we sought a scientific excuse for our digression, there was the least saxifrage along the borders of the stream and *Ranunculus aquatilis* in its very current.

Our first exploring trip ended with this adventure and the homeward climb began—a climb enriched by finding two varieties of polemonium, and on the higher rocks, golden sedum and a new white stone-crop, which we proudly identified later, as Gormanus, a genus discovered in Alaska by M. W. Gorman, and bearing his name.

There were thousands of cat's ears, too, *Calachortus Lobii*, and it is worth noticing, as additional proof of the value to science of these annual Mazama outings, that while this calochortus is common in this part of the Cascade range, and the flower just mentioned is not uncommon, the former has heretofore been reported to the State University only from Mount Jefferson and the latter not at all. The sheep have hitherto come in before the botanists.

We climbed over vast beds of heather, true and false, red, yellow and white; and from that first Lost Creek valley expedition we brought in seventy-seven distinct varieties of flowers, not mere botanical varieties, but such as the amateur might easily distinguish.

On another day, ten hours of fog spent in this charmed valley yielded less of the glory of the surrounding ranges and for that very reason showed us more intensively the flowers that bloomed at our feet. Already the glacier lilies were beginning to droop and the marsh marigolds scattered their petals as we touched them. This time we were following the lure of Lost lake and as we went along the creek, it began to fall precipitately, the valley narrowed to a bare channel between the mounta ns and small bushes marked its course—vaccinium in all its puzzling varieties, and thimbleberry, and farther down the creek, a perfect foam of elk grass. We had been told of the falls we should find by following this course, and very soon the silver of the stream had turned to milk-white and it foamed and fretted and plunged. We thought the first leap must be the falls, but another was from a greater height, and at last it leapt downward in a cascade of many veils, silver and swift and shining.

It is not an easy walk. At one point it is necessary to cross the stream at a dizzy height, on a slippery log, and prayer and alpenstock alike are needed to get you safely across. Much of the way is a scramble among huge boulders close to the water's edge, with saplings to catch at—and sometimes the sapling proves to be a devil's club. Aside, however, from the beauty of the falls, and the whole course of the creek itself, this part of the valley is deeply interesting to the botanist, for at the cascades we come into another life zone. A thousand feet above, we found the strictly alpine flowers, phloxes, polemonium, crucifers, and soft, shining pussy-paws (*Spraguea multiceps*). Here orchids and lilies multiplied. Here, too, we found betony, and elephants' heads (*Pedicularis Groenlandica*) and chimaphila and pyrola. A belated cluster of rhododendron shone among the trees. Undiscouraged by the bigger, brighter flowers, the delicate tway blade held up its tiny spike; the long-spurred green orchis was everywhere beneath our feet. We found valerian, too, and wild roses, and saw the clear shining of white bunchberry blossoms. On the face of a cliff, we found Romanzoffia, and it was like recapturing April in mid-August.

Then we came upon a marshy flat—surely the beginning of Lost lake. We even saw Indian pond lilies in bloom. A heavenly sweetness beguiled us, and we plunged boldly into the marsh and for a glorious half hour waded in water above our boot tops. The perfume was that of limnorchis, the great white bog orchid. It is not an uncommon flower, and every summer in the mountains yields a few, but it is a fortunate flower-finder who finds half a dozen in a season, and here were hundreds, thousands, tall and gleaming white and sweet beyond all other fragrances. There were multitudinous spikes, too, of tall pedicularis, and best of all as a discovery, the beautiful, floating flower-stalk of the buckbean, once so common in northwestern America, and now comparatively rare. Its flowers are not only exquisitely white but they gleam as with hoar frost at the center.

While we were still in the marsh, the fog closed down, swathing Lost lake somewhere in its folds. It was a disappointment and yet it seems in memory part of the white magic of that valley that we never found Lost lake.

Between the summit of the Middle Sister and the marsh below the cascade,—is it to be called Duncan Falls?—we found not less than a hundred and fifty varieties of plant life, not counting trees, ferns, mosses or fungi.

In our fancy, there is a certain fitness in our last vision of Lost Creek valley, with the ghostly whiteness of the fog and the heavenly perfume of the palely shining spikes of the white orchis. But when we talk together of Lost Creek we do not think of fog, but of August sunshine and a wide flower-filled valley, after miles of snow, and Katherine Tynan's verse comes to our minds:

"Not one flower, but a rout,
All exquisite, is out.
All white and golden every stretch of sod.
As though one flower were not enough!
Thank God!

Upper—South Sister from Lost Creek meadows. Middle—McKenzie river, one mile above the bridge. (*Photographs by R. L. Glisan.*) Lower—North Sister, after the snowstorm. (*Photograph by Ella P. Roberts.*)

Photograph by A C Shelton.

Winter view o Three Sisters during a heavy storm

Upper Lost Creek Meadows

By J. Duncan Spaeth

(Dedicated to the Mazamas at Camp Riley)

Up with the morning, and off o'er the snow fields
Merrily marched we away on the trail;
Bright in the sunlight, brown in the twilight
Our path wound along in search of the vale.

'Twas the valley of promise, the valley of flowers
Hid by the ranks of the sentinel hills;
Gaily the waters of Lost Creek played through it,
Fed by the snows in silvery rills.

Darkly the precipice towered above it,
Thunder-split pinnacles wrinkled and old.
Higher still, snow peaks gleamed in the distance,
Pure as the dead, white, silent and cold.

But down in the valley green meadows were smilin'
Bright with the blossoms of earliest spring,
Buttercups, marigolds, lilies and windflowers
Fluttered and laughed and danced in a ring.

Columbines swinging,
Bluebells a-ringing,
Humming-birds winging,
All of them singing,
This was their song:

Back again, back again,
Back to the springtime—
Back to fresh Aprils of first love and truth;
Forget that life's morning is gone past returning,
Forget that the suns of mid-summer are burning,
Forget and come down to the lost stream of youth.
So we dipped down the hillside
And tripped down the valley,
And danced with the flowers,
And joined in their song:

Back again, back again,
Back to the springtime—
Back to fresh Aprils of first love and truth;
Forgot that life's morning was gone past returning,
Forgot that the suns of midsummer were burning,
Forgot all the sadness of sober September,
Forgot we were nearing the nights of December,
Forgot—for we'd found the Lost Stream of our youth.

August 11, 1916.

Wild Life of the Three Sisters Region

By ALFRED C. SHELTON

The summer of 1916 found the Mazamas encamped high on "Old Obsidian," at the west base of the Middle Sister, among the snow peaks of the Cascades of eastern Lane county, Oregon. While rich in its diversity of climatic phenomena and affording a wealth of never-to-be-forgotten mountaineering experience, it was a summer in name only, and many of the birds and beasts, driven from their haunts by the extreme severity of the preceding winter, were forced to forsake their mountain homes and seek milder regions elsewhere. Winter ended, but no warm balmy days of spring blanketed the glistening peaks with sunshine nor brought forth the wild flowers to run riot with color over the soft green blankets of the mountain meadows. This year spring was but a myth in the Oregon mountains, and summer entered, cold and dreary, flashing at times an angry display of electrical wonders, raining or snowing fitfully from a leaden and murky sky, settling down with calm deliberation to day after day of cold driving fog. It enshrouded all with a chill and a murkiness which penetrated to the utmost depths of the thickest and warmest blanket in old "Camp Riley," but it could not dampen the ardor, dispel the enthusiasm nor quench the camp fires which were the center of such jolly good-fellowship. And so, on days when this heavy shroud hung like a pall over the cliffs, when weary climbers vainly shaded their eyes for the first dim, hazy outlines of little white tents in the fog, and hungrily sniffed the air for the first welcome whiff from Doc's steaming pots of oyster beverage, it seemed that this tiny colony of primates were the sole representatives of the animal world in their mountain fastnesses of snow peak, dark, sombre forest, and high cliff of black obsidian. But on days when rain and snow ceased to fall, when the cold driving fog lifted and was swept away, the sun broke forth in a blaze of splendor. In all directions stretched a wonderful vista of mountain peak and forest, with meadows, carpeted not with grass and wild flowers, but with snow bank and snowdrift, extending from the very doors of our tents, up and down the mountain sides. Such tiny areas as were bare of snow were so cold and saturated with icy water that plant life could not sprout and flourish if it would, and there was an utter absence of the wonderfully beautiful, though small, meadows so abundant throughout the region during the average summer. Only in one favored spot in the Lost Creek valley had Nature's garden burst into bloom, and high on the peaks, far above timber line, a few hardy alpine flowers, mainly perennials, sought sustenance in the vast expanse of ledge and shelf of crumbling lava. Heather, ever ready to add its cheery presence, flourished everywhere along the receding edges of the snow banks, nourished by the icy waters from the melting snow. Squirrels and chipmunks had not emerged from their winter quarters, a few gophers nosed about in the exposed patches of earth, a cony or a marmot whistled from some ledge of lava, and on bright sunny days, Oregon jays, "camp robbers," "whiskey jacks," or "summit birds," as they are variously known, came foraging around camp to pilfer and steal, or honestly search for discarded scraps from the commissary. A solitaire occasionally mounted to the topmost spray of some tall tree and favored those who were so fortunate as to hear him with a burst of melody surpassed by none save possibly the water ouzel, which in the wild fastnesses of his mountain home, perched on some spray-dashed rock in mid-stream, ran up and down his scales of wondrous music, carrying them far above the reach of noisy waterfall and roaring cascade of the icy, glacial streams which raced madly along to tumble down precipitous cliffs to north, south, east and west of the little white tents of Camp Riley. So we were not absolutely alone in the mountains, though as it was a hard year for the birds and mammals, scarcely any save a few trained observers found anything to interest them in the wild life of our mountain camp. If we could go through this Three Sisters region again, on a warm balmy day of a normal summer, when Frog Camp meadow is a riot of blue, yellow and red, when White Branch tumbles along through banks of heather and beautiful flowers, when meadows large and small dot Obsidian cliffs and wander at random up the slopes of the mountains, when chipmunks and golden-mantled squirrels scamper away from every log and rock, when the trees along the trail resound with songs, calls, and cries of mountain birds, or when from some distant ridge comes the slow dull booming of a grouse, we could in a single day find a profusion of wild life sufficient to gladden the heart of the most unobservant, and provide a wealth of unceasing joy to the naturalist.

Geographical ranges of plants and animals are expressed mainly in terms of "life zones." In the region of the Three Sisters mountains, three life zones may be clearly and easily recognized. First and highest is the Alpine-Arctic. This is the region of glacier and snow fields and rocky lava ledge, high above timber line, at the utmost summits of the highest peaks, where the only plants are a few hardy alpine forms, and where bird and animal life seldom wanders.

Below the Alpine-Arctic zone we come to the Hudsonian, which may be said to be the timber line zone, that belt extending around the bases of the higher peaks, beginning at a point near the base of Obsidian cliffs, below Camp Riley, and extending up the mountain sides to timber line. Only two trees are found in abundance in this zone, the Black Alpine hemlock (*Psuga mertensiana*) which formed the major portion of the heavy forest around camp, and the White-bark pine (*Pinus albicaulis*). Along the upper edges of this belt, at extreme timber line, these trees become dwarfed and stunted, prostrate and often grotesque in form, wind-whipped and distorted, dying out entirely along the upper edges of the Hudsonian zone where it merges into the higher, treeless, Alpine-Arctic above. Along the timber line border is a crawling prostrate form of dwarfed and stunted Alpine juniper, also characteristic of the Hudsonian zone.

Below the Hudsonian, stretching in a vast expanse along the entire summit of the Cascades, below the higher peaks, is the great belt of Jack-pine timber (*Pinus contorta*), and this is known as the Canadian zone. This is also the belt of the Noble fir (*Abies nobilis*) and the Engelmann spruce (*Picea Englemanni*). It is the zone found typically exemplified by the jack-pine forests at Frog Camp meadow and other points along the summit at a corresponding altitude.

These three zones, Alpine-Arctic, Hudsonian, and Canadian, constitute the Boreal zones, the zones of the arctic and semi-arctic regions of far northern latitudes, and the higher snow-capped peaks and summit country of our Cascade range at high altitudes, in southern latitudes.

Below the Canadian, on the slopes of the Cascades, reaching down into the western valleys, are the vast forests of Spruce (*Pseudotsuga taxifolia*), Hemlock (*Tsuga heterophylla*) and Cedar (*Thuja plicata*), and on the eastern slopes are the magnificent forests of Yellow pine (*Pinus ponderosa*). These belts constitute the life zone which we call Transition, but with this zone we are not concerned, so let us go back to the bird and animal life of the Boreal regions above, below, and around Camp Riley.

The Alpine-Arctic fields of snow and ice are inhabited by but a single bird, the Hepburn rosy finch, or Leucosticte (*Leucosticte tephrocotis littoralis*). In the ledges of lava, jutting from the snow fields far above timber line, this bird makes its home. It wanders widely through eastern Oregon in winter, but only on the Middle Sister has it ever been found nesting in this state. Three years ago we found it nesting in the lava ledges near the summit of the Middle Sister, and again this summer I found the bird in its arctic haunts. One was in a lava ledge near the extreme summit of the Middle Sister, and another was watched as it flitted in and out of crack and cranny in the edge of the Big Crevasse, near the upper end of the Collier glacier, between the North and Middle Sisters. This bird is doubtless to be found at other points at a corresponding altitude, but so far authentic summer records from other arctic regions in Oregon are lacking.

Occasionally other forms wander up into this higher zone, to the regions of ice and snow, often to meet a tragic end. This summer, in a lava ledge near the summit of the Middle Sister, we found the skeleton of a skunk, and his story was easily told. During the preceding summer he had wandered up the mountain, far above his realm, and there, possibly crippled, weak from hunger, or simply lingering too long on the summits, he was caught in a storm, buried in the snow, and perished.

Again, we read the story of another of nature's tragedies in the crater of the South Sister. A tiny Audubon warbler, leaving the protection and sheltering warmth of the forested slopes below, had flown up and up the mountain side, till he found himself, possibly at nightfall, in the crater at the utmost summit. Weary and exhausted, perhaps, from his long flight upwards, he settled himself in the snow to rest. The warmth from his body thawed the icy crystals around him, and formed a little cup-like depression into which his body gradually settled, and there he slept, and there, his tiny feet drawn in among his feathers, his head drawn back between his shouldrers, in an attitude of perfect peace and slumber, we found him—frozen.

On the lower slopes of the higher peaks, where the hemlock forests and white-bark pines form the Hudsonian zones, alpine birds may be found in abundance. Here are Clarke nutcrackers, among the largest of the mountain birds, conspicuous by their harsh, noisy calls, and the striking black and white plumage of their wings. Feeding on the soft green cones of the white-bark pines, the gray feathers of their heads and throats are often stained a brilliant purple by the sap from the cones. Here, too, flocks of crossbills feed in the conifers, wrenching the seeds from the cones with their peculiar crossed mandibles which nature has so admirably adapted to this purpose. This too is the true home of the Townsend solitaire, most beautiful songster, and yet possibly the least known, of the West. From some lofty tree or tall dead snag in the wild fastnesses of his mountain home, he pours forth his wondrous song where there are but few to hear. Varied thrushes (or Alaska robins) may be found in the deeper recesses of this belt of alpine timber, though very rarely. Most of these birds, which come south in such vast numbers during the winter, return north in spring to nest, but a few remain with us and these seek seclusion in the dark sombre forests of the higher mountains. Here they may be found feeding their young in some bright sunny mountain meadow or calling out in their shrill mysterious tones from some deep ravine. From the lower edges of this Hudsonian belt of alpine timber, down through the great belt of jack-pines and scattering firs which go to comprise the Canadian zone, bird life may always be found in abundance. Here large flocks of pine siskins are always on hand as they nest everywhere throughout this region, and the same may be said of the Audubon warblers, mountain chickadees, robins, red-breasted nuthatches, chipping sparrows, stellar jays and Oregon jays. The latter are probably the most characteristic birds of the summit, and their local name of "summit bird" is indeed appropriate. They are the first to discover camp, and in their greed for plunder will often enter the tents and pilfer from every nook and corner. In the winter months, when snow lies deep over all this summit country, these birds become a source of endless trouble to the high mountain trappers, robbing the bait from the trap sets, and often springing the traps. At this season of the year, driven by hunger they become very gentle, and will eat crumbs from the hand of any one who will feed them. Hermit thrushes may be found also throughout this summit country, but very rarely, for they are wild and shy, and seek seclusion in the deep ravines and heavy forests. All through these belts of heavy timber along the summit, woodpeckers of various species are more or less plentiful. On the lower slopes red-shafted flickers, Harris woodpeckers, and red-breasted sapsuckers are fairly abundant. In the timber on the higher slopes may be found a few Williamson sapsuckers, alpine three-toed woodpeckers, and very rarely a large flame-crested pileated woodpecker, or cock-of-the-woods, as he is often called. Game birds are rare at these high altitudes. Mountain quail seldom range above the jack-pine Canadian belt, and the same is true of the ruffed grouse, or native pheasant, but at certain seasons of the year sooty grouse (blue grouse, or hooters), may be very abundant along the summit.

Birds of prey are often seen, and those most in evidence are red-tailed hawks, as they wheel and circle high in the air. Now and then goshawks may be seen along the higher ridges, or a turkey vulture circling slowly high overhead. Cooper and sharp-shinned hawks may often be encountered among the heavy timber, where they are a source of constant terror to the smaller birds. The owls are represented only by the dusky horned owl, the largest of the tribe, found everywhere in the mountains.

In the dusk of evening, nighthawks may be seen sailing and swooping back and forth over the meadows, in their endless quest of insect prey. Along the lava ledges of the higher ridges, and the edges of the meadows, one may be so fortunate as to find the true white-crowned sparrows, but very rarely. Hummingbirds range up the mountain sides to the limit of trees, and may often be seen over glacier and snowfield far above timber line. The commonest one, and probably the only one

present, is the rufous hummingbird, though the Calliope hummer, a typical high mountain, alpine bird, found at many points along the crest of the Cascades, may be seen in the Three Sisters regions.

Mammals are plentiful in the high Cascades. During a normal summer the meadows are fairly alive with alpine chipmunks, golden-mantled ground squirrels, and gophers. The rank growth of grass, flowers, and other alpine vegetation in the meadows harbors an abundance of meadow mice, jumping mice, white-footed mice, shrews, water-shrews, and other small, night-roving mammals which are, as a rule, overlooked by any save the most observant. One of the most interesting mammals of the high mountain regions is the cony, or pika, or Little Chief hare, as he is variously called. This little beast spends the greater part of his life beneath the lava, coming out for a short period during the summer months. He is the very soul of industry. No sooner does the snow melt than the little fellow is out looking for the first tender shoots of alpine plants or grass. These are cut and piled in the sun to dry. As soon as this "hay" is thoroughly cured, it is carried beneath the lava ledges and stored for use during the long months of winter.

Marmots, cousins of the woodchuck, and often so called themselves, may be found in any of the large lava flows along the summit, where they forage through the summer, and spend the long months of winter in hibernation in the depths of the snow-covered lava. Porcupines, with their many good points, may also be found throughout this region, especially in the jack-pine forests, where they feed on the bark, especially of the saplings and younger trees.

Oregon is justly noted for the wide variety of her products, and not the least of these is the splendid quality of her fine furs. The summit country in the region of the Three Sisters is the scene of extensive trapping during the dead of winter, when small mammals have sought safety in hibernation, and when most of the birds, save a few hardy forms, have been driven to milder climates, and when snow, ten to twenty or more feet in depth, covers the summit for mile upon mile in an unbroken blanket of white. The streams produce beautiful mink and otter, their glossy and lustrous pelts cherished by all lovers of fine fur. The forests along the crest of the range furnish excellent marten, and some fisher, both of these being very valuable fur. This, too, is the home of the snowshoe rabbit, which, like the Arctic hare of the north, turns pure white in winter, to resemble more closely the snow in which he lives. Here too are found weasels in some abundance, and these, like the rabbits, turn pure white in winter and are then known to the wearers of fur as ermine. Bob-cats, or lynx cats, are among the most abundant fur bearers of the region, and hundreds of the "kittens," as the trappers call them, are taken each year. Among the larger predatory animals, wolves still range as scattered individuals or small packs through the high mountains, and cougar are widely, though sparingly, scattered throughout the region. Both of these are a serious menace to the game of the region, destroying annually many deer and elk. Bear may be said to be plentiful still in certain remote sections of the high Cascades, and coyotes are of wide-spread and universal distribution, though by no means so abundant as in eastern Oregon. But the toll of the hunter and trapper is heavy. State regulation of the trapping season has proved necessary, and has done much to afford protection to the fur bearers, whose numbers are rapidly diminishing under the persistent persecution to which they have been subjected in recent years. Mink, otter, beaver and marten are rigidly protected by law; bear go unprotected, making their way as best they can, while government trappers wage continual warfare on such predatory beasts as cats, coyotes, cougar and wolves.

This brief account is not to be considered a complete story of the wild life of the region of Three Sisters, but simply an introduction to some of the birds and mammals of the high Cascades as one would find them in their mountain haunts through winter and summer. It is to be hoped that when the Mazamas again visit the scenes of this summer's outing, they will find birds and mammals as plentiful as they always are in the high mountains, when the summer is warm and sunny, and weather conditions are more favorable than they were throughout the summer which has just passed.

Think of those wooded spaces,
Think of the campfire's cheer!
The sound, sweet sleep, the lisp
Of the leaves in the wind, the crisp
And cleanly smell of the pines.

—*Bliss Carmen.*

Birds of River, Forest and Sky

By Florence Merriam Bailey

In recalling some delightful hours once spent among the birds of McKenzie Bridge in the heart of the Cascade range, pictures of three birds, the water ouzel, the nighthawk, and the winter wren, come most vividly to mind; for they are creatures of the roaring, sparkling, glacial river, of the still, dark, coniferous forest, and above the encircling mountains, of the peaceful sunset sky.

An Hour with the Dipper

What rare memories of the Sierra Nevada and Rocky Mountains the name of dipper or water ouzel recalls! Memories of moving shadows along the dusky banks of cascaded forest brooks; memories of a gray, short-tailed bird standing alertly, questioningly, on a rock in the river as your pack train appears, and at your advance, with short-winged flight buzzing over the dashing spray to another rock farther down the river; memories of a small gray figure on the edge of a quiet lake in a lofty glacial amphitheater.

Good hunting grounds are offered the dippers by the McKenzie river when, after rushing down from the glaciers of the Three Sisters to the middle reaches of the Cascades, it has lost most of the glacial silt that made it milky above, and after plunging headlong over rapids, spreads out green in the level stretches where jutting rocks rise above the surface and shingly bars lie along the shoals. Here, unlike the river at the foot of the glaciers where the birds are found in summer, the ouzels' hunting grounds are never closed to them, for although the water from the icy peaks warms very slowly—giving a record of 41° in June—between rapids the current is too swift to admit of its ever freezing over.

A familiar chattering call and glimpses of a small gray form disappearing over the water now up and now down stream, together with an occasional whitewashed rock along the bank, has told us of the presence of this lovely bird which adds a charming touch of life to the wild mountain torrents of the west. A devoted habitue of the river who had wandered in happy exploration far up its forested banks told of repeatedly finding the ouzel near a cabin overlooking the river, and led me up along a woods trail to the spot. A long bar at a turn of the river was said by the friendly observers on the high bank above to be the favorite feeding ground of a pair of the birds, and whitened stones at the end of the shingle attested the fact. So, sitting down in sight of the spot, we prepared to watch the birds dine.

After waiting a long time for their appearance, my friend suddenly pointed to the water on the edge of the bar, where a charming little creature, looking very bluish gray in the strong sunlight, with motionless wings was daintily paddling about in the shallow water as much at ease as if it had been a web-footed duck. Memories of ouzels that hunted and swam among the cascades of Squaw creek on Mt. Shasta returned to me at the sight. In the strong sunlight, gradations of color unseen in dark woods were observable, the head of the bird looking almost purplish, the tips of its wings and the short tail dark slate, while the lighter under parts were faintly scored.

When the ouzel started to swim, it would put its head under the water as if locating something, and then, quivering its wings, disappear altogether, coming up soon after with a long, black-shelled caddice fly larva, the shell of which, as we proved later, is a remarkable mosaic of minute stones. Known locally as grampus, the larva serves as bait for many of the visiting fishermen along this far-famed trout stream. When the bird brought up a grampus, it would shake the long shell till it finally broke open, and, pulling out the yellowish brown larva, quickly swallow it.

After bringing up and eating several of the larvae, the ouzel picked about in the submerged green mats that suggested sea-weed. Once it stood on a stone green with moss long enough to bring out the strong color contrast of the green and the gray. When walking about over the rocks it would make its droll little courtesies—dip, dip, dip—till you were constrained to speak its name—dipper.

When it had had a satisfying meal it flew across the river to a stone on the shaded bank, where, in terms of protective coloration, it perfectly pictured its background, for its gray upper parts disappeared in the dark shadow, the lighter shade of its breast toned in with the sun on the rock, and only its light-colored legs were left as slender sticks quite foreign to any bird-like suggestion. But when its profiled bill and head projected into the sun, the bird form was restored. When it moved to a branch hanging over the water and the sun touched up the branch, the plump gray form became a mere knob on the limb, so perfectly did it again picture its background. But when it dipped with the sun on it, on the instant the illusion was lost, the knob became a bird again—which, to my reading, explains the dipping, for how else could shadowy intangible family forms keep together?

After the ouzel's meal out on the bar it sat quietly for some time on the shaded bank, doubtless enjoying the spray that occasionally dashed over it from the small rapids. While it was resting, I had time to look about and observe its setting—over its head a band of low green alders, above, a high dark spruce wall with glimpses of mountain tops beyond, and up the river, the rushing, roaring, glacial river, patches of white rapids enclosed by converging lines of green, swiftly running water.

A metallic chatter below us made me look down just in time to see a second bird come flying in and alight in strong sunshine where, in wren-like pose, with head raised and short tail up at an angle, it stood out clear and dark. Water-wrens, the people of the camp above well called them. As we looked, the new arrival disappeared along the bank below us.

Meanwhile its mate, if as was doubtless the case they were a pair, having at last digested its grampus dinner, stretched and dipped and started out after another meal; this time hunting over the rocks along its strip of shaded shore. Back and forth it went over a short beat, hunting till it found a larva, when, flying or swimming back with it in its bill to a certain small stone that it seemed to have selected as dining-room, it would stand and shake and shake till, as I could see through my field glass, the yellowish brown larva came out of its stony case. Once, as the ouzel, in shaking, threw up its wing to keep from slipping, I caught a flash of white under wing coverts.

While we were watching the bird, a little girl from the camp above came down the bank for a pail of water, and when I urged her to be quiet exclaimed that the water-wrens paid no attention—"even when we're hollering" up at camp. Under the bank below us the child caught sight of the second dipper, which was hunting as the first one had by the bar, going down under the water in search of grampus. But as we watched it flew across, low over the white crests of the green waves, to hunt in the shade of the opposite bank, where its mate, if it were he, was getting his supper.

At one time both birds stood on the same branch, the new comer on the low swaying tip where the water was dashing up. Its mate had found so many larvae that its crop fairly bulged and it had to rest between feasts. The newcomer worked up along the bank, getting me excited over a dark object that for a few moments I fondly imagined was a mossy nest.

But although I had to leave for another year that most interesting of discoveries, I had seen one of the choice hunting grounds of the water ouzel in the heart of the Cascades. And right royal water sprites they seemed, to have chosen this wonderful mountain river for their home. For it was a rushing, roaring river with its "wild white horses" hurrying to the sea, the great brown boulders in its bed that tried to stay their course making them stamp the green depths of the swirling current and "foam and fret" till they tossed their snowy manes! What a fascinating, beautiful river, with its rushing water, its green depths, its rolling stones, its deep-voiced plunge of rocks! How the gray water sprites must love its life and sparkle! How joyously must they sing their love songs over its dashing, glistening spray!

The Glowing Nighthawk Prairie

From the upper piazza of the Log House at McKenzie Bridge we looked across the road, with its screen of low trees behind which was a mountain park, part of a strip of original prairie, now yellow with blooming St. Johnswort, and later, when the flowers had dulled, fairly twinkling with small yellow butterflies, as if the golden flowers had taken wing! At the back of the park an enclosing wall of fir carried the eye up the timbered slopes of the mountains to Horse Pasture and the bare rocky peaks of the ridge above, said to command a wonderful view of the snowy peaks of the Three Sisters at the head of the McKenzie.

Crossing the park one morning early in July, I roused a Pacific nighthawk sleeping on the ground, and the short mottled stick, unfolding long, white-banded wings, rose high in air calling *Pe-uck* and flying about with the expert tilting, rolling flight that characterizes the aeronaut. After that, at sunset, when the prairie floor glowed a dull orange the birds could be heard from the house and I often went out to watch them. Occasionally a few Vaux swifts or a passing flock of swallows served to give scale to the four large long-winged nighthawks which were said to be feeding on winged ants high in the sky, and which between times indulged in aeronautic feats of courtship display.

Back and forth over the prairie, calling continuously *Pe-wick* or *Pe-uck*, they flew, sometimes so low that I could see the white of their under parts, but generally too high to see even the wing bars, and at times so high that their long wings became mere thread lines and almost disappeared beyond the field of vision. In feeding they flew rapidly head on until presumably they came to quarry, when suddenly putting on brakes they would almost halt, and act as if snapping up insects with their widely gaping mouths.

When not absorbed in catching insects, two birds would often fly near each other in courtship play, and sometimes three flew together, as if the matter of mates were not yet settled. Frequently one of the suitors would come swooping down close to the ground or sometimes only to midair, when he spread his wings wide and the air boomed loudly through the quills—a familiar performance indicating a peculiar fondness for pyrotechnics on the part of feminine onlookers.

One sunset, while the four long-winged birds were cavorting about the sky, as I walked across the prairie floor a soft low chorus arose from little choristers, presumably Western chipping sparrows, hidden among the glowing weeds,—a chorus so subdued and sweet that it went well with the soft evening light.

And still overhead the nighthawks beat back and forth through the sky, till their breasts grew ruddy in the sunset light, till the notch of the McKenzie leading toward the Three Sisters, earlier filled with radiant cumulus clouds, softened to rose; till, as the sun lowered, the timbered ridge at the back of the park, earlier vitalized by light and shade, had dulled, and above it the gulch of the bony lava ridge had filled with shadow, while the Saddle and Baldy had flushed with rose. Back and forth they flew till the mountains themselves grew cold and there remained only rosy streaks on Castle Rock and salmon clouds in the sky above; until at last they sailed around over the glowing field with its encircling black conifers as the evening star came out clear and bright above the golden afterglow; when down the road the light of a camp fire showed in the deepening shadows, and it seemed time to leave them to their night watch in the sky.

Forest Homes of the Winter Wren

The jolly little brown Western winter wrens, with their short tipped-up tails, enliven the humid coast belt from Alaska to California and are met with at McKenzie Bridge in the heart of the Cascades, where they are as cheering as the occasional arresting red sprays of barberry.

One that we surprised when it was hunting over a fallen tree-top stopped to look at us, and when I gave a poor imitation of its *te-tib* stretched up on its wiry legs in listening attitude and then bobbed on its springs. Its droll courtesy, much like that of the dipper, would certainly help to keep families together in their shadowy haunts, though both dipper and wren have acquired such a nervous habit of bobbing that they do it when it might attract the attention of unfriendly observers.

The little brown wrens who chatter and babble and pipe so gaily were met with in some of the choicest parts of the forest. On one of the fishermen's trails along the river where stumps bearing the marks of beaver teeth led to the discovery of hemlock poles dragged down the bank for their toothsome bark, and where a brotherhood of forest giants stood with the sun slanting in on their mossy sides, we surprised a family of wrenkins swarming over the bank like so many brown bumblebees; but they were quickly suppressed and spirited out of sight by their efficient parents.

Up the river near the dippers' home the wrens were found in one of the best tracts of timber in the whole region, where the wood road wound among trees five to eight feet through, their bare trunks rising a hundred to a hundred and fifty feet to the first branch; one tree that was measured after being cut reaching a hundred and seventy-five feet to the first branch.

In this stand of timber the spaces between the great boles were filled in only with cedars, from whose smooth flat leaves the sunlight seemed to slide off, and with low deciduous maples, both the vine and the Oregon, that caught the sun in their vivid green tops. In this wonderful forest the cheery bubbling song of the winter wren which crept around over stumps and logs was very grateful, harmonizing well with the patches of vivid sunlight.

In still another part of the forest, reached by a trail from the blooming prairie park where the nighthawks boomed at sunset, the friend who had shown me the hunting grounds of the water-ouzels up the river took me to see what was known as "Uncle's Woods," a stand of timber owned by an old man, one of the most respected and best loved characters of the region.

As we wandered about among the great Sitka spruces we heard the familiar voice of the little brown woodlander. Following him across a green carpet, where at each step we sank deep in the moss, we came to a big log covered with moss and ferns leading to a tree whose great base was heavily cushioned with the brownish green moss for which these humid forests are famous. The branches were hung with bulging pockets that suggested one of the canopied winter wrens' nests found in another part of the woods, a green nest made wholly of the fresh moss except for its reinforcement of springy twiglets that made an especially good frame for the round doorway. Instinctively I started to examine the bulging pockets, but as so many offered good nesting sites I soon realized that the search might be endless, and the woods were already dusky.

Meanwhile, wherever his house was hidden, the brown mite of which I had caught only aggravating glimpses, suspiciously refused to do more than answer me from the dark recesses of the woods, retreating as if to lead me away. As I peered vainly through the shadows in his direction under the high dark conifers a low deciduous tree stood out, fairly glowing green as if it had focused all the light now entering the darkening forest. Beyond it stood green-leafed alders and maples draped heavily with the golden brown moss. A wondrous forest home the little wren had chosen for himself! As we started away and once more crossed the mossy carpet the bird, so suggestively silent before, burst out into a low, exquisite song of happiness that could but have celebrated the escape of his loved ones.

Beyond the home of the Wrens we came to the best part of "Uncle's Woods," a stand of giant cedar. "The old man says he comes to walk among them every Sunday," my friend said gently. And no wonder, for cut off as they are from all but devout lovers of the forest, their ways are ways of quietness and in their paths are peace. Well might the old man, used to listening low to the voice of nature, stand reverently at the foot of one of these giant cedars that, with straight clean bole towering skyward confronts one with its challenge: "The place whereon thou standest is holy ground."

As we turned away from the noble brotherhood in silence and followed the narrow trail back toward the edge of the forest, the voices of the woodlanders were stilled, for the dark organ pipes stood out against the quiet light of the yellow sunset afterglow.

△△△

The Electric Storm on Middle Sister

By G. W. WILDER

The official trip of the Mazamas to the summit of Middle Sister on the morning of August 12, will long be remembered by those participating in this memorable event. The electrical storm which occurred a short time after the party reached the summit and the consequent hasty descent and return to camp have furnished topics for conversation far out of the ordinary. The personal experiences related by individuals and the several press reports seem to vary greatly in many details as might be expected in an exciting adventure of this kind, although everyone agrees substantially to the main facts which occurred. Incidentally, many inquiries have arisen as to the causes and actions of electrical storms and the dangers to human life on such exposed places as the summits of high mountains.

The party, consisting of over fifty persons, left camp early in the morning and under the very able and competent guidance of the leaders all reached the summit a few minutes after noon. At that time the air was clear and a good view could be had in all directions except to the southwest where clouds were apparently forming. A slight wind was blowing from the east and the temperature was very agreeable. The party assembled at the topmost point and were quickly busied with the

Mazama book, lunches and cameras. Those who knew pointed out the several mountains of the Cascade range, the location of various cities and otherwise informed the small groups of interested listeners.

Meanwhile the clouds in the southwest were rapidly forming and advancing towards us, notwithstanding the fact that the wind, which freshened a bit, was blowing from the east. In a few minutes South Sister was shut from view, the western sky was blackened and soon an ominous grumble of distant thunder sent a little thrill of fear through the minds of the more timid ones. However, as the conditions seemed to indicate a passing flurry and all were anxious to see more of the wonderful views, the party huddled around in groups watching the big black cloud which resolved itself each moment out of the darkening mists behind. Swiftly it appeared and soon enveloped us and then came rain, hail, and wind. At moments the clouds were rolled back by the eastern wind, and one could catch glimpses of the great wheat fields of central Oregon spread out in the glorious sunshine far down below. The sun and wind gave promise of dispelling the clouds, but the next moment we were covered again with dense banks of fog. It was a battle royal between the winds and the clouds. At other times we could catch glimpses of the clouds forming all along the west side of the Cascades and advancing as if in battle array. It was an interesting sight up there watching the struggle for supremacy. Our hopes and desires were with the eastern wind and many spoke with optimism declaring that the storm would soon pass.

An occasional flash in the south or west followed by rolls of distant thunder warned us that more was coming and the prospect of an early passing of the storm seemed disappointing. Prepared and determined to weather it out for a little while, the party nestled among the rocks and waited. The lightning flashes grew closer and sharper, the thunder became more violent and then it seemed as though the whole heavens above were charged with electricity for one could hear the crackling and snapping of discharges among the clouds. Suddenly we experienced a most astounding sensation. The rocks gave forth a hissing sound from every edge and corner, the alpenstocks crackled and hummed and one felt a peculiar sensation from the finger tips and the hair. Speechless for a moment the crowd gaped, then burst out in cries of astonishment. Fear mingled with wonderment were plainly visible on the faces of all.

This phenomenon lasted about five seconds and then ended as suddenly as it began. While members of the party were recovering from the surprise and some were attempting to explain that it was a discharge of static electricity from the clouds, the phenomenon began again and continued for several seconds. This was repeated at irregular intervals and with varying degrees of intensity. Sometimes the noises became so great as to cause anxiety as to our safety and several moved to lower rocks less sheltered from the wind and hail. Many were reminded by the sound of the discharges of the operation of a wireless telegraph plant and it did seem as though someone were working a switch, so suddenly did the phenomena begin and end each time. After a particularly long and severe experience the leaders fell to discussing the advisability of returning to camp, the ladies were struggling with disheveled hair and the young men were trying to shock each other with alpenstocks when a sharp flash of lightning, followed by a loud peal of thunder in the near south, abolished all thought of remaining and the party started hurriedly down.

Six of us elected to remain, for it looked as though the clouds might be driven back, so even was the contest between the winds. Also the suddenness of the storm led us to believe that it would soon wear itself out. While waiting we could see at intervals the sun shining on the east side of the range, although the storm had traveled all along the western side far to the north. After the main party had left, those of us remaining experienced longer and more severe electrical discharges and after a half hour of rain and hail we began to lose hope of the storm's clearing. The discharges became so violent and continuous that a prickling sensation was caused on the crown of the head and our hats joined in the chorus of hissing and crackling noises. This sensation turned to a sort of headache and the ears began to ring with high pitched noises. Even at this time no sparks were visible although the corners of rocks and the ends of our fingers and the alpenstocks were closely examined. Some rocks seem to hiss more than others and the hissing noise could be traced to definite corners and sides although there seemed to be no connection between the amount of noise and the size or location of any particular rock.

We were reluctantly thinking of leaving, when a sharp crack overhead and a flash of lightning at the same instant left no doubt in our minds, and we began a hurried descent by the short route on the north side. While going down this steep and slippery trail it rained and hailed as never before, and if we had any notions of catching up with the main party they were doomed to disappointment for all were in camp long before we arrived. We have often wondered how they made such good time. Arriving at camp we learned that the stay-at-homes were not without their exciting experiences too, and the storm was made more general by the stories of charged ironware in the cook's quarters. The gyrations of Doc. Weston and his assistants have caused many a laugh and will be equally well remembered.

Since this interesting experience on the summit of Middle Sister many have wondered how often such phenomena occur and whether the danger is as great as one would naturally fear at the time. The absence of more abundant and complete data on the subject is undoubtedly due to the rare intervals when high mountains are visited by human beings. Several of the older and more experienced members of our party declared that they had never seen nor heard of an occurrence like this one, and it is rare indeed that a whole party should encounter such a storm. To those of an inquiring mind the study of such a phenomenon offers many inducements, and although our present knowledge of atmospheric electricity is meager, it is very useful in helping us to understand some of the more common of lightning discharges.

We know from experience that bodies may become electrified by friction and that when two different kinds of materials, as glass and silk, are rubbed together the electrification may be imparted to pieces of tissue paper and other light bodies. These will become charged by coming in contact with the glass, and will be repelled and otherwise show the effects of being electrified. Many interesting experiments are well known to all students of physics as well as several important laws governing electrical action.

It is thought that the particles of water vapor in the air become charged with electricity by being blown across the rocks and trees on the earth's surface and that these particles carry their charges thus gained with them up into the higher layers of the atmosphere. When the particles assemble into a cloud the charges are under a greater strain or pressure than when the particles are in the form of vapor. In this way the clouds are electrified with a considerable quantity of electricity which is stored up under a great pressure tending to escape. Ordinarily the air is a good insulator and will not allow the electricity to escape to the ground, but when subjected to too great an electrical pressure it allows a gradual escaping of the charges and at a still greater pressure it breaks down completely and becomes a first rate conductor, allowing a complete discharge from the clouds to the earth. This in brief is the present theory of lightning discharge. Dr. Lodge, of England, divides lightning discharges into two general classes; one in which the discharge takes place slowly, and the other in which it occurs abruptly. The first he calls the steady strain discharge and the second the disruptive discharge.

The steady strain discharge is the result of the clouds slowly drifting over a mountain top under conditions which allow the discharge to take place fast enough to prevent the building up of a great pressure. The disruptive discharge takes place when the conditions are such that the pressure is built up very quickly as when one cloud suddenly approaches or meets another one. In such a case the air is unable to discharge the electricity as fast as it accumulates and breaks down, resulting in a distinct flash accompanied with a peal of thunder. The steady strain discharge takes place quietly as a rule, although in many cases it produces a hissing noise such as experienced in our adventure.

On the plains and in the valleys the lightning discharges are usually of the disruptive type, for the clouds float along at a considerable distance above the ground and the intervening air prevents any gradual escape of the electricity. When clouds come near each other, however, the pressure becomes so enormous that the air then breaks down and a flash is the result. On the mountains the clouds come nearer to the earth since many of the peaks in our Cascades rise far above the level of clouds during storms, and consequently the conditions for a gradual escape of the electricity are better. Whether the mountains are always able to relieve the clouds fast enough by the steady strain method to prevent an accumulation of electricity and so prevent a violent discharge of the disruptive kind, is the much argued question among the students of science. There are evidences of violent strokes of lightning, although they are quite rare as regards the extreme tops of mountains. There are plenty of evidences of such strokes on the sides and at places farther down. Most of such evidence is in the form of accounts of persons who have witnessed the phenomena. Other evidence is found in the great heating effect which such lightning strokes produce, such as the fusing of the rocks or of mineral veins.

It will also be noticed that the average storm forms and starts operations quite high up in the atmosphere and gradually descends to the lower levels. It was this knowledge that induced the six to remain behind the main party. We hoped that the storm would soon break or settle down to lower levels and leave us above in clear air. Although we did not remain long enough, the storm did descend as was shown by the several violent flashes that occurred far down below in the valleys, some time after, while we were nearing camp. One of the effects was noticed the next day by one of the members of the party, who discovered a large tree freshly shattered by lightning.

In order to seek evidence of lightning strokes on the summit, the writer made a second ascent of Middle Sister one week later, accompanied by Mr. B. W. Griffith of Los Angeles. A careful search under the finest of weather conditions failed to reveal any conclusive evidence of lightning, although it is doubtful if one could tell from the rocks whether they had suffered from lightning strokes or not, as their natural weathering might be confused with fusing effects. It will be remembered that the rocks forming the summit are igneous in character and would not reveal such effects as readily as other kinds. The result of our trip left the matter in doubt.

Since our eventful trip, attention has been called to a Mazama record box which was formerly on the summit of this very mountain and is now on exhibition in the rooms of the Chamber of Commerce. This box is flat and made of zinc. The corners are pierced with holes and one of them shows unmistakable evidence of having been fused as if by a lightning stroke. This would indicate that violent and unsafe lightning discharges have occurred on the summit of Middle Sister and that it is not a safe place for human beings to be during a storm.

It will be seen that the Cascade range plays an important part in the weather conditions of central Oregon. The valleys between the several mountain ranges from the coast to the Cascades experience very few violent lightning strokes on account of these ranges, which keep the air rather well discharged of electricity. It is usually the great plains of the Middle West of our country where there are no mountain ranges for hundreds of miles that very violent storms are expected every season.

△△△

The Jaunt of the Four

By JOHN A. LEE

An organization is often at a loss to know what to do with its ex-presidents. Seeming to forget that they no longer are in authority they manifest at times a decided penchant for "butting in." The result is embarrassment for the existing administration. With a consciousness of this fact more or less clearly defined, four ex-presidents of the Mazamas decided that, for their summer's trip of 1916, they would not enroll themselves as members of the Club's outing but would organize a little expedition all their own. They were influenced to this decision by the further fact that, though having scaled nearly all the snow peaks in Oregon and Washington and many of those in California and British Columbia, there was a particular section of their own state that they had never visited. None had ever caught more than a passing glimpse of the many scenic attractions to be found along the Deschutes, and this region, it was agreed, was to be their main objective. Incidentally they planned to drop in upon and pay their respects to the Mazama camp at the Three Sisters. Then they would hie themselves away again before their presence could be considered in the least objectionable.

C. H. Sholes, M. W. Gorman, R. L. Glisan and the writer made up the party. Sholes has so far fallen from grace as a hiker as to have become an earnest devotee of the joys of motoring. Gorman is a tree expert of national repute, has a general knowledge of botany that is exceeded by few, and is an ardent collector of botanical specimens. Glisan, as an amateur photographer and tireless globe trotter, has in his collection, probably, more complete sets of views taken on different trips than is possessed by any other amateur in this particular line of accomplishment. The writer can boast of no special outing talent, except that he is supposed to develop a mild form of insanity whenever a trout stream is known to be in striking distance. These were the avocations of the respective members of the party, but the vocation of all for the two weeks at their disposal was to enjoy life to the fullest in the great out-of-doors.

Sleeping bags, provisions, camp equipage, pack-sacks for side trips, botanical presses, and fishing tackle were all quickly gotten together and loaded into Sholes' "Hup," and on the morning of August 10 we bade goodbye to Portland and were away. Speeding rapidly up the west side of the Willamette through Newberg, Dayton, Hopewell, Independence, Corvallis, and Eugene, thence turning eastward up the McKenzie, we made camp the first night on the banks of this delightful stream, twenty-four miles from Eugene and one hundred and sixty miles from Portland. The roads thus far were good, the "Hup" was working beautifully and no stops were made except for lunch, to add a few articles to our commissary, and to collect an occasional plant specimen that Gorman "simply had to have." On the McKenzie the first Incense cedars (*Libocedrus decurrens*) were noted, a tree that is readily distinguishable from the more common Western red cedar (*Thuja plicata*) in that the leaves on the branchlets are tipped up vertically instead of extending out horizontally as in the case of the Western red cedar. Here the two frequently were seen growing side by side. The writer gazed longingly at the pools and rapids of the McKenzie, having in mind the far famed piscatorial reputation of this stream, but the others said no, and early morning found us on our way.

Until after noon we continued along the McKenzie, making slower progress as the road was not so good, and passing through some fine groves of Oregon's most valuable and prolific tree, Douglas fir (*Pseudotsuga taxifolia*). Western hemlock (*Tsuga heterophylla*) was also common and an occasional Western yew (*Taxus brevifolia*), was observed. Crossing Lost creek at 2:30 p. m. we soon were on the heavy grade that leads up into McKenzie Pass. Steep as was the grade and rough the way, the auto never flinched and 4:30 p. m. found us in Frog Meadow, at an altitude of 4,750 feet. Here, at this sub-alpine elevation, we noted a variety of coniferous trees: Western hemlock, which farther north in the Cascades is not found so high above sea level, Noble fir (*Abies nobilis*), Lovely fir (*Abies amobalis*), Western white pine (*Pinus monticola*) and Lodgepole pine (*Pinus contorta*). This last-named tree also has a wide altitudinous range, unless, as Gorman strongly contends, a distinction is to be drawn between the *Pinus contorta* of the coast and the *Pinus contorta* of the mountains and elevated plains. The latter he would designate as *Pinus contorta murrayana* or *Pinus murrayana*, a distinction which Sudworth, however, does not recognize. Thus do great minds differ.

At this point we left the auto, made up our packs, taking provisions for four days, cached all surplus outfit and soon were "hitting the trail" for the Mazama camp. As we made elevation up the trail the tree life rapidly changed. The various species before mentioned soon gave way to Mountain hemlock (*Psuga mertensiana*), Alpine fir (*Abies lasiocarpa*) and White-bark pine (*Pinus albicaulis*), which are the three characteristic trees of the alpine regions of the Cascades. Our packs were heavy, and soon perspiration welled from every pore. By dint of steady plugging, however, we placed the lava beds behind us, clambered up Obsidian cliff through drifts of snow and hove into camp just at 7 p. m., being received with vociferous acclaim. At the camp fire that night we were each formally presented by Prexy Riley in his happy and inimitable style. So fulsome was he in extolling our supposed virtues and achievements as Mazamas that we were glad the shadows were deep so as to hide our blushes.

We will not dwell at length upon the doings of the Mazama camp during our four days sojourn there. They are fully recounted elsewhere in this number of Mazama. As members of one of the official climbing parties, we experienced the severe electric storm on the summit of the Middle Sister, when "each particular hair stood on end like quills on the fretful porcupine." One very pleasant day was consumed in botanizing in Lost Creek valley, another in ascending to the north summit of the North Sister. We regretted that time did not permit us to essay the climb to the south and somewhat higher summit of this peak, immortalized by the late and much lamented Prouty. Our bivouac among the stalwarts on "Punkin Ridge" proved to be well chosen. In the preparation of "chow" each of the four was chef, with ever ready and voluble defense of his prerogative. As a result a highly edified audience was always present at meal time "to see the animals perform."

Photographs by R. L Glisan

Upper left—Mr. Glisan and his "Dolly". Upper right—Paulina falls, below West lake. Lower left—Ranger's trail to rim of crater through lodgepole pine Lower right—Lava inferno, near road f E C

Photographs by R. L. Glisan.

Upper left—Road across lava bed beyond Frog Camp. Upper right—"The Four" leaving Mazama camp. Middle—Panorama of Paulina lakes and crater basin. Lower left—Canyon of the Deschutes. Lower right—Among yellow pines near Sisters.

It was with a feeling of genuine regret that we paid our final respects to the camp, shouldered our packs and descended to the auto at Frog Meadow.

The air of early evening was clear and crisp as we crossed McKenzie Pass. Anxious as we were to make the town of Sisters on the eastern slope before dark, we could not refrain from lingering in the pass to survey the scene about us. To the south loomed the Three Sisters, pure in their mantle of snow; to the north were Washington and Jefferson; while all between was lava, lava, lava—lava that had flowed from countless different cinder cones in streams that crossed and criss-crossed in inextricable confusion. Piled up into huge windrows, contorted and twisted into all sorts of fantastic shapes, forming huge caverns, large enough to take in a good sized house, black and forbidding in aspect, these lava beds were withal most interesting. In one place a large lava stream had encountered a large mound or butte in its flow. Separated into two branches, it swept down on either side of the butte; then the two, again uniting, formed a veritable island in a river of lava. Fascinated and awed, we were loath to depart, but the shades of night were falling and this was no place to camp.

Descending rapidly, we were soon among the yellow pines of central Oregon—Western yellow pine (*Pinus ponderosa*). Having sufficient water for cooking, we made a dry camp a few miles west of Sisters, in a vast open park, absolutely devoid of undergrowth. The deep and beautiful coloring of the tree trunks seemed intensified by the moonlight and the effect was almost spectral.

Passing through Sisters by noon, we soon were at the Metolius, a cold, crystal-clear stream that pours out of the lava from a number of great springs and becomes a mighty river in a stretch of seven miles. It is some trout stream, too. Thirty beautiful red-sides fell victims to our lure. Here we found the first Western larch (*Larix occidentalis*) observed on the trip. Locally it is known as tamarack. The leaves of many of the specimens seen had already (August 17) turned a deep yellow and were beginning to fall, for this tree, though a conifer, is annually deciduous.

After a day and a night spent on the Metolius, we returned through Sisters and our route now took us to Bend, where we went into camp on the banks of the Deschutes. As we left Sisters the yellow pine disappeared and scattering junipers—Western juniper (*Juniperus occidentalis*)—dotted the landscape, seeming to delight in a soil that was little more than lava. Along this route, however, some fine farms were noted and irrigation ditches appeared to be everywhere. During the night spent in Bend the temperature fell sharply and we snuggled deep in our sleeping bags. It was no surprise the next morning to find that ice had formed in the water pail. We could not help wondering how our friends were faring up yonder in Mazama camp. Sad to say, this frost destroyed many of the gardens in the Bend section.

Bend is the metropolis of the Deschutes country and is a thriving little city, prettily located, and has two of the largest lumber mills in the West. They cut yellow pine exclusively, obtained from the great belt of timber extending south from the town. Both above and below Bend great irrigation projects have been developed, taking their water from the Deschutes.

The forenoon of August 18 was spent in a visit to the ice cave eighteen miles southeast of Bend. This cave is not large, perhaps sixty feet long, thirty wide and fifteen high, and the floor of the cave a hundred feet below the earth's surface. The interesting thing about this and similar caves found elsewhere in the lava regions of the Pacific states is, that the ice remains through the hottest seasons and seems to renew itself, even in mid-summer, when portions have been removed. On this side jaunt we had our first accident, fracturing the crank case of our auto in fording an irrigation ditch. The repairs in Bend delayed us only a few hours, however, and by nightfall we were in camp amid the pines sixteen miles south of Bend. The sharp staccato of a coyote echoed through the forest as we dropped off into slumber and this also was our matin call.

Planning an early start, the "Hup" went on a strike and refused to budge. It was probably frozen up, for the night was even colder than the one spent in Bend. Oregon jays, perched on limbs nearby, guyed us as Sholes and the writer tinkered at the auto. As we were about to despair, the engine finally relented and again we were off.

We were now bound for Paulina mountain, which, as Prexy Riley would say, was to be the *piece-de-resistance* in our feast of sight-seeing on this portion of the trip, with perhaps a day to be spent in fishing on the upper reaches of the Deschutes. The substantial was first to be partaken of, as is proper in any well ordered meal, but we missed the turn off road to Paulina and brought up in LaPine, thirty-three miles south of Bend. We were now only eight miles from Pringle Falls, a point on the main branch of the Deschutes held in high esteem by the disciples of Izaak Walton. So we decided to reverse the order of our scenic repast and first have the dessert. A slight indisposition on the part of Glisan, causing him to feel like delaying the strenuous pull up to Paulina, coupled with a little diplomatic maneuvering on the part of the writer, influenced to this decision. Hence, to Pringle Falls we went. Here, as along the whole route from Bend, we were in the forest, yellow pine for the most part but occasional stretches where the lodgepole pine prevailed. The fishing was a little disappointing, as the season was too late for the best results on the Deschutes, though Glisan became the proud possessor of a seven-pound Dolly Varden. Just how this prize came into his possession is another story and one we will leave to him to relate.

After spending one night at Pringle Falls and the whole party now feeling fit as fiddles, we started for Paulina lakes and mountains. Returning through LaPine and reaching it at noon, we varied our camp fare by partaking of a splendid home-cooked luncheon at a modest restaurant kept by a good housewife. Then the climb up to Paulina began, following a route over which we were directed at LaPine. The road was narrow, just a way cut through the pines, but the grade, while very steep, was generally even, and Sholes negotiated the sharp turns with the hand of a master. Finding that our gasoline was getting low, we stopped just short of the lakes and pitched camp on the brink of the canyon above Paulina creek, two hours out of LaPine.

A book might be written of Paulina, of the mountain proper and the view from its summit, of its lakes, hot springs, flora and peculiar geologic formations. For though hitherto little advertised, it is properly a rival, among the scenic wonders of Oregon, of Mt. Mazama and Crater lake. Only a few of its prominent features can here be considered.

Paulina mountains lie about thirty miles southeast of Bend. They are a part of no chain or system but stand out alone. The summit is a huge crater, circular in shape, some six miles in diameter from rim to rim and hence nearly twenty miles in circumference—almost equal in area to the crater of Mt. Mazama or Crater lake. The rim rises from one thousand to two thousand feet above the floor of the crater. Unlike the rim of Crater lake, however, it has been broken away for an interval on the west and its inner as well as outer slopes are generally clothed with timber. The highest point on the rim, which has been named Paulina Peak, is on the south and has an elevation of 8,475 feet. It is used by the Forest Service as a fire observation station, and a cabin has been provided for the observer. Extending down into the crater from Paulina Peak is a well defined lava stream, especially interesting from the fact that Professor Russell, of the U. S. Geological Survey, pronounces it the most recent lava flow in the United States. Within this lava stream is a dike of obsidian, much resembling the obsidian at the base of the Middle Sister, where the Mazamas were encamped.

On the floor of the crater are two beautiful lakes, both of great depth. The name Paulina lake has been given to one and East lake to the other. They vary little in size and each is approximately two miles in diameter at its widest point. They are a short distance apart, but are not connected, at least not by any surface flow. Paulina creek is the outlet of the former and plunges over a splendid fall just below the lake. It flows westerly into the East Fork of the Deschutes between Bend and LaPine. Four years ago both of these lakes were stocked with steel-head trout. They have appeared to thrive and already have attained a weight, some of them, of as much as fifteen pounds.

Our party spent two days at Paulina. The first morning we were there, we climbed Paulina Peak to obtain the view, which had been described to us in glowing terms. We were not disappointed. All of central Oregon lay at our feet. To the north, east, and south extended the high, treeless plateau of this section of the state. Silver lake could be seen and Steens Mountains showed in the distance. Looking westward, the Cascades stretched out through three great states—Shasta, McLoughlin or Pitt, Mazama (with its crater plainly visible, though the surface of Crater lake was hidden), Scott, Thielsen, Diamond Peak, Maiden Peak, Bachelor Butte, Broken Top, The Three Sisters, Washington, Three Fingered Jack, Jefferson, Hood and Adams, seventeen snow peaks, all clear-cut on the horizon. Fred Childers, the ranger, was all hospitality, and full of information. Also, during the few hours that we were his guests, he noted and reported by phone to the central station two different forest fires, which goes to show the great utility of maintaining such observation stations.

In the afternoon Glisan and the writer tried trolling on the lakes, but the trout were sluggish from spawning and refused to be lured. Despairing of success, Glisan returned to camp while the writer decided to investigate the creek below the falls. Zip! Bing! Almost at the first cast a fourteen inch rainbow rose to the fly and was duly landed. Three more beauties furnished trout enough for supper. The angler's mania now possessed him, and the next forenoon, while the others were paying a visit to the hot springs at East lake, he determined to drink deep of the joys of this, the most completely soul-absorbing of all outdoor sports. The conditions were ideal—a cold, clear stream, not too small or too large, descending in a succession of pools and rapids, its banks free from undergrowth, and, as he knew, the fish were there. Temptation, as it is, to tell how this big one was landed and that one lost (the largest, of course), he will refrain and be content in saying that twenty-two fine fat rainbows were the product of the half-day's sport.

Amid the distraction of his quest for trout the writer was able to note some beautiful specimens of White fir (*Abies concolor*) growing on the banks of Paulina creek. Gorman had already discovered one other specimen, which was the first that any of us had ever seen. This is a very different tree from the White fir of the lowlands, (*Abies grandis*) or Grand fir. The bark of the former is deeply furrowed and in this respect its trunk bears a resemblance to Douglas fir.

The period of our outing had now almost reached its limit and likewise has the space allotted for its recital. After a trout feast such as Lucullus might have envied, reluctantly we started on our homeward way. Bend was reached by nightfall and camp pitched under the junipers by the Deschutes, where we had bivouacked before.

Resuming our journey the next morning, we followed down the Deschutes, crossing the river twice, and passing, during the day, through the towns of Redmond, Culver, Metolius and Madras. From Redmond north the country is open and for the most part rolling, with little natural growth except sagebrush and juniper. Mt. Jefferson is the most conspicuous object in the landscape and is an impressive sight as viewed from almost any point along this route. There is some fine farm land in this section, notably about Redmond, Culver, and on what is known as "the plains," lying northwest of Madras. In some places irrigation is resorted to, but dry farming is more common. The farmers were busy harvesting and the crops appeared to be good.

This country impressed us visitors from the Webfoot section as having one serious drawback. Water can be obtained in many places only by boring to great depths or by hauling from considerable distances. In the beautiful plains section, level as a floor and checker-boarded with roads at each mile interval, a long barrel-shaped water tank set on wheels was noted in almost every farm yard. It struck us that to remedy this situation, a pipe line system might be brought down from a point farther up the Deschutes.

We were now making for the Mecca crossing of the Deschutes and, as we stopped to inquire the way, the famous Mecca grade would be referred to in awesome tones. We could well understand this feeling as we approached the head of the grade and gazed down into the frightful canyon of the Deschutes. There was the river, a full eleven hundred feet below us, and it looked as though we might throw a stone into its current. Three miles of descent with careful driving placed us safely at the foot of the grade and we went into camp at Mecca, just opposite the Warm Springs Indian Reservation. Numerous Indians were gathered about the store which, with the railroad station and warehouse, constitutes the sole enterprise of the place. The river here, augmented by Crooked river and the Metolius, is a swift and powerful stream. The fishing was tried but with slight success. The stratification in the sides of the canyon is of decided interest. One is indeed lacking in imagination who cannot picture the successive and mighty changes of nature that served to pile up the stratum upon stratum of lava, silt and other formations exposed in the walls of this wonderful canyon, through which the river has gradually worn its way.

We were now on the last leg of our homeward journey. A day spent in the hard and tortuous climb up through the reservation and on up the slope of the Cascades, a night which we hoped to spend at Government Camp just south of Mt. Hood, a quick descent on the west side, and our outing would be but a memory. As was the plan so was the execution, with little of incident to be recorded. Our impression of the reservation was, that it may have been well selected as an abode for the Indians from the standpoint of fish and game but poorly chosen when agriculture is considered. Two coyotes trotted away from our approach as we passed through the reservation. Glisan sought to bag them with his little "game getter," but his prowess was no match for their cunning. By noon we had crossed the reservation and come to the little town of Wapinitia, set on a high open table land of varying fertility.

The afternoon's pull up to Government Camp was full of botanic interest, as well as being a tax on the nerves of Sholes, for the road is far from good. Here on this easterly slope of the Cascades, leading up to one of the lowest passes crossing the divide, are assembled nearly all the species of coniferous trees to be found in Oregon. It seems that the low elevation of the pass (only 3,880 feet) has enabled most of the species peculiar to the humid western slope to cross over and neighbor with their cousins of the more arid belt. In addition to all of the various species already mentioned we noted the Engelmann spruce (*Picea engelmanni*) and Alaska cedar (*Chamaecyparis nootkatensis*).

After a pleasant night spent at Government Camp, where old Mount Hood looked down upon us, where the grandparents of one of our party had camped many years before in their toilsome ox-team journey across the plains, and where Glisan rounded out his set of almost eighty views taken on the trip, we descended the westerly slope and were home. We had covered 704 miles by auto, and had completed one of the most varied and interesting circuits to be made in the West.

In making frequent reference to the coniferous trees seen on the trip, it must not be thought that Oregon does not possess also many broad-leafed species or is not rich in other flora. It was thought well, however, not to burden this narrative with mention of too many forms of botanic life. It was thought, too, that such reference as has been made would be justified, or at least excused, from the fact that just now the Educational Committee of the Club, with the able help of Mr. Gorman, is conducting a course of study of the coniferous trees. Gorman collected many plants on the trip, which he is now busy classifying.

Slight reference has been made to the birds and other fauna of the sections visited. We did not go prepared to hunt, the season was too late for the best bird study, and besides there was no member of our party who felt equal to any adequate identification of the various forms that were seen. In the high mountain regions numerous Oregon jays, Clarks crows, and magpies were observed and we had the satisfaction of seeing one specimen of the famed Deschutes raven. Strange as it may seem, not a single jack-rabbit was noted on the trip.

△△△

Mt. Hood to Mt. Jefferson on Foot

By ALFRED F. PARKER

For months our "Mt. Hood to Mt. Jefferson trip" had been uppermost in our minds. Whenever any of us happened to meet, conversation invariably drifted to that subject, and by the time "next summer began to be generally known as "this summer," we had thought about it so much that every detail of our equipment and itinerary had been figured out to a nicety. How enjoyable it is to look ahead and plan, secure items of useful information here and there, decide what and what not to take, and look after all the little details which make for convenience and comfort in the wilds!

The "mobilization" was set for Saturday, August 7, 1915, and on the afternoon of that date, the six of us, amid much hilarity, strapped our packs on the outside of the auto stage, tumbled in, and headed for Government Camp, at the south base of Mt. Hood. Our party consisted of Messrs. George H. Young, Arthur D. Platt, John W. Benefiel, Charles A. Benz, Jamieson Parker, and the writer.

After a dusty but interesting ride across farming country and up the canyon of the Zigzag, we arrived at our destination at about sunset. Here we stopped only for a few moments to adjust our packs, and admire the beautiful spectacle of the upper slopes and summit of the mountain illuminated by the crimson rays of the setting sun. We then pushed on a couple of miles southeasterly to Summit House, where we pitched our first camp. It was now quite dark, but we had our fire going in a few minutes, and our broiling beefsteak soon emitted a pleasant aroma in the still night air.

By eight o'clock next morning we were on our way down the road to the south, and the day being Sunday, we passed numerous automobile parties out for the week-end. At noon we halted at Clear lake, a small sheet of water which belies its name, where we had a swim and lunched. A good road brought us a few miles farther on to the Clackamas Meadows ranger station, where the ranger and his wife received us hospitably. A telephone message for Mr. Platt had been received there from his home, requesting his immediate return to Portland. He did not know the reason, but determined to hurry back at once, and telephoned to Government Camp, asking that an automobile be sent out to meet him. Insisting on none of us accompanying him, he set out afoot and alone, and we bade him goodbye with regret. We did not learn until nearly a week later the tragic cause of his recall—the death of his sister, by drowning, at Clatsop beach.

We camped that night at a pleasant spot near the ranger station, and all of us slept soundly, as our first day, although not particularly hard, had tired us, on account of our being not yet thoroughly broken in to traveling with packs averaging thirty-five pounds in weight.

The road which we had followed terminates at this point, but an excellent trail commences; and next morning, after a conference with the ranger, we proceeded on our way. We traversed dense forest until nearly noon, when we emerged into a burn and had a fine view of some of the lower peaks along the range to the south, notably North Pinhead Peak and Olallie Butte, the latter capped with snow. The trail descended for a mile or two to a little stream, where we made our noonday halt. We then began a long gradual ascent which brought us to the summit of the ridge which unites North Pinhead and West Pinhead peaks. The day was warm, and as the forest at this point had been entirely destroyed by fire, the sun beat down upon us unmercifully. It seemed as if we never would cease to ascend, but we kept going until at last we reached the crest of the ridge, where we had a fine view of Mt. Hood to the north. Then, as we began to descend from the divide, a shout from the one who happened to be in the lead called our attention to our first view of Mt. Jefferson. It seemed surprisingly near already, with its clear-cut outlines illuminated by the late afternoon sun against a background of deep blue sky. The sight put new life into us, and we pushed on a few miles further to Lemiti creek, our third camp. No water was in sight, the creek being practically dry, but we dug a hole among the boulders in its bed and thus obtained enough for our limited needs.

We held a conference around the camp-fire that night, and decided that on the following day we would proceed only as far as Olallie meadows, a distance of about three miles, leave our packs there, and spend the day on Olallie Butte. This peak rises more than 2,500 feet

Photographs by Charles A Benz

Upper—Olallie Butte from Monon lake   Middle—Mt. Jefferson from Olallie lake
Lower—Mt Jefferson from Olallie Butte.

Photographs by Charles A. Benz

Upper—Mt. Jefferson from canyon of Whitewater creek. Middle—Summit of Mt. Jefferson, taken from an altitude of about 9000 feet. Lower—Mt Jefferson from Jefferson Park.

above the meadows to a height of 7,243 feet, and we had heard that its summit afforded one of the finest views in Oregon.

We accordingly tramped the next morning to the meadows, where we selected a pleasant camping spot and left our packs, taking with us only what we required for our lunch. We then followed a beautiful trail up the mountain in a zigzag course at an easy grade, and in a couple of hours reached the summit.

The view almost beggars description. It is as comprehensive as the panorama from Hood or Jefferson, but owing to the lower altitude of the viewpoint, has not that flattened-out appearance so noticeable from the higher peaks. The sudden sight of Mt. Jefferson, looming up as it does only a few miles southward, almost took our breath away at the first glance. It is easily the dominating feature of the landscape, with the Three Sisters and Broken Top appearing over its shoulder to the left. Westward the Cascades lay in a great rolling mass as far as the eye could reach, and northward we could trace the crest of the range to Hood and Adams, looming up on the horizon. We were so much closer to the eastern side of the divide that we seemed to look down almost perpendicularly upon the wheat fields of the Deschutes valley. The country between us and Jefferson looked enticingly beautiful, diversified as it was with numerous lovely lakes, with rocky shores. Two of these, Olallie and Monon, were of considerable size.

We spent most of the day on the summit, reveling in the view and playing in the snow, which still lay in a huge drift on the north side of the mountain. We also conversed for some time with the ranger who was stationed there on the lookout for fires. It was with much regret that we finally descended late in the afternoon, and made our camp for the night at the place we had chosen in the morning.

The part of the trail which we followed the next day was more varied in its attractions than any we had yet traversed. It led through dense forests to Olallie lake, where we halted for some time, admiring the magnificent sight of Mt. Jefferson, directly across the lake, reflected perfectly on its still surface. The trail from this point skirted the west side of Olallie and Monon lakes and then rose abruptly up a steep slope nearly a thousand feet to Breitenbush lake, a beautiful little tarn surrounded by a grassy meadow, with groves of trees scattered here and there, and altogether a most alluring camp-site. Here we had a long rest and a swim—our first for three days. It was such a restful spot that we cast many a "longing, lingering look behind" as we left it. But we had made up our minds to reach Jefferson Park that night; so about two o'clock we packed up again and resumed our journey. The trail terminating at this point, we picked our way through the open forest up a long ridge running north and south, which joins another ridge running east and west. The woods here were beautiful, with no underbrush, and with grass and flowers growing profusely underfoot. As we rose higher and higher, we gradually left almost all vegetation behind us and struck great snow fields. Our view was very similar to the one from Olallie Butte the day before. It was interesting to note the difference in weather between the eastern and western regions of the state. Over the entire western side of the range and the Willamette valley hung great masses of dark cloud, while on the eastern side the sky was blue and the sun shone brightly.

When we finally struck the highest ridge we looked down on the other side and saw Jefferson Park nestling at our feet. From where we stood it looked like just what it is—a mountain paradise—so green and park-like, with its miniature lakes and picturesque little groves. After resting here for a short time we scrambled down the thousand-foot divide to the floor of the valley, the beauty and attractiveness of which came up to our highest expectations. Our camp that night was in an ideal little nook on the north side of the valley.

A better camping spot than Jefferson Park, or Hanging Valley, as it is sometimes called, I am sure does not exist. It has all the regular requirements, in wood, water, grass and trees, and so many little lakes that each member of a party, be it ever so large, can have his own private bathroom. And wherever you go, old Jefferson towers majestically above you, with its eternal snows stretching downward almost to your feet.

Next morning we started on our "home stretch," as it were—a swing around the mountain to Pamelia lake, on the southwest side. Here we had friends encamped who, we knew, would give us a royal welcome. We had planned our final camp at this spot, on account of its accessibility to the mountain, and had shipped in from Detroit a quantity of supplies which we knew would be awaiting us. One of our favorite methods of torment had been for two of us to discuss in the presence of a third all the luxuries of our future commissary, at a time when we were subsisting for the most part on a diet of rice and bacon.

In going down to the west end of the valley we came across Mr. Owens and his wife, of Corvallis, who had been camping quite near us, without either party knowing of the other's proximity. We chatted with them for awhile, and then proceeded a short distance to the south and west, scrambling across the canyon of Whitewater creek just below the snout of its glacier. There was no trail here, and the traveling, with our packs still heavy enough, was very hard. The thick growth of underbrush was particularly exasperating, especially when we tried to hold an even elevation in working our way along the side of a steep slope. We crossed ridge after ridge, each one of which we thought would surely be the last. Finally we swung around so far to the south that we could recognize Three-fingered Jack and Grizzly Flats, and then we knew that we were nearing our destination. One more long, hard pull through the densest kind of brush, and a wild leap and slide down a steep, dusty slope brought us to the waters of Milk creek, from which it was only a few minute's walk to Pamelia lake. We were greeted by an enthusiastic reception committee, to whom we gave an account of our trip, and then the delicacies contained in our fascinating commissary demanded our attention. By that night we were comfortably settled in our new camp, after five glorious days en route.

Mr. Young was obliged to leave us on the second day after our arrival; but the rest of us lingered for over a week after his departure, loafing about and taking numerous side-trips into the wonderful surrounding country. After climbing without mishap the topmost pinnacle of Mt. Jefferson, we broke camp on August 20 and returned to Portland by way of Detroit and Albany.

Will any of us ever forget the joys of those two short weeks—the wonderful scenery and healthful exercise, and every evening, at "the end of a perfect day," that fitting climax—the camp-fire? Sitting under the glorious stars around our cheerful little blaze, with that indescribable feeling of good-fellowship which always exists among congenial companions amid such surroundings, who would ever think that he had a care or trouble on earth? As we remember such experiences, we are reminded of Service's lines:

"The freshness, the freedom, the farness—
O God! How I 'm stuck on it all!

Mazamas Among the Canadian Peaks

By W. E. STONE

The Columbia river, flowing northward from its source, reaches a point some two hundred miles above the international boundary and then, turning south, calms its turbulent glacial waters in the Arrow lakes before crossing the line into the United States. Enclosed in this "Great Bend" is found some of the grandest mountain scenery of the northwest. Exploration of this vast region has scarcely penetrated beyond a few miles on each side of the Canadian Pacific Railway until recent years. In the north the Sir Sandford region has been recently visited and its wonders made known through the work of Palmer and his colleagues. In the south the first climbs in what is known as the Purcell range were made only six years ago and while the more accessible valleys of this region have now been visited, a great part of the country remains unexplored and scores of first class virgin peaks invite the mountaineer.

In the summer of 1915 the writer accompanied by Mrs. Stone visited this country for the first time and with those intrepid climbers, Mr. and Mrs. A. H. MacCarthy of Wilmer, H. O. Frind of Vancouver, and that prince of guides, Konrad Kain, penetrated into new regions and made several first ascents.*

In 1916 after a week in the annual camp of the Alpine Club near Simpson's pass, we again turned to the Purcells and an expedition was organized to go to the headwaters of Tobey creek and its tributaries, comprising one of the important feeders of the upper Columbia. In this region we expected to connect on the north with the sources of Horse Thief creek, where our climbs of the previous year had been made.

With a pack train of thirteen ponies (which proved indeed an ill-omened number for one unlucky cayuse) in charge of two Shuswap Indians, and a cook, we set out on July 28th, anticipating a two weeks' trip. The call of the mountains proved too strong, however, and twenty-five days had elapsed before we returned to our starting point. During this time we encountered no human being except of our own party; we made ten camps and reached the sources of six streams, traveling upwards of one hundred and fifty miles, some of the way over trails and occasionally cutting our way through the primeval forests. This is a game country and we frequently saw deer, goats and plentiful signs of bear. Many high points and several passes were occupied and eight

*Described in the Canadian Alpine Journal and Appalachia for 1916.

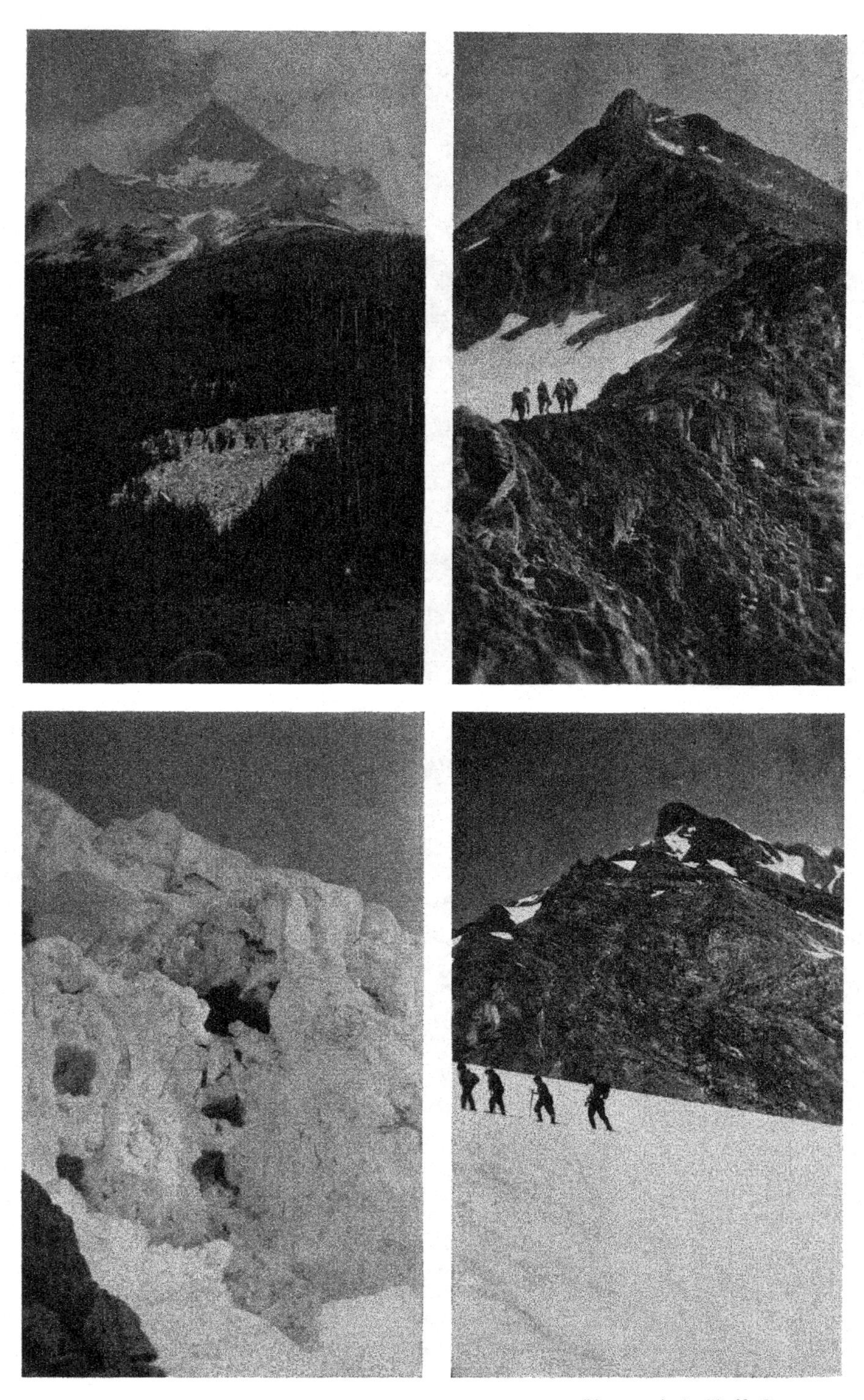

Photographs by W E Stone.

Upper left—Unnamed peak east of Jumbo Fork. Upper right—Peak 11,000 feet high east of Jumbo Fork. Lower left—Ice fall, glacier west of Glacier Peak. Lower right—Black Diamond, head

Photographs by W E Stone

Upper le —Jumbo range from the southwest Upper right—Uto and Sir Donald Lower left—Mt Monica
Lower ght Range west of north fork of Tobey creek

ascents of ten thousand feet or over were made, of which seven were virgin peaks. With the exception of the loss of one pony which rashly left the trail on a steep grass slope covered with fresh snow—a most dangerous combination—the party experienced no mishap, although about all of the fundamentals of our food supply were exhausted before we returned to the delectable fruits and vegetables of the ranch.

The trip was full of interesting and exciting experiences and from the mountaineering standpoint was unusually successful. The ascent of Mt. Nelson was made from a high camp on Nelson creek where ice formed over night and a snow squall whistled about our ears. The climb was without difficulty, at first over a small glacier and then up a long ridge of rotten rock. On account of its outlying position in the range Mt. Nelson affords a wonderfully extensive view, particularly to the eastward over the Columbia valley and the Rockies.

At the head of the north fork of Tobey creek we camped three days and from the high points in that region renewed acquaintance with all of the peaks climbed the previous year about the head of the south fork of Horse Thief creek—Jumbo, Commander, Farnham, St. Peter, Delphine, Spearhead and Peacock. One day's climb at this station was memorable. Starting out to traverse a range bounding the west side of the valley and including three peaks nine to ten thousand feet in altitude, we found ourselves at 4:30 p. m. on the summit of the farthest and highest point. To retrace our route during daylight hours was impossible while further progress along the ridge was cut off by an impassable cliff. The alternative was a descent direct to the valley down the east rock face of the peak and across an unknown glacier. For five hours we hung upon that cliff like flies, making the glacier just at dusk. An hour more in the semi-darkness brought us to the head of a steep scree and rock slope which disappeared below into impenetrable darkness. Midnight finally brought us into the glow of the welcome campfire after a day of sensational experiences.

A few days later saw us at the head of the south branch of Jumbo Fork, where we camped long enough to make four ascents and to explore thoroughly the surrounding region. The outstanding event at this station was a back-packing expedition over the divide to the headwaters of a branch of Glacier creek, beyond which a splendid snow peak had attracted our attention on numerous climbs during the past two years. From our bivouac the mountain stood up against the western sky, a beautiful snow-white mass at the head of a great snow field from which three glaciers descended to the valley at our feet. Four hours were required to cross the glacier working our way up a convenient ridge past a wonderful ice fall on to the snow field, where, just as we arrived, a thunder-storm with snow enveloped us. Having taken our bearings as a precaution, we kept on, although unable to see more than a few feet ahead, and at the end of an hour the clouds broke and revealed before us the object of our quest, a tent-shaped peak completely covered with snow. The ascent was made up a sharp corniced snow ridge at the north, the barometer indicating a little over eleven thousand feet. Traversing the mountain, we found time to climb a bold rock peak to the south of over ten thousand feet. Here we encountered another snow storm with freezing wind and were glad to avail ourselves of a long glissade down to the snow field and then to retreat to the valley where heavy packs were waiting to be carried over an eight thousand foot pass to the base camp.

From another camp at the head of the north branch of Jumbo Fork we climbed to the snow pass at the north overlooking Star Bird glacier and Lake Maye with a phalanx of great peaks (Monica, Bruce, Jumbo and others unnamed) arising on all sides. The first ascent of a fine pyramidal peak on the east side of the valley, a part of the Jumbo range, with an altitude of over eleven thousand feet was accomplished.

Finally we reached the head of Tobey creek and Wells pass where another splendid range of high snow mountains challenged our efforts. But the uninterrupted spell of fine weather broke and after a steady storm of two days which left the whole country buried in snow far below the timber line, we realized that climbing was impossible for a time and in a three days march covered the distance back to the beautiful MacCarthy ranch.

After more than three weeks on the trail it was a great contrast to enjoy the comforts of Glacier Hotel in the Selkirks for a few days and with Mrs. Stone, to make the ascent of Sir Donald by the northwest arete, regarded as one of the "sportiest" climbs in the Canadian mountains. Spectacular and interesting it is but not so difficult as many others, and, having accomplished it as a climax of the season's efforts, the next day saw us homeward bound.

The Mountain's Boast

By Charles H. Sholes

I lift my peak to the sun, I challenge its fire;
    To hurricane's fury and roar I never have bowed.
The terrors of night may appal, I lift my head higher,
    And gaze o'er the world more kingly and proud.

Upreared in dim ages, sculptur'd through seasons and years,
    I'm monarch of clouds and despoil them with gales;
Rejoice in the flight of the night-blooming spheres,
    While thunder below wild storms in the vales.

Implanted am I on the rock-based ribs of earth,
    Far down where creation's ceaseless life-throbs jar·
When restive I rumble and roar in my mirth,
    Till incense curls upward from peak unto star.

The glory and grandeur of empire are centered in me;
    Man's soul I inspire wherever his vision is clear;
Whom my majesty awes from vain-glory is free,
    And heroes are born where my pinnacles rear.

Eons of creation are preserved in me,
    While man's epitaphs are as tracings on snow
Dissolved by my breath, o'erwhelmed by the sea,
    Vain records outworn by swift waters that flow.

Capitalizing Scenery

Scenic Appreciation by the Nation and by the Individual

By NATHAN A. BOWERS

Jules Martin, a figure prominent in the development of British Columbia since the days of the Hudson's Bay Company factors, recently made his first visit to Europe. On his return he was eagerly questioned as to how he compared the famous Alps with the Canadian Rockies. His reply was a most indifferent negation. He hadn't even seen any mountains over there.

"But, surely, you were at Martigny and Tasch—and you saw not those mountains—ah! those most wonderful mountains?"

Oh, yes," he said, "now that you speak of it, there was some rising ground thereabout, but all of it could be put in one of our valleys. Why, man, there just can't be nothing that could hold a candle to the Selkirks and the Rockies."

Old Jules was so glad to be home again that he could not be quite reasonable in making such comparisons. But in his view, absurdity though it was, there lay a certain note of truth that is destined, in time, to be native-born as a sentiment in the heart of every true American. Other countries have known their snowy heights longer than we have known ours and they have surpassed us in the appreciation of the mountains. Indeed, their loyalty has attained world-wide fame. But we have, in our own right, mountain ranges that inspire a love and a loyalty second to none on earth and the spirit here and in Canada should be—will be—such that we shall not take second place in the appreciation of our "woods and templed hills." This will come to be true of our nation as a people because we are awakening to the possibilities that have heretofore remained unknown and undeveloped.

The evidence in federal policy is the opening up and exploiting of national parks; with individuals it is the rate at which love of the outdoors and action afield are taking precedence over passive pleasures and pastimes indoors. Always there has been more of the outdoor spirit in the west, but with the progress of development America is becoming more closely knit. There is more of the west in the east and more of the east in the west than ever before.

The National Aspect

Our national parks, having always been directly under the jurisdiction of Congress, have long been looked upon as a liability and treated almost as an evil incidental to national affairs. Such appropriations as have been allotted to them were spent without explanation of the return on the investment and there was a popular feeling that it was money squandered. The parks were inaccessible; there were no accommodations or transportation facilities and naturally enough their attractions were unknown. The appropriations for parks were too often in proportion to the influence of those who represented the district—or, rather, of those who represented the voting power of the district.

Then a keen-minded economist pointed out that Swiss scenery had been so advertised that it attracted tourist trade amounting to $250,-000,000 annually. About the same time the general passenger agent of one of our large railway systems estimated that the sum of $500,-000,000 was annually spent abroad by American tourists.

There were a few Americans who knew that our scenery is not in-inferior to that found in Europe. They knew that with proper development and advertisement the natural attractions of our parks could be made to divert much of this overseas tourist tide into our western states. So the slogan, "See America First," went forth and earnest nature lovers urged a sane policy of handling national park affairs.

This movement met with prompt response. Last year Mr. Lane, Secretary of the Interior, appointed a General Superintendent of National Parks and told him to "do things." This man took hold in earnest. He estimated that within a very few years an annual total of at least $50,000,000 could be saved to this country if the parks were properly opened up and adequately advertised. It was pointed out that the expenditures for such development would be good business ventures because the funds would be used chiefly for substantial improvements, such as roads, bridges, trails and chalets. Also concessions could be granted which would, under government supervision, be at least self-supporting. So the policy was accepted as good, and the national parks suddenly ceased to be liabilities and became assets.

Under the new policy, roads and trails are being built that will make the points of interest accessible, and an entirely new plan of operating concessions within the parks is being inaugurated. Heretofore a hotel keeper, for example, could get only a short term lease and on this basis of course he could not afford to spend much on improvements. Now lessees are to be given long term contracts and will be required to construct only such chalets and hotels as shall be designed in accordance with the general plans of the Superintendent of National Parks. These permits are to be issued for a twenty-year term at the end of which time the chalets will become the property of the government. Meantime the concessionaires are to pay the government fifty per cent of their net revenues. The books are to be at all times open to government inspectors to insure fair play.

An important feature of this scheme is that the lessee becomes virtually a co-partner with the government. When financial or operating difficulties arise it will be mutually desirable for lessee and government agent to study the matter together and co-operate in finding some solution. This plan eliminates at the outset the source of trouble with lessee which is now most common, and at the same time it places at the service of each, through the government agents, the benefits of experience at every other concession in the park system. This will not only make for economy and profit, but will insure the visitor the very best service feasible and an opportunity to see more of the park than would be possible with a less efficient arrangement.

National Parks of the United States

Name	Location	Area in Square Miles	Roads and Trails	Hotel Accommodat'ns
Sequoia Park..........	California....	2,525	....	....
General Grant Park....	California....	40	....	Several hotels
Yosemite Park	California....	1,124	75 miles r'ds. 700 mi. trails	For 3,500 guests
Crater Lake Park......	Oregon	159,360	....	....
Mt. Rainier Park	Washington ..	324	34 mi. roads	Several hotels
Glacier Park	Montana	1,400	....	....
Yellowstone Park......	Wyoming	33,480	Sufficient r'ds and trails	72 hotels
Mesa Verde Park	Colorado.....	640	....	....
Hot Springs Reservat'n.	Arkansas	14	....	Sufficient
Platt Park............	Oklahoma....	47	....	....
Wind Cave Park.......	South Dakota	312	....	....
Rocky Mountain	Colorado.....	358	....	Several hotels

The first construction work was undertaken in Yosemite Valley in California. The first chalet, a $150,000 structure, is now nearing completion on the floor of the valley and a smaller hotel is under way up on the heights at Glacier Point. Other mountain inns are to be built in this park at the rate of three each year until the chain of twenty-three is complete. About seventy-five miles of road will be constructed to bring the park up to the standard of accessibility.

The success of the first step in California is being watched closely so that later construction may profit by the first venture on the new basis. Several inns and about seventy-five miles of road will be required in Mt. Rainier Park, while Glacier Park requires forty-three inns and sixty miles of road. If success attends the present plan to extend Sequoia Park so as to include Mt. Whitney and the Kings River canyon, this enlarged area will be allotted forty-seven inns and an extensive road system.

Thus elaborate plans have been developed and the first work to be carried out according to the new scheme is now well under way. It is very desirable that this program so well laid out should have the support of the next Congress. Each of us can help by evincing interest and helping to advertise the parks. If there is a popular demand there is no doubt that full development will follow.

The Viewpoint of the Individual

So much for the change in our attitude as a nation. As for the individual each is "a law unto himself" for rarely are two personalities in perfect agreement on temperamental matters. However, even in our interpretation of Nature's message there are some fundamentals which must be common ground—among Mazamas surely there is much more than a few fundamentals that will be found in common.

But to get directly at the heart of the matter, do we fully improve our opportunities; do we make the most of our association with the snow peaks and the forest silences? Are we not apt to accept the healthy exercise and the sport of outdoor life as representing practically all the value of our trips, regarding as rare occasions, or "soul feasts" those times when we really sense a message or even a meaning in Nature's revelations?

In the rut and habit of daily routine most men lose their perspective; because of too close scrutiny we fail to grasp the general scheme of the picture. To be truly broad minded we must get out of ourselves, as it were, and analyze conditions from an impersonal viewpoint. The minds of great men are said to be able to do this frequently, but most of us need some help in getting a new angle of things. Oftimes this comes to us. For example, we renew acquaintance with an old friend, find him full of ambition and fairly bubbling over with enthusiasm, and we say the meeting with him was "refreshing.' In reality it was that he brought us a new viewpoint, and the reaction upon our minds was a stimulation.

In defining "inspiration" Webster speaks of a "stimulating influence upon the intellect" and he associates this with "high artistic achievement." If our minds are capable of receiving beneficial influence of this sort let us not leave the matter wholly to chance occurrences. Rather let us go over it most thoughtfully and in seeking our inspiration aim at the highest sources. Truly in this sense we may hitch our wagon to a star. We find a certain mental stimulus in contact with humankind, but this is not the highest Occasionally, in the quiet contemplation of some phenomenon of Nature, we sense the broadening uplift of a glimpse of the divine. This is at once our invitation and our opportunity. We should be quick to realize that it is well worth while for us to study and cultivate, each according to his tendency and his capacity, the conditions that make for such glimpses. We need clearer vision—a view often enough so that it becomes an influence on character.

Our study and progress along this line is dependent altogether upon our method of thought. It is something wholly within ourselves. It is not easy to discuss it freely, and in fact so far as interchange of experience is concerned there is not a great deal to be gained thereby. Companionship is an essential to our greatest development because of the opportunities it affords—and it gives us much pleasure and comfort beside—but we cannot leave even to those nearest to us any share of the thinking that determines individuality. We do our really deep thinking alone. Alone we win and lose our greatest struggles. One of the foremost mentalists of the age has even written, "I am not alone if I read or if I write." At these times, he reasons, he is listening to or speaking to his friends. He goes on to say "but if one would be alone, let him look at the stars." It is out in the silent places that one may hear the most and there he may come to know himself best.

It is this one feature, the fact that it is a strictly personal experience, that has reduced almost to a myth the inspirational element of life in the lands that remain as God made them. Scorned and scoffed at by those of us who do not sense it, those of us who think we do cherish it secretly, without a clear understanding of whence it comes or how. So often we hear it said of some wondrous scene, "No use trying to describe that to those who haven't seen it and those who have need no description." It is largely so in reducing to words impressions of the inspirational uplift that comes from our trips in the open.

But at least we can start with those fundamentals common to many of us. On this basis we may even discuss ways and means that may help in our searching for the truth. Two essentials there are: one who looks outward and upward for that "stimulating effect on the intellect," to begin with, must be "friendly" with himself and he must be honest with himself. Next comes the attitude or frame of mind. Gauge this by two units of measurement—first, time, or rate of thought, and second, degree, or extent of familiarity with that about which the thought centers. The thought that shapes itself slowly, analytically, is the thought of the listener; it is in accord with the "passive" or receptive mind. Then, of course, one can see farther and hear more if he draws upon a larger store of knowledge and experience.

To illustrate the time and degree elements, I cite first the tourist sated with travel, who led his party through a famous art gallery almost on a run and, Baedeker in hand, looked back as they neared the exit to call out encouragingly, "Fine, we did that last mile in nine minutes. His haste excluded Rembrandt and Angelo. Again, John Muir's intimate friend has said that while others heard the music of the wind in the pines, Muir heard a song whose words swayed his mood and awakened a response within him. John Muir knew the woods.

Not much has been published that deals at all specifically with ways and means whereby the individual may capitalize scenery. Perhaps it is better so because each can best work out his own interpretation. So if the foregoing comment serves to encourage the reader in further thought on the subject on his own account it shall have achieved its highest aim—unless it may, perchance, lead to a more able discussion of the matter in a later issue of "Mazama.

Although specific comment is not common, some beautiful thoughts have been expressed in more general terms. Among these is the "Creed" of Walter J. Sears, in which he urges mankind "To seek in Nature the meaning of the infinite truth; to understand that the laws of growth are the laws of God; to believe that the melody of birds, the laughter of children, the unmeasured sacrifice of motherhood, and the ceaseless yearning of all men for a wider outlook and a nobler existence are prophetic of the perfect joy and love and life of another world; and so believing to find rest as in the shadow of a great rock against all the storms that beset us; to look out upon the silence of the starlit nights, the peace of autumn days and the solemnity of the boundless seas and feel the regnant spirit of hope, that, soothing the hurt of grief, healing the wound of wrong and calming the fever of doubt, fills the soul with the faith that transforms the shadows of earth into the spl endors o Heaven.

They beckon from their sunset domes afar,
Light's royal priesthood, the eternal hills:
Though born of earth, robed of the sky they are;
And the anointing radiance heaven distills
On their high brows, the air with glory fills.

—*Lucy Larcom.*

Hunting the American Chamois Without a Gun

By WALTER PRICHARD EATON

No guns are allowed in Glacier National Park, which was once the great hunting ground of the Blackfeet Indians, and later a part of their reservation. Although the park is but half a dozen years old, already the bighorns and the goats are coming back. The winter brings them down the slopes, and Walter Gibbs, the ranger at Many Glacier, took a photograph of a flock of 134 sheep close to the chalets last April. (It is still winter in April in the northern Rockies.) As the snow melts and the verdure begins to reappear on the slopes, they retire farther and farther up, till by tourist time you have generally to watch them with field glasses if you wish any view more intimate than that of a fly crawling on the forty-ninth story of the Woolworth tower.

By climbing yourself, however, you may sometimes get a closer view, and ultimately you can reach the ledges where the goats travel and actually find their trails—well-beaten little paths that lead over the spines of the Divide or zigzag down dizzy ledges. There is something endlessly fascinating in watching the big white Rocky mountain goats (which are really a kind of antelope, we are told) feeding on the side of a 3,000-foot precipice, which looks from the base absolutely unclimbable. The other day we sat on the shore of Iceberg lake (or on the edge of the snow field under which we were assured the lake lay) and watched a herd of twenty goats nearly at the top of Castle Ridge, a vast rock cliff which rings Iceberg lake like a Titanic stadium. The goats were tiny white specks at first, and only to be detected when in motion. Gradually, however, they worked down the face of the cliff, one behind the other, and a prospective climber could have mapped out his route by following their trail. They worked down to a big snow patch which rested on the top of the steep debris slide of shale—a characteristic of all Rocky Mountain cliffs in this region—and at that point they were near enough to us that we could distinguish the kids. There were three or four kids in the herd, which gambolled and frisked out on the snow exactly like small boys. When the herd started up, a goat chased the kids back into line, and the procession reclimbed the rock wall, by exactly the path they had taken down. We watched carefully to see, and we were able to predict, from the descent, the exact points at which they would make their switchback turnings.

But from the bottom of the cliff it was impossible to see what they could find to eat on those bare, forbidding precipices of stone, to which not even a stunted tree could cling. The only way to find out was either to climb a cliff ourselves, or else reach some summit and look over. We elected the latter method as less likely to cause bereavement in our families.

The easiest way to reach a summit in Glacier Park is to go up to the top of one of the passes, by horseback, and then follow up the shoulder to right or left. For the most part, the Continental Divide in the Park is a knife blade ridge of shale stone, from 8,000 to 9,000 feet high, with peaks that are almost rock chimneys on this blade. The passes go over at points where the blade is lowest and widest, reaching it by climbing the side of one of the numerous peninsulas which thrust out to the east. An ascent of one of these passes in July—July of 1916, at any rate—is a seven or eight-mile trip out of midsummer into early spring, with each stage of the season proclaimed by marvelous gardens of wild flowers.

As we were working up a ravine toward a pass yesterday, for instance, through dense forest at times where the trail was a path of black mud in the leaf mould, at times through open glades beside a tumbling green stream, we saw goldenrod on one side of the trail, flanked by pipsissewa and self-heal, while directly opposite was a twenty-foot square bed of deep blue annual larkspur, larger and richer in color than any we can grow from "store" seed in our own Massachusetts garden. Down in the hollow the stately blooms of the Indian basket grass, or bear grass, as it is variously called, were fading. But beside the tiny brooks which crossed the trail were masses of monkey flower, a striking plant about sixteen inches high with deep red blossoms like small petunias. It ought to live, one would suppose, in New England, at least in the hills, and would be an addition to any waterside garden. There were also pretty borders of the aster-like flea-bane and beds of vetch and others of bunchberry blossoms, like little dogwood blossoms on the ground. But the prettiest of the wood flowers were the little pink bells of the twin flowers.

Presently our attention was diverted from the flora to the fauna by the sight of a fresh beaver dam close to the trail, in a rich meadow bottom fringed with willow between the great fortress walls of the canyon. The beavers had raised the level of water nearly two feet, with a dam at least sixty feet long, and the flooded ridge above was full of their canals. But, unfortunately, no beavers were in sight.

As we began to climb, the earlier flowers appeared in greater and greater abundance. In one little hollow, a dozen feet wide and not over thirty long, we counted eighteen varieties while sitting in the saddle. There were, among them, pink spirea, tall blue false forget-me-nots, beautiful bushes of golden shrubby cinquefoil (the curse of the Berkshire farmers, who call it hardback), lavender wild onion, several stately clumps of yellow long-spurred columbine, much wild valerian, larkspur, paint-brush and head after head of the bear, or basket grass.

This peculiar grass, which grows in clumps and is so coarse and sharp the horses will not touch it, bears the most conspicuous flower in the Rocky Mountains. It sends up a fleshy stalk half an inch in diameter and from two to four feet tall. This stalk, in turn, bears a raceme of cream-white blossoms which is often a foot from base to crown, and so thickly clustered that it looks from a distance like a great white Bartlett pear, wrong end up, on a stick. The red hot poker plant is the nearest analogy in our gardens, but it lacks the delicacy of individual blossom in the cluster, and the peculiar shape. These bear grass blossoms are everywhere in the upland meadows, growing even in among the stunted spruces or under the trees. They wave their white plumes in the wind against the background of the firs like little forest armies on the march.

We ate lunch surrounded by a score at least of fat ground squirrels which became almost as tame as gray squirrels on the Common, and ate out of our hands. They are about the size of a gray squirrel, rufous and greenish-gray in color, and they are omnipresent in the northern Rockies. They sit up on their haunches and look at you, like prairie dogs, press their front paws against their stomachs, and with each pressure emit a peculiar, bird-like peeping squeal—exactly as if they were mechanical toys, with a noise-maker in the middle.

At the top of the pass we left the horses and began the ascent of the shoulder, toward a peak of piled boulders not unlike the summit cone of Washington, though much less of a dome, and about 9,000 feet high. It wasn't a long climb, but almost every foot of the way was a revelation of tiny alpine gardens in among the rock crannies. Of these small alpine plants the most beautiful was easily the purple saussurea (or so we judged it), but running it a close second was the moss campion—its little masses of pink flowers on their velvet green mat adorning both the crannies and the tops of the rocks. The true forget-me-nots, a true, pure blue, persisted a long way up, and sometimes we found shrubby cinquefoil dwarfed to a few inches in height, exactly as a spruce is dwarfed to a ground shrub. The blossoms, however, remained their full size. The beautiful pink heather also reached altitudes far above timber. Grass persisted in the crannies at the very top, and in the grass several varieties of tiny alpine plants.

Almost to the summit, too, we found the whistling marmots, big fat fellows, looking much like their eastern cousins, the woodchucks, in all but color. Their front quarters are almost cream color. These mountain fellows were very tame, letting us get near enough to photograph them, and, after we had passed on, coming out again on the same rock and resuming their sun baths.

On the top of the Divide, from the chimney we ascended, we could see plainly a game trail running along for two miles. On one side the precipice dropped down abruptly more than a thousand feet. On the other side, in the northern hollow of the curve described by the ridge, was a high glacier, mostly snow covered, with the snow coming up in places to the very ridge itself. The ridge was perhaps thirty feet wide, and the game trail ran right along the middle of it. At the farther end is ascended a considerable summit, and was dispersed in many very faint trails which led out and down upon the lower ledges. Looking over the ledge anywhere you could see a dozen possible ways down, even for a man, and you could see that the ledges were much wider than they appeared from below, many of them bearing little hidden gardens of alpine plants, moss and grasses—poor enough picking for a horse, but evidently sufficient for the goats. Goat signs were frequent, and even up here, at the top, was plentiful evidence of the sheep. From our perch we saw two goats rounding a ledge a mile or so away, and a lone sheep outlined on a snow field far below us. We also saw two of the chief foes of the goats—the bald-headed eagles. (The golden eagle is equally a foe.) Our guide told us he had found eagle's nests surrounded by the carcasses of kids and lambs, and affirmed that the eagles pierce the body and eat out nothing but the lungs.

We had a rope along, and at a favorable spot went far out on the steep glacier, which was poised on its high shelf far above the canyon below. It was still covered with snow, however, and only at the extreme lower edge had begun to develop any crevasses, so that it was no more interesting than a mere snowfield. We finally descended to the pass by a shale slide, which is the quickest way, but calls for cast iron boots. Any climbing in the Rockies, in fact, calls for cast iron boots. Not only should big hob nails invariably be worn, but the soles of the boots should be twice again as thick as you think they ought to be. Only thus can you walk in comfort over the omnipresent sharp shale stone.

As we dropped into camp in one of those beautiful upland meadows which are characteristic of the lower end of the green lakes in the glacial cirques at the head of the Rocky Mountain canyons, the shadows were filling the bottom of the vast rock amphitheatre, while the peaks stood up in full sunlight, their snowfields glistening, their strata upon strata of vari-colored ledges making a pattern of rich warm color. Two large Clarke nutcrackers (a handsome black and white bird characteristic of the high Rockies) were calling in crow-like tones, and in the firs behind our tent a hermit thrush was singing, but with less clarion timbre than in Franconia. The daylight lasted on the peaks long after the canyon meadow was in shadow, the upper ledges and snowfields and glaciers growing more and more ethereal. Finally the light faded from them, too, and the western sky alone held light, against which the ragged Divide stood up in silhouette like a stage set by Urban or Gordon Craig. The world grew still. Only a porcupine rustled near camp, and a coyote barked in the distance.

Perhaps it is only fair to add that in the morning we found that the porcupine had eaten the entire sleeve out of a sweater and consumed the better part of two stout rope halters. But he never will eat anything else. The Government cannot forbid clubs in the park.—*From the "Boston Transcript."*

Partner, remember the hills?
Those snow-crowned battlements of hills
We loved of old.
They stood so calm, inscrutable and cold,
Somehow it never seemed they cared at all
For you or me, our fortune or our fall
And yet we felt their thrall;
And ever and forever to the end
We shall not cease, my friend,
To hear their call.

—*Berton Braley.*

Photographs by Charles A Benz

Upper left—Unnamed falls on Greenleaf creek. Upper right—Twahalaski falls, Upper Multnomah creek Lower left—Wahclella falls, Tanner creek Lower right—Unnamed falls on stream east of Lindsey creek.

The Lesser Waterfalls Along the Columbia

By H. H. Riddell

The features of natural scenery along the course of the Columbia river from Portland to the east through the Cascade range, are so many and varied in form and contour, that but few have become well known to the public. The bold escarpments that overhang the river, and the peaks that rise along the rim of the canyon are known to all who have traveled along the river, as are the major waterfalls. Multnomah, Latourell and Horsetail falls are familiar to all wherever the Columbia is known. It is but few, however, that know of the many lesser falls and cataracts that cascade down the rocky beds of the many creeks and streams flowing from the highlands on the north and south to a confluence with the Columbia. These are mostly tucked away out of sight from the river and from roads and trails and in places where climbing and hard walking are necessary to find them.

Perhaps thirty or more beautiful waterfalls, each possessing a beauty all its own, are situated within easy walking distance of the banks of the river.

The clear overleap of Latourell falls calls for no description. It is the first of the waterfalls to be seen as one journeys up the river. The cliffs of basalt that form the southern wall of the Columbia gorge are extremely hard and difficult of erosion. Latourell creek has been unable to eat its way into the mountain side, and the fall is clean cut and leaps free from the wall, falling to its basin without obstruction.

Something more than a mile east of Latourell, Young creek comes dashing down in a tortuous twisted cataract into the depths of Shepperds Dell. The stream is not large, but the several drops, and white cascades in which the water plunges down the steep slopes give one an idea of the many aspects that a stream can assume when rock walls and gravity have an opportunity to work their will.

Under the highway bridge the fine Bridal Veil falls pours over a dyke of black basalt in a filmy white ribbon that readily suggests the name. It is unfortunate that the stream is diverted into flumes so that the stream bed is often dry.

Just west of Angels Rest a small creek drops over a precipice forming Coopey falls. It is a pleasing sight, especially when the stream is flowing bank-full after a storm. A half mile to the east, Dalton creek comes down from the heights in a succession of abrupt falls, invisible from the river or road, but well worth the effort necessary to see them.

Directly above Multnomah Lodge, Mist falls comes in a film of cloud-like spray from the high cliffs, wasting into a cloud of mist in its thousand-foot descent, and gathering its waters at the head of the talus slope to cascade down in a dash of foam to the river level.

It is but a short distance further to the beauteous Wahkeena that bursts full blown from under a mountain mass to shoot down in a continuous dazzling fall and cascade for hundreds of feet; the last mad dash being in sight of the highway, and forming one of the finest of its sights.

Multnomah is of course the most impressive of all the waterfalls, its height and volume combining to make it a masterpiece of nature.

Before the construction of the Larch mountain trail but little thought was given to Multnomah creek above the great fall. It had been practically inaccessible; but the new trail has made it easy to visit the upper courses of the stream. Just above the brink of the fall is a pretty cascade where the waters drop into a basin to gather themselves for the great leap into the river canyon. A short way above, a beautiful cascade is caused by a dyke of black basaltic rock that crosses the bed of the stream, and just above this the superb Twahalaskie fall is a thing of beauty. Other cascades and falls abound along the course of the stream, all of them unnamed.

A splendid fall of over three hundred feet marks the southern end of the wonderful Oneonta gorge. Here Oneonta creek makes a wild plunge into the deep recesses of the gorge, which is so narrow that the pool into which the water dashes fills the chasm from wall to wall. This fall is easily visited from the highway, a short walk along the bed of the gorge bringing one to its foot in a few minutes. Above this fall are others, not so high but of surpassing beauty, marking the wild course of the stream from its birthplace on the slopes of Palmer Peak and Larch mountain.

Horsetail creek, so called from the splendid fall which the stream makes in its final plunge to the river level, offers a bewildering maze of cataracts throughout its length from Nesmith Point to the highway bridge. These are unnamed, mostly unknown; and like the flowers of Gray's Elegy, their beauty wasted with no human eye to see or ear to hear, the water comes down in myriad fall and cascade, a lash of foam from the snow-clad summit of Nesmith to the quiet level of the Columbia.

During the winter months a number of waterfalls, fed by the copious rains that drench the crests and basins along the escarpment near Cathedral rock and to the east, add a touch of beauty that is difficult to describe, and which is absent in the summer and autumn. McCord creek is marked by the superb Elowah fall. This, like Latourell, drops sheer from a basalt rim into a basin. The overhang of the rim is such that one can pass entirely behind the column of descending water, and can find a dry seat amid the saxifrages and wild asters, from which to view the uplifts of Hamilton mountain through the mists of the fall as they drop curtain-like in front. McCord creek from its sources in Latourell Prairie lies in so steep a bed that it is a continual cascade, lashed into foam by the huge blocks of andesite over which the creek pours.

A half mile or more back in the mountains from the massive arch that spans the gorge of Moffett creek, the pretty Wahe fall drops for almost a hundred feet into a deep chasm that has been eroded into the mountain. Above this is yet a second fall, down the sheer cliff of a rock-walled canyon, that is one of the finest of the many cataracts of this region. It is inaccessible, except to the experienced mountaineer, as it requires a difficult climb along the steep sides of the canyon wall.

Wahclella falls marks the upstream limit of the trout in Tanner creek. Here the stream has eroded its way deep into the mountainside. This fall and its companion cataracts are easily visited by a trail up the creek from Bonneville. It was named by a Mazama committee after the tribal name of the village which occupied the site of Bonneville.

On Eagle creek are the Metlako falls, so called from the mythological Goddess of Salmon, and above are yet other falls unnamed and but seldom seen. The rushing torrent of the Eagle from its birthplace in Wahtum lake on the southern slopes of Chinidere is a succession of falls, cascades and cataracts.

From Eagle's crystal flood to the twin falls of Gorton creek are no falls along the river, except a winter cascade on Ruckel creek; but as one rounds the talus of Shellrock mountain and passes the lower lifts of the approaches of Mt. Defiance, a beautiful sequence of falls appears. Along Lindsey creek from the highway crossing are several falls, and farther to the east six falls of unequaled beauty drop into their several basins dashing their spray in a misty cloud. These are without names, save "Starvation." May anathema be the lot of him who imposed this malphonious and unsuitable name on this beauty spot of creation.

On the north side of the Columbia are a number of waterfalls as beautiful in their way as those that have been mentioned. On Rock creek less than a mile above Stevenson, the stream passes over a ledge of sedimentary rock, and drops in a broad sheet for almost a hundred feet. A short distance above is a similar fall of lesser height. These falls differ from the others in spreading their flow and pouring over the brink in a broad sheet, and not in a concentrated mass, as do the streams on the Oregon side.

Greenleaf creek out of its sources in the basin of Table mountain pours its floods over a rim, the broken fragments of the huge buttresses of the fabled "Bridge of the Gods," and drops two thousand feet in a quick succession of beautiful falls and cataracts. This stream, when swollen by heavy winter rains, presents a wonderful appearance, each fall a gem of purest ray, clean from the hand of the Master Artisan.

On Cedar creek in a little nook shortly above its confluence with Hamilton creek is a pretty fall that makes a long drop down a perpendicular wall of rock. Seldom seen, this beauty spot but awaits the time when a trail will be opened to render it more accessible.

Lest a horror similar to "Starvation" be perpetrated it is to be hoped that the Geographic Boards of Oregon and Washington will assert an activity that will give to each waterfall, mountain peak, and natural feature a fit and appropriate name.

With the construction of the roadway along both banks of the Columbia river and the trails that are being built by the Forest Service an increasing public interest is being manifested in the beautiful scenery of our mountains. It is a matter of increasing importance that the natural features be aptly named. Serious attention has not been given to this in the past. Inattention has caused a varied nomenclature as incongruous as it is inapt, with its reduplication of such commonplaces as "Cedar creek," "Eagle creek," and similar names. An inspection of the topographic sheet of the Mount Hood quadrangle where three "Cedar creeks," three "Bear creeks," and others appear, serves to illustrate this anomalous condition.

In our regional nomenclature place names of Indian origin are many. They are of special aptitude in most instances, and particularly to the natural objects about which hover a wealth of traditional reminiscence and mythical lore. This, it seems, should not be entirely lost or forgotten. The ancient name "Woutoulat" is preferable to Rooster rock. The one is redolent of the old time myths, and the religious symbolism that reaches back to a time anterior to the traditions, while the other serves no purpose, outside of being a mere name, the unsuitableness of which invariably calls forth the query, "Why?" The Beacon rock of Lewis and Clark sits with better grace on the great andesite monolith than does Castle rock. Multnomah falls, narrowly escaped being named after the earlier name of the creek and dubbed "Coon Creek falls."

Mazama activity has been exerted in naming several of the falls, peaks and creeks along the course of the river. It should be carried further and fit names given to the many unnamed mountains, waterfalls and creeks in our nearby mountains.

Perhaps the most noticeable change effected by Mazama influence was the substitution of Wahkeena, for Gordon falls. This singularly euphonious name came instantly into popular favor, and is now universally used. The graceful fall that marks the plunge of McCord creek to the river level was named "Elowah" after an old Indian name of the place. The falls on Tanner creek also were given the tribal name of "Wahclella," after the Indians who formerly dwelt amid the groves on the present site of Bonneville.

For a like reason the local name of "Wahe" was given to the falls on Moffett creek. The turbulent summer rivulet, and fierce winter torrent which descends in a dash of spray and foam from the steep slopes of Yeon Peak has been termed "Tumalt creek" in place of "Devils Slide" creek. Tumalt was an Indian of the Wahclella tribe who dwelt on the beach near the mouth of this creek. He rendered material aid to some of the white settlers in escaping from the savages at the Cascades massacre in 1856. He was so unfortunate as to return, and in the heat of the excitement following the defeat of the Indians by the troops, was executed as a participant in the uprising. The high rock pinnacles midway between Multnomah falls and Oneonta gorge have been named "Winema," a Lutuamian term meaning chieftainess, and applicable because of a mythical tale in which a maiden rallied her tribesmen and inflicted defeat on a band of invaders. She fell in the battle, and Talapus raised the pinnacles where she fell. The mountain between Tanner and Eagle creeks has been called Wauna Point, after the legendary bridge which was at one time thought to have spanned the Columbia at this point. "Wauneka point" is the appellation given to the mountain between McCord and Moffett creeks. It means "Place of red sunsets."

In keeping with the general plan of preserving pioneer names as well as those of Indian origin, the mountain directly south of Warrendale on the divide between the Columbia and Bull Run, has been named Nesmith Point, after Oregon's pioneer senator. The name Deadman's creek, an inappropriate and grewsome title, has been eliminated, and the pretty mountain torrent, which in less than four miles descends some 3,500 feet from the edge of the Benson plateau was renamed Ruckel creek after J. S. Ruckel, who first settled on the flat at the creek's mouth, and constructed the portage around the Cascades on the Oregon side. Shellrock mountain near the sources of Tanner and

McCord creeks was renamed Mt. Talapus after the legendary deity of the Indians. The summit southwest of Nesmith Point was named Palmer Peak, after Joel Palmer, who in 1843 was the first known person to attempt the ascent of Mt. Hood. He succeeded in reaching Crater rock. The huge mountain mass to the south and east of St. Peters Dome was named Yeon mountain, in honor of J. B. Yeon, as an expression of the public appreciation of his work in building the Columbia river highway.

To the south of Cascade Locks is a massive mountain pile rising to an altitude of 4,000 feet, and culminating in Benson plateau, which has never received a name. It is suggested that it be called Wallala mountain. It is the aboriginal name for the village that in former days occupied the site of Cascade Locks. The Indian treaties caused the removal of the remnant of this tribe to the Warm Springs reservation. With the removal of the Indians the old tribal name has been nearly lost.

Not all of the mountains adjacent to the Columbia, in the Cascade range have been named. Within the area covered by the Mt. Hood quadrangle, along the drainage slope of the river, are some twenty summits exceeding 3,500 feet in altitude, that are nameless. Of this number ten are of greater altitude than Larch mountain. Across the Columbia, the companion peak to Table mountain is without name. A dozen others lie back in the ranges to the north.

Some one has called the mountain between the forks of Herman creek, Mt. Woolyhorn. For want of anything better the members of the State Agricultural College Geologic Survey have so termed it.

The topographic work for the Troutdale quadrangle has been in progress during the past summer. It is to the interest of the people of Portland to see that appropriate names are included for all the prominent natural objects. This will insure a satisfactory map. As an example we have Vancouver point, named by Lieutenant Broughton on the first attempted exploration of the Columbia in October, 1792, as the point where he terminated his work and returned to the sea. This point, which juts prominently on the Washington side a short distance above the mouth of the Sandy, was definitely located by Mr. T. C. Elliott, George H. Himes and the writer, by aid of a photographic reproduction of a map drawn by Lewis and Clark. This point has been called Cottonwood point. Other places of historical interest have been almost forgotten and the historic or aboriginal names have disappeared, while local names that bespeak the misdirected intelligence of surveyors and map makers have obscured the work of the earlier explorers.

Two Useful Botanical Manuals

By M. W. Gorman

Our local walks have now become so well established and so important a part of our training for the more serious work of the annual outing that any aids to the interest of the members who take part are to be hailed with pleasure. The three scientific subjects to which attention is directed on these trips are geology, botany, and ornithology, and we derive pleasure and improvement from these outings in proportion to our powers of observation. While the presence of someone versed in these subjects is always of great assistance to the learner in answering questions that arise and settling matters of doubt, yet the instructions given, to be effective and permanent, should be supplemented by the use of a good manual on the particular subject studied.

On the subject of botany two very desirable manuals for our region, in addition to those mentioned in our last bulletin, have recently appeared, viz.:—*Northwest Flora by Professors T. C. Frye and G. B. Rigg.*

This book is a small quarto volume of 453 pages, written primarily to supply students of the University of Washington and incidentally of the other universities, colleges, and high schools of the northwest with a reliable textbook on the flora of this region. The area covered is identical with that of Howell's "Flora of Northwest America," viz:—Oregon, Washington, Idaho, and southwestern British Columbia.

The cryptogamous plants (ferns and fern allies) are omitted, only the true flowering plants being included, and the large scope of the work is indicated by the number of species and varieties described, viz:—3,412, of which 3,166 are native and some 246 are introduced plants, chiefly European. It is somewhat to be regretted that these introduced species are not indicated by small caps in the text as is done in "Gray's New Manual of Botany," a distinction that is of great help to the student and beginner.

Complete, concise, and very usable keys to the families, genera, and species have been compiled, a useful glossary added, and a most commendable effort has been made to give the common names of practically all the species described. The latter feature at once makes the book highly usable to the layman and general reader, and to still further recommend it in this respect, a description is given of the origin or meaning of the generic name in each case.

This is followed by a conscientious effort to indicate the distribution of each species by adding after the description, the following letters, viz:—the letter C for plants occuring in the Cascade Range, E for plants east of the Cascade Range, U for plants of southwestern Oregon, and W for plants west of the Cascade Range. That errors should occur in this feature is, perhaps, inevitable, and, as the authors are both men actively engaged in teaching, they probably never had an opportunity to collect plants in Multnomah and Columbia counties, Oregon, or in Clarke and Cowlitz counties, Washington. In any event the fact remains that numerous species credited to the region east of the Cascade Range and naturally belonging there have long since found their way through the Columbia gorge and are now well established on Hayden island and along the borders of the Columbia river in the above named counties. Among these we might mention a few, such as:—

Chenopodium botrys	Jerusalem oak
Corispermum hyssopifolium	Bug-seed
Mollugo verticillata	Carpet-weed
Saponaria officinalis	Soapwort
Lepidium apetalum	Roadside peppergrass
Convolvulus sepium	Hedge bindweed
Melissa officinalis	Garden balm
Nicotiana attenuata	Narrow-leaved tobacco
Aster frondosus	Short-rayed aster
Erigeron divergens	Spreading fleabane
Bidens vulgata	Tall beggar-ticks
Artemisia biennis	Biennial wormwood
Artemisia gnapholodes	Lobed wormwood
Artemisia dracunculoides	Linear-leaved wormwood
Artemisia Lindleyana	Lindley's wormwood

while a few species, chiefly weeds, such as the common St. John's-wort, credited to west of the Cascade range, can also be found east of it, and a very small number, such as *Euphorbia crenulata* (Wood spurge), credited to southwestern Oregon, can be found on Rock island and some other islands in the Willamette river.

That some species have been omitted is not surprising in a first edition. Among these we might mention a few introduced plants that are now fairly well established about Portland and in the Willamette valley generally, viz:—

Chenopodium glaucum	Oak-leaved goosefoot
Galenia secunda	One-sided galenia
Paronychia Chilensis	Chileno whitlow-wort
Ononis arvensis	Rest-harrow
Conium maculatum	Poison hemlock
Foeniculum vulgare	Fennel
Lysimachia Nummularia	Moneywort

Lysimachia terrestris	Bog loosestrife
Phlox paniculata	Garden phlox
Myosotis versicolor	Varied scorpion-grass
Mazus rugosus	Mud-flower
Sherardia arvensis	Field madder
Echinops Ritro	Porcupine flower
Senecio sylvaticus	Lowland groundsel

These and other omissions and corrections, together with such new species from this region as may be described in the meantime, can be correctly included in the next edition.

When a second edition is being prepared, we cannot too strongly urge upon the authors the advisability of including the cryptogams, as everyone, student and layman alike, is interested in our ferns. This is all the more necessary as Howell's Flora also omits the ferns and fern allies, the author's intention having been to include them in a second volume which it is to be regretted he did not live to complete. One other improvement might here be suggested. The importance of a good index in any book, particularly in a scientific work cannot be overestimated, and if we are not asking too much of the authors, we would suggest that in the next edition an index such as that in Robinson and Fernald's "Gray's New Manual, 7th ed." be used. The facility and time-saving in using such an index compensates the student tenfold for the trifling additional cost or added bulk of the volume.

Flora of the Northwest Coast by Professors C. V. Piper and R. Kent Beattie. This, the most recent of our manuals on northwestern botany, is a handsome octavo volume of 431 pages. The region covered includes the area west of the summit of the Cascades from the 49th parallel south to the Calapooya mountains on the southern border of Lane county, Oregon.

As the book is chiefly intended for the use of students in our high schools and colleges, the treatment given the subject is strictly scientific, but a careful perusal of its pages will show that it is also available for the use of the layman and general reader who are interested in botany. The number of technical terms used is not very great, but for the benefit of those to whom they are not familiar, a useful glossary is appended.

The total number of species and varieties described is 1,619, of which 61 are ferns and fern allies and 1,558 are flowering plants. Among the latter, 176 species are introduced plants, chiefly European, and 1,382 are natives of the Northwest. Of the above 1,558 species, 54 are trees, 100 are shrubs and 1,404 are herbaceous plants.

The keys to the families and genera are exceptionally good and as concise as the subject will permit. The generic descriptions are followed by a short key to the species in all cases except those consisting of a single species, a feature of great assistance to the student and the amateur, while the specific descriptions are models of clearness and conciseness.

The specific description of *Betula glandulosa*, "Low shrub, about one meter high," will have to be amended if *B. Hallii* of Howell is to be included therein, as specimens referred to the latter, can be found 12 to 15 feet high on Mt. Hood and Mt. Jefferson.

In the description of *Rosa gymnocarpa*, page 205, the statement is made,—"leaflets 5–9." So far as our observation goes, not a single full-grown bush of this rose occurs in this vicinity that does not show eleven leaflets to be common, and a reference to "Douglas Journal," Page 113, shows that as long ago as 1825 that keen observer states "Rosa sp., leaflets 9–11."

The authors were evidently actuated by a conscientious desire to include only such species as were positively known to occur in the region covered. To do this consistently, it became necessary in nearly all cases to describe only such plants as were collected within the above limits and to be found in the national herbarium at Washington, D. C. The result of such precaution is that about 150 species of native and probably 25 species of introduced plants that actually occur in this area have been omitted. This of course can be corrected in a second edition. The fact that such well known shrubs as:—

Cytisus scoparius	Broom
Garryia Fremontii	Quinine bush
Baccharis pilularis	Groundsel bush

such common weeds as:—

Agrostemma Githago	Corn cockle
Saponaria Vaccaria	Cow herb
Melilotus officinalis	Yellow sweet clover
Conium maculatum	Poison hemlock
Bellis perennis	Daisy

and such well known native plants as:—

Delphinium Oreganum	Oregon larkspur
Ranunculus Bolanderi	Bolander's buttercup
Therofon majus	Large-leaved saxifrage
Solanum triflorum	Cut-leaved nightshade

and many others were omitted, should be a strong incentive to botanists in western Oregon to see that full and complete collections of our plants are sent to the National herbarium in the near future.

Left—Midsummer snow scene in Mazama camp (*Photograph by N. A. Bowers.*) Middle—Lost Creek falls. (*Photograph by R. L. Glisan.*) Right—Crevasse in Lost Creek glacier, South Sister. (*Photograph by Winter Photo Co.*)

Photographs by Charles J Merten

Scenes on midwinter skiing trip, south base of Mt. Hood

The Skiing Trip

By Margaret A. Griffin

On Thursday, December 30, 1915, at half-past one o'clock, thirteen curiously clad young people were to be seen loading themselves, their packs, snowshoes and skis into the waiting automobiles in front of the Northwestern Bank Building, the home of the Mazamas. Sedate townsfolk passed by with looks of inquiry. Others paused to watch, and to them it soon became plain that these were not Eskimos—only parka-clad Mazamas, off to the hills for a holiday.

Our destination was Government Camp, which is located at the base of Mt. Hood, on the south side. We were to make as much of the journey as possible by automobile and to walk the rest of the way on snowshoes. We were soon on our way, and, being pleasure bent, made no complaints when the road grew rough and an occasional mudhole delayed our progress. What is a mudhole when broad Mazama shoulders are put to the wheel? We were in search of Winter and hailed with delight the first snow. We arrived at Rhododendron Inn about six o'clock in good spirits and ready for our ten mile walk. At Toll Gate we left the automobiles, which could not well go further on account of the snow, donned our snowshoes, shouldered our packs, and set out for Government Camp.

Many of us were inexperienced in the use of snowshoes and consequently there was some variation in speed; but with a leader ahead and a guide in the rear, each felt free to choose his own gait, and the party soon broke up into groups. The road, so well known to us in summertime, was lost to sight under many feet of snow; and familiar landmarks were either snowed under or had assumed strange disguises. Nevertheless, those ahead led us straight through to Government Camp, unerringly.

I shall not soon forget that weird journey. The hills seemed to tower higher and higher as the night closed in upon us. Steadily we crunched over the snow. White shrouded trees awed us. Strange sounds broke the forest hush and checked us, listening. Bears? Cougars? Only the wind. So the miles slipped behind us. Then the lights of Government Camp twinkled reassuringly, foretelling the welcome of blazing fires, a hot supper and the hand clasp of our genial host, Mr. Pridemore.

We found snow in great banks, some over twenty feet deep, surrounding the buildings, completely covering the fences, lying four to six feet on the level and so changing the general appearance of the place as to make a new country of once familiar ground.

We arose early the next morning, that we might be about the serious business of learning to ski. Many of us had much to learn—how to slide, turn, sidle up a hill; and above all, how to fall, that we might unassisted rise again; for skis have a queer way of tying themselves into knots. A very small hill satisfied the requirements of most of us that first day and, selecting our skis, we went to work. When we were thoroughly tired with this unaccustomed sport, we were glad to return to the warm fires, the bountiful supper table and the cheer of the evening hours. The big hotel had been turned over to us and we made ourselves comfortably at home. I think there was never a group of Mazamas that gave itself over so completely to the holiday spirit or crowded so much fun into a few short days. There was not a dull moment, and when the entire party gathered around one long table in response to the dinner bell, good spirits ran riot.

We had a New Year party Friday evening and watched the Old Year out and the New Year in before we slept the sound sleep that we deserved.

Saturday was a day of uninterrupted pleasure—if you call it pleasure to tease long forgotten muscles by laborious climbs and swift descents, often enlivened by a sudden passage in the air or a still more sudden burrow in the snow. The weather was perfect—everyone in good spirits. Feeling confident that our practice of the previous day had not been in vain, we selected a much steeper hill than before, and all day long we played, sliding and climbing and tumbling in the snow. Sometimes we did unexpectedly clever things—on one foot perhaps. Sometimes we started six abreast, with disastrous results at the foot of the hill. But, on the whole, we made good progress, by practice and the assistance of those experts and near-experts who were included in the party.

Skiing is an exercise which brings every muscle into play and calls for considerable skill in balancing. Some of the steeper hills made demands upon our nerve, for though the descent was swift and certain, our ultimate destination and our method of arriving were most uncertain. There was an exhilaration in the rapid downward flight and a fascination in executing a long curve, balancing and guiding and gradually coming to a stop, still erect.

In the afternoon we climbed for an hour or more far up the side of Mt. Hood and reaped the harvest of our morning's work in long swift slides, with no tedious climbs between. Down one slope and then another we went, often without pause, curving around trees, swooping down over sudden drops and rising over mounds of snow.

It was a day plucked from the calendar of childhood's happy years, but even such days must end. Reluctantly we turned our faces homeward. The sun dropped behind the hills and the jagged outlines of the ranges sharpened against the afterglow. Swiftly the winter twilight fell over a white world. Gently and sweetly we were drawn back into the dignity of our grown-up spheres by the hand of Him Whose world is given us.

Sunday dawned clear and bright—our last day. We had our first good view of Mt. Hood that morning. The atmosphere was unusually clear and the mountain seemed close enough to touch, and was magnificently big and white and still. As if in farewell, she swept aside the curtains of fog and we looked for a long time, fixing in memory that vision of massive beauty.

There followed a few more hours on the hills; then a chill wind blew down from the mountain and dismissed us, whispering something that we could not understand. We returned to the hotel, put away our skis, fortified ourselves with another of those hearty dinners. We were reluctant to go. Throughout our stay fortune had seemed to favor us. The weather had been crisp but not too cold for skiing. Our parkas had kept us warm. The snow had been in good condition and there had been no wet clothing to fret us. But the time for departure had come. We said goodbye to our host and hostess—indulgent to the last—and to the hills, the mountain and the snow. Then came the bustle of departure. Very soon after dinner we were on our way. Once more the soft "crunch, crunch" of the snowshoes over miles of snow, the waning day, the twilight, and the homeward ride.

Then, as we glided noiselessly into Portland, that whispered message of the wind was made clear to us, and we laughed to see our city blinking sleepily under her new blanket of soft white snow. Her hills called us to come with our skis, and in fancy we could hear the tinkle of sleigh bells and the "Track! Track!" of the coasters. Our fun was not over; it had just begun.

It is easy running from a mountain's top
down to the valleys at its foot,
But difficult and steep the laborious ascent, and
feebly shalt thou reach it.

—*M. F. Tupper*

Geographical Progress in the Pacific Northwest

By LEWIS A. MCARTHUR

Excellent progress has been made during the past year along geographical lines in Oregon and Washington, and many new maps and publications have been issued that contribute materially to our knowledge of areas that heretofore have been but little known. These maps and publications are valuable alike to those interested in development work of all sorts, including commercial enterprises, and to those who care for the outdoors.

The maps issued by the United States Geological Survey include the following quadrangles: Tualatin, Albany, Estacada, Chehalis, Coyote Rapids in central Washington, Arlington, Condon and Willamette valley, sheets 7 and 8 in the vicinity of Corvallis, Albany and Lebanon. These and previously existing maps published by the U. S. Geological Survey may be secured in Portland for ten cents each, with the exception of the Willamette valley sheets which are more expensive.

It is understood that the following sheets will soon be off the press; Diamond lake, north of Crater lake; Priest rapids in central Washington, Pasco and Willamette valley No. 9. Advance sheets of the following maps may now be secured from the Geological Survey, and the engraved editions will be issued next year: Hillsboro, Salem, Kerby in southwestern Oregon, Wallula and Pasco.

Efforts are now being made by the Oergon Geographic Board working with the Forest Service to secure the addition and revision of names on a number of the older Geological Survey sheets in Oregon and Washington, and already considerable data has been submitted. It is believed that as the older maps are reprinted much new information will be included on them. In fact, a new printing of the Mount Hood sheet is just off the press with a great deal of revised nomenclature, much of which originated with the Mazama special committee on names along the Columbia river highway. This new edition shows the highway and all the changes recommended by the Mazamas. and it should be used in all cases instead of the old edition published in 1913. The Oregon Geographic Board has submitted a large number of corrections to the U. S. Coast and Geodetic Survey charts in the way of new names and correct spellings, and these data are being incorporated in new editions as fast as they are issued. The Coast Survey has been very willing to co-operate in this work.

During the past summer field work by the Geological Survey was completed on the Mt. St. Helens, Connell in eastern Washington, Walla Walla and the Oregon portion of the Troutdale quadrangles. In addition about half of the Twickenham quadrangle in the John Day valley was mapped.

In addition to the taking of topography as outlined above, there were three extensive triangulation schemes finished in this territory last summer and the computations are now under way in Washington. The U. S. Coast and Geodetic Survey completed the extension of its primary arc from Utah across Idaho, over the Blue mountains and down the Columbia river to a connection with the coast arc near Portland. This new arc located a large number of geographic positions in Oregon, made more accurate many old positions, and will put all the Oregon triangulation heretofore completed, on the standard North American base. It will furnish control for many additional topographic maps along its route.

The Geological Survey extended triangulation from the John Day country northeast of Prineville southwestward to the summit of the Cascade range, locating about 30 new positions, and giving control especially to the Bend quandrangle, which will probably be mapped next summer. The Geological Survey also completed an extensive triangulation net north from Diamond peak along the western flanks of the Cascade range to the McKenzie river, which will be used as a base for many new maps, including the Diamond peak sheet which is on the program for completion next summer.

Last year the Geological Survey completed a very accurate survey of the Skagit river in northern Washington, mapping a section that has been but little known. The advance sheets have been issued. An interesting volume on the water powers of the Yakima valley has been issued by the Geological Survey, entitled Bulletin 369. This book contains many maps of the eastern slopes of the Cascade range.

During the past two or three years fine soil maps have been issued by the Department of Agriculture covering practically all of western Washington. They may be obtained from the Bureau of Soils in Washington.

The Forest Service has issued a very interesting map of the Columbia Gorge park, showing contours and a view of the Cascade range north of Mt. Hood, and in addition excellent maps have been issued of the Oregon, Deschutes, Siuslaw, Siskiyou, Ochoco, Umatilla, Whitman, and Malheur National Forests in Oregon and of the Rainier, Chelan and Washington National Forests in Washington. New maps have been completed for the Crater, Wenaha, Okanogan, and Wenatchee National Forests, and these will be published during the coming year, and in addition, work is being completed on the Olympic and Snoqualmie National Forest maps, and also on a fine topographic map of the Columbia National Forest, compiled by the photographic process.

During the coming spring it is planned to complete new maps of the Santiam, Cascade and Umpqua National Forests, as these districts are becoming more popular as summer camping grounds, and maps are needed for proper fire protection.

The Reclamation Service in co-operation with the state has issued valuable reports on the Deschutes, Ochoco, John Day, Malheur, Harney, Warner, Rogue river, and Silver lake projects, and these contain new maps. The map of the Warner lakes region is exceedingly interesting, as it represents a contribution to the geographic knowledge of the state that is very valuable. All previous maps of this section have been of little value.

During the past summer the United States Geographic Board compiled a volume of decisions of that body, including some 335 pages. This volume is highly valuable as a reference work, and it supersedes all previous volumes of decisions of the U. S. Board. There are over 1,000 of these decisions that apply to names of features in Oregon and Washington.

Bureau of Associated Mountaineering Clubs of North America

During the summer of 1915, I visited the mountaineering clubs and geographical societies of the country and suggested the formation of an Association for the furtherance of common aims, and for the establishment of headquarters in New York where mountaineering information might be collected and made available. The plan was outlined as follows:

It was proposed to form an association of clubs and societies, each of which shall co-operate through its secretary, and transact its business by correspondence with the general secretary. Each club shall send its printed matter which will be added to the collection of mountaineering literature established in the New York Public Library. An annual bulletin of information on the membership, officers, and activities of the leading organizations shall be issued. The secretary of each club will notify the general secretary of the movements of local members who have interesting slides, and who can address the members of the Association at such times as they may be in different parts of the country.

One of the most important features of a club's activities is that of its library. Members should be encouraged to read what is being done in the mountaineering world, for education in this direction is as essential to a true appreciation and enjoyment of mountaineering as is the work in the field. Copies of many of the new books in mountaineering will be sent to each club for review in its annual publication and bulletins, thereby materially assisting in the growth of its library.

It is believed that the existence of this Association will have a valuable influence in many directions, and, occupying the field, its activities may expand as experience and occasion make desirable.

Meeting with a favorable response to the above ideas, I sent out a preliminary letter and received unofficial replies in approval of the plan. At the annual meeting of the American Alpine Club, held at the New York Public Library on Jan. 8, 1916, I presented these letters and asked that the Councilors of the Club be instructed to consider the plan and to send out an official letter to each club inviting it to become a member of the proposed association.

After due consideration, the Councilors of the American Alpine Club sent such a letter in March to the leading clubs, asking them to join in a Bureau of Associated Mountaineering Clubs of North America. Securing a majority of acceptances, they declared the plan in operation on May 2, 1916.

The first official act of the Bureau was the publication in May of a bulletin containing statistics of the membership, officers, and activities of the leading mountaineering clubs and geographical societies of the continent. The present membership of the Bureau comprises the following organizations. (Some others await the annual meeting of their directors.)

American Alpine Club.
Appalachian Mountain Club.
British Columbia Mountaineering Club.
Colorado Mountain Club.
Geographic Society of Chicago.
Geographical Society of Philadelphia
Hawaiian Trail and Mountain Club.
Mazamas.
Mountaineers.
Prairie Club.
Sierra Club.
United States National Parks Service.

A valuable reference collection of mountaineering books has been formed by the New York Public Library in the main building at 476 5th Ave., and we have secured the deposit of the library of the American Alpine Club. The combined collection promises to become one of the most important in existence. A collection of photographs and enlargements of mountain scenery in all parts of the world is also being made, and contributions of mounted or unmounted views will be appreciatively received.

LEROY JEFFERS,
General Secretary,
Librarian American Alpine Club.

Photograph by Winter Photo Co

A 1916 view taken on site of Mazama camp of 1910, showing large quantity of snow unusually late in the year.

Harley H Prouty.

In Memoriam

Harley H. Prouty

On September 11, 1916, at St. Vincent's Hospital, Portland, Oregon, Harley H. Prouty, after an illness of a few weeks departed this life. Mr. Prouty was born in Newport, Vermont, June 26, 1857, and was in his sixtieth year at the time of his death. He was the son of John Prouty, and was descended from one of the oldest New England families, its records dating back to 1667. His brothers, Charles A. and George Herbert Prouty are well known public men, the former having served for many years as Interstate Commerce Commissioner, and the latter having been governor of Vermont.

John Prouty, Harley's father, was one of the pioneer lumbermen of Vermont and in this business Harley spent his early manhood. After graduating with honors from St. Johnsbury Academy, Vermont, he spent several years in the lumbering business in Canada, subsequently selling out his interests in the Prouty Milling plant and coming to Seattle, Washington. From Seattle he came to Portland, residing here for many years.

Through habits of thrift and close attention to business, Mr. Prouty a number of years ago accumulated a competence, and for the last ten years had retired from business and devoted his time to traveling in this country, Europe and the Orient. He was greatly devoted to mountaineering and kindred sports. He was an active member of the Sierra Club, the Alpine Club of Canada and an ex-president of the Mazamas. As an alpinist he had few superiors in this country. Prouty Peak, a summit peak of the North Sister, in the Cascades, was first ascended by him and has been so named to commemorate his splendid work on the mountains.

Mr. Prouty was modest and retiring but exceedingly interesting as a conversationalist and effective as a writer. His sympathies were with struggling humanity—evidenced by his munificent testamentary gift to the Salvation Army for carrying on their work in his adopted city.

In the passing of Mr. Prouty, the out-of-doors life has lost a most ardent devotee, the Mazamas an honored and useful member, and the State a worthy citizen.

Hart Keokuk Smith

Hart K. Smith was born September 15, 1875, in Wayne County, Iowa. His early education was received in a public school in Hume, Missouri, where he lived until he was twenty years old, and later on it was supplemented by a course at a college in Macomb, Illinois.

His boyhood days, like those of so many others of his countrymen, were spent on a farm until his favorite vocation impelled him to travel over several of the middle and Pacific states in his desire to study and collect Indian tools and implements.

He arrived in Portland early in 1904, and for about ten years was in the employ of the Pacific Telephone and Telegraph Co. as store-keeper. In this vicinity, with its wealth of aboriginal village sites, kitchen middens, fishing grounds, shell mounds, and ancient implement factories about St. Johns, the Peninsula, Sauvie Island and Oregon City, he found a field with unlimited possibilities, and for some years all his spare time was spent in the collection and study of these interesting relics of a bygone age and a vanishing race.

His keenness in the pursuit of this study is well exemplified by the following incident. An old farmer on the Peninsula on whose place it was well known that an old village site existed, obstinately refused to allow anyone to search for relics thereon. A freshly plowed field offered such an irresistible temptation to Mr. Smith that he calmly awaited for a very foggy day and then diligently walked over the field, furrow by furrow, and was duly rewarded by finding five primitive implements left or mislaid by the early inhabitants.

His powers of observation were of a high order and his keenness in detecting an artifact and in determining its use were such that the writer never knew him to make an erroneous decision on these subjects. He was the first person in Portland to purchase and read the invaluable work of W. K. Moorehead on "The Stone Age in North America" and of Frederick Smith on "The Stone Ages in North Britain and Ireland." The writings of these enthusiasts only stimulated him to further efforts and incited greater zeal in this absorbing study. That he accomplished so much in this branch of science with such limited leisure and means, may be a surprise even to his fellow members, when it is learned that he collected almost 3,000 specimens of the red man's handicraft, and that these collections extended over eleven states.

Coupled with his archaeological bent and keenness of observation were a high regard for the rights and feelings of others, and a spirit of altruism of which only those who knew him intimately were aware.

His entrance into the fold of the Mazamas opened to him a new field of good-fellowship and gave him an opportunity for study in the great outdoors that he appreciated and enjoyed to the fullest extent.

A severe cold, contracted in the autumn of 1915, gradually but surely led to tuberculosis, and on October 26, 1916, he passed away, mourned alike by relations, fellow members and friends.

M. W. Gorman

Emil Franzetti

In the fullness of his powers, in love with life, and actively engaged in the carrying out of many plans whereby others might derive pleasure and health, Emil Franzetti, an honored member of the Mazamas was called by death on Saturday, the nineteenth of November, last.

His record is an open book. As proprietor of Rhododendron Tavern, close to Mt. Hood, he had come to be known, loved and respected alike by both travelers and neighbors.

To be intimately acquainted with Emil Franzetti was to get a new vision of life in its best aspects. No one who has known him can ever forget that wondrous smile, or the warm clasp of that strong right hand, which was in deed and in truth a right hand of fellowship.

His fellow members can recall only with pleasure the many happy and profitable hours spent when on one occasion or another, they were the guests of this true friend. He never failed so to give of himself and of his interest that the success of the outing was assured from the beginning.

Not only under his own roof was he known to Mazamas, for occasionally he would join in the outings, and so he came to be recognized as one of the organization's foremost mountaineers and nature lovers.

Emil Franzetti was born thirty-five years ago in Osmete, a border town of the Italian Alps. He came to America about eleven years ago. His occupation was that of a chef, perhaps his most prominent position being with the staff of the Waldorf-Astoria in New York City. Moving to Portland eight years ago, he at once established himself at the head of his art in this city. Four years ago he purchased Rhododendron Tavern, where he resided until the time of his death.

"I climb the hill; from end to end
Of all the landscape underneath
I find no place that does not breathe
Some gracious memory of my friend."
—*Tennyson.*

Osmon Royal

Mazama Outing for 1917

The Mazamas have decided to visit Mt. Jefferson, located in the Santiam National forest, August 4-19, 1917, for their twenty-fourth annual outing. The camp will be either at Pamelia lake or Hunts Cove at the southwest base, or in that veritable paradise on the north side, Jefferson Park.

Pamelia lake, although much lower than Jefferson Park, makes a good camping place where a plentiful supply of trout may be obtained at our very door. From either Pamelia lake or Hunts Cove, the mountain is easily accessible, with an elevation of 6,000 or 7,000 feet to overcome in ascending to the summit. The route into Pamelia lake, after a railroad journey via Albany and Detroit, Oregon, at the terminus of the C. & E. branch of the Southern Pacific railroad, leads one up the beautiful North Santiam river through a primeval forest, over a trail of some fifteen or twenty miles, with very agreeable grades. From Pamelia lake, many side trips entice one to the numerous waterfalls and cataracts, and afford varied views of Jefferson and other prominent peaks in the vicinity.

If Jefferson Park should be chosen for the campsite, a different route would be followed from Detroit, the trail leading up the Breitenbush a distance of twenty or thirty miles, permitting a visit to the famous Breitenbush hot springs. Once established in Jefferson Park, the scenic attractions on every hand are unbounded. There are many crystalline lakes ever reflecting Mt. Jefferson on their placid surfaces. This park occupies a hanging valley nestled close to the side of the mountain, at an elevation of about 6,000 feet, to reach which the trail leads over a ridge 7,000 feet in elevation, affording a glorious view of Mt. Jefferson and Jefferson Park.

Jefferson Park is about three miles in length by one mile in maximum width, and is at the crest of the Cascades, its waters plunging down at each end of the valley in a succession of cataracts and draining into both eastern and western Oregon. Camp in this spot will be of interest to the botanist on account of the great variety and profusion of the flowers.

A prospectus giving more specific details will be issued early in 1917.

L. E. Anderson,
Chairman Outing Committee.

Mazama Organization for the Year 1916-1917

OFFICERS

William P. Hardesty (418 City Hall, Portland)........................*President*
A. Boyd Williams (131 East Nineteenth St., Portland)............*Vice-President*
Miss Beulah F. Miller (629 East Ash St., Portland) ....*Corresponding Secretary*
Miss Jean Richardson (888 East Washington St., Portland) ..*Recording Secretary*
Miss Martha E. Nilsson (320 East Eleventh St. N., Portland)..*Financial Secretary*
Miss Mary C. Henthorne (Library Association, Portland).... *Historian*
Roy W. Ayer (131 East Nineteenth St., Portland).....................*Treasurer*
Leroy E. Anderson (213 N. W. Bank Bldg., Portland).*Chairman Outing Committee*
Robert E. Hitch (602 Fenton Bldg., Portland)..*Chairman Local Walks Committee*

COMMITTEES

Outing Committee—Leroy E. Anderson, Chairman; Francis W. Benefiel, Miss Martha E. Nilsson.

Local Walks Committee—Robert E. Hitch, Chairman; Charles J. Merten, W. W. Ross, Miss Agnes G. Lawson, Miss Margaret A. Griffin.

House Committee—E. C. Sammons, Chairman; O. B. Ballou, Miss Pearle E. Harnois, Miss Anna Bullivant, Miss Charlotte M. Harris.

Entertainment Committee—Miss Nettie G. Richardson, Chairman; Arthur S. Peterson, C. V. Luther, Miss Alice Banfield, Miss Minna Backus.

Publication Committee—Miss Mary C. Henthorne, Chairman; Alfred F. Parker, Miss Beulah F. Miller.

Educational Committee—Arthur K. Trenholme, Chairman; John A. Lee, A. M. Churchill, Miss Lola Creighton, Miss Jean Richardson.

Library Committee—Miss Mary C. Henthorne, Chairman; Leroy E. Anderson, Charles A. Benz, Miss Beulah F. Miller, Miss Ella P. Roberts.

Membership Promotion Committee—Roy W. Ayer, Chairman; T. Raymond Conway, Miss Harriet E. Monroe.

Membership Committee—A. Boyd Williams, Chairman; Miss Beulah F. Miller, Miss Martha E. Nilsson.

Auditing Committee—Robert F. Riseling, Chairman; B. W. Newell, Miss Martha O. Goldapp.

Constitution and By-Laws of the Mazamas

as approved at a special meeting
held in Portland, Oregon, June 29, 1916

ARTICLE I.—NAME

The name of the organization shall be "Mazamas."

ARTICLE II.—OBJECTS

The objects of this organization shall be the exploration of snow peaks and other mountains, especially of the Pacific Northwest; the collection of scientific knowledge and other data concerning the same; the encouragement of annual expeditions with the above objects in view; the preservation of the forests and other features of mountain scenery as far as possible in their natural beauty; and the dissemination of knowledge concerning the beauty and grandeur of the mountain scenery of the Pacific Northwest.

ARTICLE III—OFFICERS

Section 1. The affairs of this organization shall be controlled and managed by a board of nine directors, who shall be known as the Executive Council, and whose duties shall include the appointment of all committees, and the filling of all vacancies in the Executive Council. They shall be elected annually and shall hold office until their successors are elected and qualified.

Sec. 2. The Executive Council shall hold regular meetings once each month. Special meetings, however, may be called at any time by the President, and, in his absence, by the Secretary, by giving 24 hours' notice, either written or verbal, of the same to members thereof. The attendance of any member of the Executive Council at a meeting thereof shall be a full waiver of all notice of said meeting by the members so attending. A majority of the members of the Executive Council shall constitute a quorum for the legal transaction of all business.

Sec. 3. The members of the Executive Council shall be nominated and elected as follows: The Executive Council shall, at least six weeks previous to the annual meeting, appoint from the members of the club at large a nominating committee of five. It shall be their duty to nominate a ticket of eighteen candidates for the Executive Council for the ensuing year; provided, however, that the name of any member proposed to the committee in writing by any ten (10) members of the Club shall be added to such ticket. Within two weeks after its appointment the said committee shall file its report with the Corresponding Secretary of the Club, who shall, at least three (3) weeks previous to the annual meeting, have printed and mailed to each member of the Club a ballot of such nominees. This ballot shall have the names of all nominees arranged with a blank space for the insertion of any additional name, the nominees selected by the Nominating Committee to be so designated, and opposite each name a space for the marking of a cross. Upon each ballot shall be the following words:

"BALLOT FOR OFFICERS OF THE MAZAMAS.

'Annual election Monday, October (Here insert date of annual election). Polls open from 1 to 4 p. m. Directions for voting."

'Vote for nine candidates by marking a cross opposite the names of the candidates selected. Vote in person at the annual election or mail your ballot; in which case indorse your name on the envelope;

otherwise the ballot will not be counted. The election is so conducted by the judges as to keep each vote secret."

With such ballot the Secretary shall mail a stamped envelope, with the following address and words printed thereon:

'Executive Council, Mazamas,
(Here insert postoffice address)
Portland, Oregon.

"Ballot from...

Sec. 4. The annual election of members of the Executive Council shall be held at the Club Rooms on the first Monday in October of each year and the voting shall be by ballot. No notice of such election, except that given by the mailing of such ballot, shall be necessary.

Sec. 5. The polls shall be open at 1 o'clock p. m. and shall be kept open until 4 o'clock p. m. on the day of election. A plurality of votes shall elect.

Sec. 6. The Executive Council shall appoint five judges of election from the members of the Club at large to supervise said election, a majority of whom shall be competent to act, and the Corresponding Secretary of the Council shall refer to them, unopened, all the envelopes containing ballots.

Sec. 7. The Judges of Election shall at the time of the annual election and before opening the envelopes, check off the names of those thus voting, and shall thereupon open and destroy said envelopes, and, without examining the ballot, cast said ballot in a box provided therefor. At the close of election the judges shall count and report to the Executive Council in writing the number of votes cast for each candidate and the names of those elected to serve as members of the Executive Council, and the Corresponding Secretary shall thereupon notify in writing the members elected.

Sec. 8. The officers of this organization shall be a President, Vice-President, Corresponding Secretary, Recording Secretary, Financial Secretary, Treasurer, Historian, Chairman of the Outing Committee, and Chairman of the Local Walks Committee, who shall be chosen annually by the members of the Executive Council from their own number.

Article IV.—Duties of Officers

Section 1. It shall be the duty of the President to preside at all meetings of the organization and of the Executive Council; to enforce the by-laws; to call such meetings as he is empowered to call and to perform such other duties as usually devolve upon the office of President. The President and the Recording Secretary shall execute deeds and other instruments on behalf of the corporation when authorized so to do by the Executive Council.

Sec. 2. It shall be the duty of the Vice-President to assume the duties of the President in his absence.

Sec. 3. The duties of the Corresponding Secretary shall be to conduct the official correspondence of the Club, send all notices of all meetings, circulars, and other information to members of the Council and of the Club.

Sec. 4. The duties of the Recording Secretary shall be to record the proceedings of all meetings; receive all applications for membership, sign all orders drawn on the treasury of the Club, and shall call attention to such business as may properly come before meetings of the Club or Council, and see that the same is properly disposed of after action is taken.

Sec. 5. The duties of the Financial Secretary shall be to collect and receive all moneys, pay over and account for, monthly, to the Treasurer, send notices of annual dues and assessments to be collected by the Club, and make a report to each meeting of the Council as to the amounts received and paid over by him to the Treasurer.

Sec. 6. The duties of the Treasurer shall be to receive all moneys of the Club, and keep a fair and faithful record of the same. He shall make a written report at the annual meeting and shall be ready to report at each meeting of the Council when called upon so to do.

Sec. 7. It shall be the duty of the Historian to keep a record of the field work of the Society, and to submit the same at the annual meeting each year; also to collect, classify and preserve in suitable form all obtainable written or printed accounts of the Society and its expeditions and any other descriptive or scientific information concerning mountaineering in general, and especially concerning the mountains, lakes, streams and other natural scenery of the Pacific Northwest.

Sec. 8. It shall be the duty of the Chairman of the Outing Committee:

1. To preside at all meetings of the Outing Committee.
2. To be the executive head of the outing.
3. To conduct all the official correspondence of the outing.
4. To make all necessary outing contracts in the name of the Club, provided that any contract involving as much as $100.00 shall be approved by a majority of the Outing Committee and by the Executive Council; all funds for the outing shall be paid into the general treasury, provided that while actually in the field the chairman shall receive and disburse all funds, keeping an accurate and detailed account of the same. All bills shall be paid by the warrant system except that bills incurred while in the field may be paid from funds in the hands of the chairman, collected while in the field as provided by the Council.
5. The Chairman of the Outing Committee shall not be required to pay any annual outing assessment.
6. Mazamas, who contemplate any trip or outing advertised as a Mazama trip or outing, must first secure the sanction of the Outing Committee or Council and make a financial report to the Council at the close of such trip or outing.

Sec. 9. It shall be the duty of the Chairman of the Local Walks Committee to supervise the local walks of the Club and remit monthly to the Treasurer all funds collected from said local walks.

Article V.—Librarian

The Executive Council shall be empowered to employ a clerk whose duties shall be to have charge of the Library and offices of the Mazamas under the direction of the Historian and Executive Council of the Club. Said Clerk shall be at the service of any of the officers of the Club to assist them in the performance of their duties, and shall perform such other duties as may be assigned by the Council. The compensation shall be determined by the Executive Council.

Article VI.—Committees

Section 1. Outing Committee: A committee consisting of three members whose duty it shall be to take charge of the annual outing and supervise all other outings of the Club except as otherwise provided for in the By-Laws.

Sec. 2. Local Walks Committee: A committee of five members whose duty it shall be to arrange for and publish a bulletin of the local walks and take charge of same, fix upon a nominal charge for those taking part in each walk.

Sec. 3. House Committee: A committee of five members whose duty it shall be to have general supervision and care of the furnishings of the club rooms, and any lodge hereafter acquired, or other property of the Club.

Sec. 4. Entertainment Committee: A committee of five members of the club whose duty it shall be to provide and arrange for all entertainments for the Club.

Sec. 5. Educational Committee: A committee of five members whose duty it shall be to provide and arrange for all educational meetings of the Club.

Sec. 6. Library Committee: A committee consisting of four members in addition to the Historian, who shall be chairman of the committee, whose duties shall be to make suitable provision for all publications belonging to the Club and to make suggestions for adding to the same; also to secure written reports and accounts of the local walks and local expeditions of the Club, together with photographs of local scenery on the walks.

Sec. 7. Membership Promotion Committee: A committee of three members whose duty it shall be to take all proper steps to increase the membership of the Club.

Sec. 8. Membership Committee: A committee of three members of the Executive Council who shall investigate the qualifications and eligibility of applicants for membership.

Sec. 9. Publication Committee: A committee of three members whose duty it shall be to supervise all publications of the Club.

Sec. 10. Auditing Committee: A committee of three members whose duty it shall be to audit semi-annually the books of the Treasurer and of the Financial Secretary and, at the close of the annual outing, the books of the Outing Committee, or of any committee required to make a financial report to the Executive Council, and report to the Council a statement of the result of the audit of the books of said officers, with any recommendations concerning said account deemed advisable. The bank book and books of account kept by each officer, shall be at all times open to the inspection of the members of the Auditing Committee.

Article VII.—Membership

Section 1. There shall be three classes of membership—active, life, and honorary.

Sec. 2. Any person who has climbed to the summit of a perpetual snow peak, on the sides of which there is at least one living glacier, and to the top of which a person cannot ride, horseback or otherwise, shall be eligible to active or life membership.

Sec. 3. Applications for active or life membership shall be made in writing to the Recording Secretary, endorsed by at least two active or life members, accompanied by satisfactory proofs of eligibility and by initiation fee in case of active membership. All applications shall be referred to the Membership Committee and then posted for two weeks prior to action by the Council or Club. Applications for membership, after posting and on report of the Membership Committee or on failure of said committee to report, may be acted on by the Executive Council at any meeting regularly called, the majority of the whole Council being necessary for election; or by the Club by ballot, at any meeting regularly called, the majority of all members present being necessary for election.

Sec. 4. Persons who have rendered distinguished services to the Club, or are eminent for achievements in exploration, science, or art, shall be eligible for honorary membership.

Sec. 5. Nominations for honorary membership must be made in writing to the Recording Secretary, at least thirty days before the date of the annual meeting, and be signed by not less than three active or life members, and shall contain a statement of the reasons why election is urged. The call for the annual meeting shall contain the names of all persons so nominated. Honorary members shall be elected by ballot at the annual meeting only, a two-thirds vote of all members present being necessary for election. Not more than one honorary member shall be elected in any one year.

Sec. 6. Honorary members shall not be required to pay dues, neither shall they have the right to vote or hold office, but shall have all other rights and privileges of active members.

Sec. 7. The Executive Council shall have power, by a unanimous vote of all its members, to expel from the Club any member for such cause as shall in its opinion justify its action; provided that due notice of the charge has been sent the member in question to his or her last address known to the Club, and a reasonable opportunity of defense afforded, and a formal statement of the case made to each member of the Council. Any member so expelled shall have the right of appeal to an Annual or General meeting of the Club.

Sec. 8. Resignations of members shall be accepted only in case dues are paid in full.

Article VIII.—Meetings

Section 1. The annual meeting of the Club shall be held on the first Monday in October, and written or printed notice thereof shall be sent by the Corresponding Secretary to each member at least ten days previously.

Sec. 2. Special meetings shall be convened by the President at any time upon written request of five active or life members, and written or printed notice thereof specifying the object of the said meeting, shall be sent by the Secretary to each member at least one week previously.

Article IX.—Financial

Section 1. Initiation fee of $3.00 shall accompany each application for active membership, provided that no dues shall be charged a new member from date of election to the first day of January following said election.

Sec. 2. The dues of all active members shall be $3.00 per annum. Such dues shall be payable in advance at the beginning of each calendar year, except in the case of new members, whose dues shall be payable as provided in Section 1 of Article IX of these by-laws.

Sec. 3. The Financial Secretary shall, prior to February first of each year, mail notices to all members stating dues are payable. Any active member failing to pay his or her dues before the first day of May, next following the date same are payable, shall be delinquent and it shall be the duty of the Financial Secretary at once to post the names of all delinquent members upon the bulletin board in the rooms of the Club, where they shall remain posted for a period of 30 days. The Financial Secretary shall also, and simultaneously with such posting, send notice by mail to all delinquent members of the fact of their delinquency and the posting of their names, and in such notice shall warn them that if their dues are not paid within a period of 30 days from and after the date of posting, their names will be presented to the Council to be dropped from the roll. The Council shall, in their discretion formally drop from the roll of members the names of all members so delinquent whose names have been so presented to the Council by the Financial Secretary, after their names have been posted and the notice and warning have been given by the Financial Secretary as before provided.

Sec. 4. An entrance fee of Fifty ($50.00) Dollars shall be charged life members; and they shall not be required to pay annual dues.

Sec. 5. The Treasurer shall pay out money of the Club only on warrant of the Recording Secretary, countersigned by the President and authorized by vote of the Executive Council.

Sec. 6. The Treasurer shall have executed a fidelity bond in favor of the Club in protection of moneys of the Club in his possession in amount to be determined upon by the Executive Council at the beginning of each fiscal year; provided that the cost of said fidelity or surety bond shall be defrayed by the Club.

Article X.—Quorum

Nine active or life members shall constitue a quorum for the transaction of business.

Article XI.—Consuls

The Executive Council may appoint from members of the Club local Consuls to represent the Club in principal cities where desirable. Their duties shall be to render assistance to the Executive Council and to perform any such duties as may be designated by said Council from time to time.

Article XII.—Honarary Members of Council

Section 1. The Executive Council may, at their discretion, elect annually, by an unanimous vote, an Honorary President, who must be a member of the Club, and who shall have pre-eminently distinguished himself in mountaineering, exploration or research.

Sec. 2. The Executive Council may also elect annually four Honorary Vice-Presidents, who must be members of the Club, and who shall be selected for such offices by reason of their prominence in matters identified with the purposes for which the Club was organized, or because of some material aid and assistance they may have rendered the Club.

Article XIII.—Official Seal

The Executive Council shall procure a corporate seal containing date of incorporation and Mazama emblem.

Article XIV.—Amendments

Amendments to the by-laws may be made at any regularly called meeting of the Club, provided that such amendment or amendments shall have the signatures of not less than five active members of the Club and are acquiesced in by two-thirds of those recording their votes.

Book Reviews

Edited by Beulah F. Miller

"The Columbia, America's Greatest Highway Through the Cascade Mountains to the Sea.

There was little left to be desired, in text or picture, in the first edition of Mr. Lancaster's classic of the Columbia canyon and the Highway; this "little," comprising pictures of the completed road and the bridges, more scenes of enchantment in the gorge, and several new pages of text in the author's eloquent and devotional style, is supplied in the second edition published in 1916.

There are two brief chapters of popular science describing the formation of the Cascade Range and Sierra Nevada and of the Columbia river and the gorge, giving glimpses of the Indian life in the Columbia basin, and narrating the adventures of the fur traders and the early missionaries.

Under the chapter entitled "The Struggle to Possess the Land," excerpts are given from the simple but absorbing diaries of the heroic pioneers who suffered harrowing hardships in the earliest journeys over rude trails through the canyon, and by crude rafts down the river. In dramatic juxtaposition with these annals of 1849 are placed beautiful pictures of our paved boulevard of today, its walls and viaducts and bridges, and the cataracts and palisades as seen from the Highway.

Thirty-five new half-tones, including striking views of St. Peters Dome, of Eagle creek and its picturesque arch bridge of water-washed boulders, are added; and the exquisite photographs in natural color are the crowning pictorial feature of this, as well as of the first edition.

As a handbook de luxe of the Columbia river scenery, and a dependable short historical sketch of early settlement and transportation in the Columbia basin, and as a graphic record of the successful achievement of a great public enterprise, the book makes a wide appeal. Frank Branch Riley.

Lancaster, Samuel Christopher. *The Columbia—America's Greatest Highway Through the Cascade Mountains to the Sea.* 1916. Lancaster. Portland. $2.50.

"Tourist's Northwest."

This is a most excellent guidebook to the Pacific-Northwest. It covers Oregon, Washington, northern Idaho, Glacier National Park, British Columbia, and Alberta. General descriptions of the country are given, interspersed here and there with interesting bits of history and legend. Accurate information concerning transportation, routes, customs, motor roads and steam ship lines, hotels, and festivals is included. The opportunities for hunting, fishing, mountaineering and other sports are outlined. The book is well written, contains good illustrations, several maps and an index, and should prove most valuable to any one traveling in the Northwest. B. F. M.

Wood, Ruth Kedzie. *Tourist's Northwest.* 1916. Dodd $1.75.
(Supplied by the Bureau of Associated Mountaineering Clubs of North America.)

"Lore of the Wanderer."

This is a collection of essays from well known English authors such as Stevenson, Hazlitt, Symonds, Ruskin, Steele, and Bacon. It is filled with the spirit of the outdoors world and should prove a most welcome addition to the small books for the camper. B. F. M.

Goodchild, George, ed. *Lore of the Wanderer: An Open-air Anthology.* Dent. 1s

"RAMBLES IN THE VAUDESE ALPS." A most readable and friendly account by F. S. Salisbury of his impressions of many extended and pleasant walks through the Vaud Canton of Switzerland.

The author describes most charmingly, and with authority, the flora and the tree-life of the region. The book is, in fact, a popular botanical guide.

In the description of the vast alpine scenes, Mr. Salisbury colors his word-pictures with an intimacy and sympathy and simplicity which constitute by no means the least delightful feature of his book. LOTTE B. RILEY.

SALISBURY, F. S. *Rambles in the Vaudese Alps.* 1916. Dutton. $1 00.
(Supplied by the Bureau of Associated Mountaineering Clubs of North America.)

"WILD LIFE CONSERVATION" This is a series of lectures delivered before the Forestry school of Yale University in 1914 It supplements Dr. Hornaday's work, "Our Vanishing Wild Life," and tells what, in his opinion, is the conservation remedy.

Graphically he pictures the destruction of wild life that is taking place. This, he asserts, is due in a major number of cases to liberal state laws and lack of any uniform national law. In the remaining cases it is carried on in the utter absence of any preventive laws.

Dr. Hornaday regards the Weeks-McLean bill as the most important legal step yet taken towards conservation. This act, passed by Congress in 1913, provides for governmental protection of migratory birds.

He says: "It is the most potent and far-reaching measure ever enacted for the protection of our native birds, and any occurrence that would impair or destroy its usefulness would be a national and continental calamity."

Again he says: "The law is necessary because of the utter inability of more than one-half of our states to protect their migratory birds by state law." In fifteen states, it is asserted, authorities have sullenly resisted appeals for legislation that would stop the slaughter. "The federal law," says Dr. Hornaday, "terminates that situation permanently."

Speaking of conservation movements, he points out as an instance, the fact that Glacier National Park, being a game refuge, probably saves for future generations, the mountain goat, which was otherwise threatened with extinction. Considerable attention is given to the detailed study of the economic value of birds. The author shows the value of these birds in the destruction of injurious insects. He is not opposed to hunting, but his protests are to save species of wild life that are being exterminated.

One chapter is devoted to a discussion of animal and bird pests that prey on other forms of wild life. Means of destroying this predatory life are given. In his final chapter he sets forth the duty of the public in the great conservation movement.

Frederic C. Walcott, in a subsequent chapter, gives a history of the development of game preserves, and shows the value of this means of artificial propagation. He summarizes in detail, artificial propagation and other protective movements now being carried out in this country and Canada. A bibliography of recent works on game concludes the book. FLORENCE J. McNEIL.

HORNADAY, WILLIAM T. *Wild Life Conservation.* 1914. Yale University Press. $1.50.

"BLACKFEET TALES OF GLACIER NATIONAL PARK."

That Glacier National Park, which has been a national park for only a few years, is rapidly becoming a popular playground, is shown not only by the greatly increased number of tourists who visit it each year, but also by the number and variety of books and magazines articles which have recently been published regarding it.

The Blackfeet Tales are written by an adopted member of the Indian tribe bearing that name, who married one of their girls, and for years lived with them as an Indian. There are stories of great chiefs and brave warriors and beautiful Indian maidens; and weird and fantastic tales of their gods, as told the author by his friends of the tribe when gathered about the lodge fire, while the pipe was going the rounds. There are legends of many of the lakes and streams and falls in the Park; and interwoven with all, the author's own adventures and old-time experiences.

Last of all, the book contains an earnest plea for the preservation of the old Indian names, for which so many times meaningless substitutes are found on the maps. MARTHA OLGA GOLDAPP.

SCHULTZ, JAMES WILLARD. *Blackfeet Tales of Glacier National Park.* 1916. Houghton. $2 00. (Supplied by the Bureau of Associated Mountaineering Clubs of North America)

"DAVID DOUGLAS JOURNAL.

The early explorations of the Oregon country in the search for the Northwest Passage, and the subsequent establishment of trading posts aroused much interest in the vast region of western America. The Royal Horticultural Society of London desired to learn about the flora of this new land, so in 1823 they sent David Douglas, a young Scotch botanist, to make a study of the plants of North America. On the first trip, made during the summer and autumn of 1823, he visited only the eastern coast. He collected a large number of American oaks, and a list of these with descriptions is published in this volume. The following year Douglas sailed to the western coast and began his work along the Columbia river and its tributaries. During the latter part of the expedition he went to California.

This diary is a most readable one, describing in quaint detail the many experiences he had, the people whom he met, the customs of the Indians, the flora and fauna he found, as well as conditions of weather. Descriptions of pines are given, a list of the plants introduced by David Douglas, and an account of his tragic death in the Sandwich Islands in 1834. The book will prove as interesting to the student of history as to the person desiring scientific information. B. F. M.

DOUGLAS, DAVID: *Journal Kept by David Douglas During his Travels in North America, 1823–27, together with a particular description of thirty-three species of pines.* 1914. London, William Wesley and Son. $5.04.

"THROUGH GLACIER PARK."

Mary Roberts Rinehart has written a most interesting account of a horseback trip through Glacier Park.

So graphically does she tell it, that she takes you along with her. You experience all the thrills of the real trip as you read, the excitement of preparation, the picturesque outfit, and the departure. You meet the leader of the party, Howard Eaton, "a sportsman and a splendid gentleman." You note the change, a few minutes down the trail, from civilization to the Great Wilds—follow the winding trails down, up, under and around—feel the peacefulness and grandeur of the mountains, valleys, streams, trees and flowers—learn the traits of the saddle and pack-horses—hold your breath as you read of her trip on the pack-horse that always walked on the edge of the precipice with two legs usually dangling over.

Her definition of a mountain pass is truly a refreshing one: "A pass is a blood-curdling spot up which one's horse climbs like a goat and down the other side of which it slides as you lead it, tramping ever and anon on a tender part of your foot. A pass is the highest place between two peaks. A pass is not an opening, but a barrier which you climb with chills and descend with a prayer. A pass is a thing which you try to forget at the time and you boast about when you get back home" She observes the change in the party a few days out: the joy of doing appeals more strongly; utter disregard is paid to complexion and sore muscles; there is eagerness to press on in any weather, over trails so mired that the horses literally fight their way through; and exultant feeling is displayed as the party stands on the Triple Divide from which water flows into the Gulf of Mexico, the Pacific Ocean, and Hudson Bay. She tells of the animals of the Park, the beaver, deer, elk, lion, coyote and bear. The visit to the garbage dump to see the bears was amusing until a grizzly "the exact size of a seven passenger automobile with a limousine top and same rate of speed," was about to give his attention to the unarmed visitors, after putting the hound and black bears to rout. As a climax, she describes a fishing trip down the Flathead Rapids and the pride with which she takes her catch aboard the train and gives orders for packing and later cooking.

Incidentally the author decries the treatment of the Flathead Indians, speaking particularly of the negligence of the government in making needed improvements and furnishing an adequate number of rangers. She also gives a little information about hotels, guides and horseback rates.

Altogether it is a delightful little book that may be read easily in an evening.

ALICE BANFIELD.

RINEHART, MARY ROBERTS *Through Glacier Park.* 1916. Houghton. $0 75.
(Supplied by the Bureau of Associated Mountaineering Clubs of North America)

"NATIONAL PARKS PORTFOLIO. This portfolio consists of pamphlets describing nine of the largest national parks. They are beautifully illustrated and contain short descriptions and brief legendary notes of the parks. This portfolio should do much to interest the people of the country in the nation's delightful playgrounds.

B. F. M.

U. S. INTERIOR DEPT. *National Parks Portfolio.* 1916. Scribner.

"BOOK OF CAMPING AND WOODCRAFT. This work is a revised edition of the "Book of Camping and Woodcraft" which was published in 1906, containing much new material. It is intended to meet the needs of the outdoor enthusiast who goes to accessible camping places as well as those of the camper who travels light. Careful details in regard to clothing, camp equipment and camp making are given. Chapters are devoted to tested provision lists and camp cookery. It is a very valuable encyclopedia of information for anyone living in the wilds. The second volume which is soon to be issued will be eagerly received.

B. F. M.

KEPHART, HORACE. *Camping and Woodcraft.* Vol. 1. 1916. Outing Pub. Co. $1.50.
(Supplied by the Bureau of Associated Mountaineering Clubs of North America)

"THROUGH THE BRAZILIAN WILDERNESS."

This is Roosevelt's account of his zoo-geographic reconnoissance through Brazil. This trip was originally inaugurated as a lecture tour through Argentina and Brazil, upon invitation of the governments of these countries. After making arrangements for this trip the author decided that instead of returning around South America by boat he would come north through the interior of Brazil and into the basin of the Amazon. He notified the Museum of Natural History of New York City of his intentions and they gladly took advantage of the opportunity to send with his expedition several naturalists who were to collect specimens of this region. On reaching Rio de Janeiro the Minister of Foreign Affairs of Brazil authorized the geographic features of the expedition and sent a native Colonel of the Governmental Service to accompany him for the purpose of mapping and exploring the little known Matto Grosso or highland wilderness of Brazil, and a river in the locality whose course and location had been but vaguely determined, giving them the official title of the Roosevelt-Rondon Scientific Expedition.

Their route was in a northerly direction from Buenos Aires, up the Paraguay River to the head of navigation, across the high interior wilderness by pack train, and down the course of the Rio Duvida. This trip was started December 9, 1913. The naturalists of the party found numerous specimens. The author shows his familiarity with tropical natural history in his descriptions, makes comparisons of various birds and mammals with those of Africa and our own country. His literary style is narrative and conversational and his subject matter is excellently arranged and complete in detail, an added interest being given by occasional abrupt terminations of certain phases of his subject with transitions to others more or less related.

The most important and most dangerous portion of his trip was that of exploring the "River of Doubt." The line of the Brazilian Telegraph Commission crossed this stream near its source but its course from there was doubtful. About two months was consumed by this trip, all possible varieties of hardships being encountered. Many days were spent in portaging around dangerous rapids, and in building dugout canoes to replace those lost from time to time. All manner of pests had to be contended with and, though there are few large animals in South America that man should fear, the smaller creatures are often dangerous and difficult to deal with, some of the most common being the Piranha, a carnivorous fish of only a few pounds weight, and ants that will attack and devour large animals or man. After following this stream down for a distance of about one hundred and ninety miles they found it to be identical with a river known locally as the Aripuanan, a large affluent of the Madeira and indirectly of the Amazon, but as to its character above the head of navigation nothing had hitherto been known. The expedition thus established the connection and identity of the different parts of the stream and at the same time collected other valuable geographical data, in recognition of which the Brazilian Government gave the river the official title of Rio Teodoro. The party reached the Amazon and disbanded early in May, 1914, having taken five months for the trip.

The book is completely indexed as to subjects touched upon and besides, contains appendices with details of equipment and commissary, and notes on the work of field zoologists and geographers in South America.

C. V. LUTHER.

ROOSEVELT, THEODORE. *Through the Brazilian Wilderness.* 1914. Scribner. $3.50.

"ALASKAN GLACIER STUDIES. This book has been supplied to the club by the Bureau of Associated Mountaineering Clubs of North America. It was reviewed by a member of the organization in the 1915 issue of Mazama.

TARR, RALPH STOCKMAN & MARTIN, LAWRENCE. *Alaskan Glacier Studies of the Nationa Geographic Society in the Yakutat Bay, Prince William Sound and Lower Copper River Regions. Ill* 1914. National Geographic Society. $5 00.

"THE ALPS." This is a very enjoyable little book of ten chapters, dealing with the Alps from the combined point of view of the mountaineer and the man of letters.

To us, surrounded as we are by mountains whose historical associations, aside from the Indian legends, are the scantiest, it is pleasing to have recalled the momentous incursion of Hannibal into the Alps, where his elephants "were ever readie to run upon their noses"; and we reverence the eternal snows more keenly after refreshing our memory with the story of the world and its making and realizing anew how much the lofty summits of the Alps had to do with the European march of progress.

Following the general discussion of pioneering in the Alps, several individual peaks, including Mont Blanc, Monte Rosa and the Matterhorn, command our attention as their dramatic stories unfold. We run in perusal the entire gamut of emotions, from amusement at the droll picture of Aggasiz in his mountain hut to horror at the grim tragedy that followed close upon the heels of the first victory of the Matterhorn.

In the closing chapters, which treat of Modern Mountaineering and The Alps in Literature, the author takes occasion to extol the joys of ski-jumping and to treat of the various advantages of climbing with and without guides; ending his little volume very happily with a tribute to the rich gifts to literature that the Alps have made, are making, and will continue to make until the appreciative human soul needs no longer to strive for verbal expression. BEATRICE YOUNG.

LUNN, ARNOLD. *Exploration of the Alps.* 1914. Holt. $0.50.

'THE MOUNTAIN." Mountain making, the foot hills, uplands, glaciers, and other phases of mountains are treated scientifically as well as popularly in Mr. Van Dyke's book. Not only is the result of his serious study of mountain masses given, but there are also vivid descriptions of the hills and mountains in their many aspects, together with an historic word pageant of the passing of the Indian and the encroachment of the pioneer upon his wild domain. Ruskin's well-known mountain passages are quoted—and also criticized from a scientific standpoint.

He sounds a needed warning to mountaineers when he says: "The average ascent-maker seems to have better legs than eyes. He sees little save the man ahead of him He is doing a stunt—not seeing a vision. People like John Muir and Leslie Stephen are rare in mountain literature.

Many people will agree with him also when he writes: "The mountain ranges have not ceased to be a source of mystery. Again and again, as we ride away, we turn in the saddle to look at their massive forms against the sky. They keep drawing us with a new look and an old lure. They are not paintable, they are not habitable, they are not wholly understandable, but, perhaps, for that very reason, they are wonderful. May they always remain so!" M. C. H.

VAN DYKE, JOHN C. *The Mountain.* 1916. Scribner. $1 25.

The Mazamas

Any person of good character who has climbed to the summit of a snow peak on which there is at least one living glacier and the top of which cannot be reached by any other means than on foot is eligible for membership in the society; the annual dues are $3.00. Following is a list of members of the Mazamas:

ABEL, A. H, 502 Worcester Bldg.
ABISHER, Marie, 335 14th St.
ACTON, HARRY W, 519 West 121st St. N. Y.
ACTON, MRS HARRY W., 519 West 121st St., New York.
ADAM, RICHARD, 344½ Alder St.
ADAMS, LORING K., 730 Chamber of Commerce
ADAMS, W. CLAUDE, 1010 E. 28th St. N.
AITCHISON, CLYDE B, Court House.
AIKIN, OTIS F., 919 Corbett Bldg
ALLEN, ARTHUR A., Care of Portland Rowing Club
ALMY, E. LOUISA, Care of H. L. Johnson, Glendive, Mont.
AMOS, DR. WM. F, 1016 Selling Bldg.
ANDERSON, DR FREDERICK, Orenco, Ore.
ANDERSON, LEROY E., 213 Northwestern Bank Bldg.
ANDERSON, LOUIS F., 364 Boyer Ave., Walla Walla, Wash.
APPLEGATE, ELMER I., Klamath Falls, Ore.
ATKINSON, R. H., City Pass. Agt. O.-W. R. & N. Co
ATLAS, CHAS. E., 423 City Hall.
ATWELL, F. C, Mallory Hotel.
AVERILL, MARTHA M, 1144 Hawthorne Ave.
AYER, R. W., 131 E 19th St.
BABB, HAROLD S, 578 Miller St.
BACKUS, LOUISE, 675 E. Alder St.
BACKUS, MINNA, 675 E. Alder St.
BAGLEY, FRANK H., Care of Lumbermens Bank:
BAILEY, VERNON, 1834 Kalorama Ave., Washington, D. C.
BALLOU, O. B., 80 Broadway.
BANFIELD, ALICE, 570 East Ash St.
BARCK, Dr C, Humboldt Bldg, St. Louis, Mo.
BARNES, CHAS. A, JR., Box 636 Tacoma, Wash.
BATES, MYRTLE, 448 E. 7th St
BEACH, LEE N, 731 East Ash St.
BEATTIE, BYRON J., 830 Rodney Ave.
BENEFIEL, FRANCIS W., 750 E. Ankeny St.
BENEFIEL, JOHN W., 110 East 20th St.
BENEFIEL, WILSON, 110 East 20th St.
BENSON, B. M, 20 Washington Bldg.
BENZ, CHAS A, 441 Eleventh St.
BERNARD, W. R, 213 N W. Bank Bldg
BLACK, W. J. Lumbermens Bank.
BLAKNEY, C. E, Milwaukie Ore.
BLUE, WALTER, 1306 E. 32nd St. N.
BODLEY, R M, 4519 Powell Valley Rd.
BODWAY, W. P., General Delivery.
BORNT, LULU ADELE, 641 E 13th St.
BOWERS, NATHAN A., 501 Rialto Bldg., S. F. Cal.
BOWIE, ANNA, 297 E. 35th St.
BOYCE, EDWARD, 207 St. Clair St
BREWSTER, WM. L., 808 Lovejoy St.
BRONAUGH, GEORGE, 355 Hall St
BRONAUGH, JERRY ENGLAND, Title & Trust Bldg
BROWN, G. T., 500 East Morrison St.
BRUCE, ROBERT C., 729 7th St, New York City.
BULLIVANT, ANNA, 269 Thirteenth St.
BULLIVANT, E H, 269 Thirteenth St.
BUSH, J. C, 88½ Grand Ave.
CALHOUN, MRS HARRIET S, 369 E. 34th St.
CAMPBELL, DAVID, 404 Boyer Ave, Walla Walla, Wash.
CAMPBELL, JOHN C, 518 E. 38th St. N.
CAMPBELL, P. L, 1170 13th Ave, E., Eugene, Ore.
CASE, GEORGENE M, 302 12th St.
CATTERLIN, LLOYD L, 762 Hancock St.
CECIL, K. P, 429 Beck Bldg.
CHENOWETH, MAY, 104 E 24th St. N.
CHRISTIANSON, WM D, 134 Colborne St., Brantford, Ontario, Canada.
CHURCH, WALTER E, Eugene, Ore.
CHURCHILL, ARTHUR M, 1229 Northwestern Bank Bldg
CLARK, C. M., 213 N 23rd St
CLARK, E. A, 1381 E. 17th St
CLARK, F. N, 1108 Westover Road.
CLARK, J. HOMER, 706 Glisan St.
CLARKE, D. D, Care of Water Board Office, City Hall.
CLARY, RALSTON J., Cor. Killingsworth & Albina Aves.
COALMAN, ELIJAH, Sandy, Ore.
COCHRAN, HARRISON H., 3424 Pillsbury Ave., Minneapolis.
COLBORN, MRS. LOIS B., 383 Summer St., Buffalo, N. Y.
COLLAMORE, MARY ERNA, Care of North Pacific College, E. 6th & Oregon Sts
COLLINS, W. G, 510 32nd Ave. S., Seattle, Wash.
COLVILLE, PROF F. V., Dept. of Agriculture, Washington, D C.
CONNELL, DR. E. DEWITT, 628 Salmon St. or Selling Bldg.
CONWAY D J, 4705 60th, S E.
CONWAY, T. RAYMOND, 4705 60th, S. E.
COOK, Arthur, 243 W. Park St.
COOK, COURTNEY C, 1220 Spalding Bldg.
COURSEN, EDGAR E, 658 Lovejoy St.
COURSEN, GERALDINE R, 658 Lovejoy St.
COWIE, LILLIAN G, Wellesley Court, 15th & Belmont.
COWPERTWAITHE, JULIA F., Station "E."
CRANER, HENRY C, Room 214, M. A. A. C.
CREIGHTON, LOLA I., 920 East Everett St.
CROSS, HARVEY E., Oregon City, Ore
CROUT, NELLE C, 1250 Hancock St.
CURRIER, GEO H, Leona, Ore.
CURTIS, EDWARD S, 614 Second Ave, Seattle, Wash
DALCOUR, NELLIE MAE Karl Hotel.
DAVIDSON, PROF. GEO, San Francisco, Cal.
DAVIDSON, R. J., 458 E. 49th St. N
DAY, BESSIE, 690 Olive St, Eugene, Ore.
DEKUM, ADOLPH A., 111 6th St
DILLER, PROF. JOS. S, U. S. Geological Survey, Washington, D C
DILLINGER, ANNA C., 121 E 11th St.
DILLINGER, MRS C. E, 121 E 11th St.
DONAHUE, TYRRELL E, 5004 33rd Ave, S. E.
DORFMAN, ANNA, 233 10th St.
DRAKE, J. FRANCIS, Pittock Block.
DUFFY, MARGARET C., 467 E. 12th St.
EARLEY, GEO C., 370½ 13th St.
EDWARDS, J. G., Agt. N. P. Ry., Vancouver, Wash.
EGGERSGLUESS, ERNEST, 170 11th St.
ELLIS, EDITH, Salisbury Apt. 244 Sandy Road.
ELLIS, PEARL, Salisbury Apt., 244 Sandy Road.
ELLISON, JAMES H, 875 Haight Ave.
EMMONS, A C, 1474 Yeon Bldg.
ENGLISH, NELSON, 267 Hazel Fern St.
ESTES, MARGARET P., 1063 E. Washington St.
EVANS, WM W., 744 Montgomery Drive.
EVERSON, F. L, 361 Tenth St.
EWELL, ELAINE, 608 E. Taylor St.

Faber, Gertrude F., 1904 East Washington St.
Fallman, Mrs. N. A, 151 Park St.
Farrell, Robt. S., 140 Front St.
Farrell, Thos. G, 328 E. 25th St.
Fellows, Lester O, 4319 74th St.
Filloon, Ray M, Troutlake, Wash.
Finley, Wm. L., 651 E. Madison St.
Fitch, R. Louise, Eugene, Ore.
Flemming, Miss M. A., 214 Post Office Building
Flesher, Nathan, Carson, Wash.
Ford, G. L., 309 Stark St, or 1101 E. 19th, N.
Forman, W. P., Y. M. C. A.
Forsyth, Mrs. C E, Wolf Creek, Ore.
Foster, Forrest Lloyd, 354 49th St. S. E., City.
Fraine, Cora D, 335 14th St.
Franck, Albert C., Box 136, San Diego, Cal.
Franing, Eleanor, 549 N. Broad St., Galesburg, Ill.
Freeman, D. C., 112 E 11th St.
Freeman, Ethel, 423 Irving Ave., Chicago.
Fries, Samuel M, 641 Flanders St.
Fuhrer, Hans, Rowe, Ore.
Fuller, Margaret E, 409 16th St.
Gaines, J. E., 482 Skidmore.
Galusha, Ora W., 30 Hillcrest Parkway, Winchester, Mass.
Garrett, Geo, 646 Cypress St
Gasch, Martha M. 9 E. 15th St. N.
Geballe, Pauline, 782 E. Yamhill St.
Gehr, Harritt B, North Powder, Ore.
George, Mrs. Mary E., 616 Market Drive.
George, Melvin C., 616 Market Drive.
Getz, Florence I., 1016 Clackamas St.
Gile, Miss E E., 770 Flanders St.
Gill, Mark, Care of J. K Gill Co.
Gilmour, W. A, Title & Trust Bldg.
Glascock, W. V., 115 N 23rd St.
Glisan, Rodney L., 612 Spalding Bldg.
Glover, Truman J, Fairview, Ore.
Goddard, E C., 491 Mill St
Goldapp, Martha Olga, 455 East 12th St.
Goldrainier, Adele, 22 W. Jessup Street.
Gorman, M W., Forestry Bldg.
Gould, Joseph L., 312 City Hall.
Graves, Henry S, U. S. Forestry Service, Wash., D C.
Gray, R. W., 724 Y. M. C. A.
Greeley, Gen'l. A. W, Washington, D. C.
Griffin, Margaret A., 303 Title & Trust Bldg
Griffith, B. W., 1736 Kane St., Los Angeles, Cal.
Guerne, Chas. A, Athena, Ore.
Haffenden, A. H S, 4236 49th Ave., S. E.
Halas, Dr Jan V., 67 Front St.
Hardesty, Wm. P., 60 E. 31st St. City, or 418 City Hall
Harnden, Edward W., 1617 Barristers Hall, Boston, Mass
Harnois, Pearle E, 1278 Williams Ave.
Harris, Charlotte M., 1195 E. 29th, N.
Harza, L F, Great Lakes Power Co, Ltd, Sault Ste. Marie, Ont.
Hathaway, Warren G, 1288 Rodney Ave
Hawkins, E. R., 655 Everett St.
Heath, Minnie R., 665 Everett St.
Henderson, G P, 1135 E. Taylor. St.
Hendrickson, J Hunt, 211 North 24th St.
Henry, E. G., 488 N 24th St.
Henthorne, Mary C, Library Assn, 10th and Yamhill Sts
Hermann, Helen M., 965 Kirby St.
Heyer, Jr, A. L, 253 6th St. or Pacific Power & Light Co, 1220 Spalding Bldg.
Hibbard, Wayne E, Pocatello, Idaho
High, A s s, 300 N. 13th St., Vancouver, Wash ugu tu
Hild, F. W., Denver Tramway Co., Denver, Colo.
Hilton, Frank H., 502 Fenton Bldg.
Himes, Geo. H, Turney Blk, 2nd & Taylor Sts.
Hine, Andrew Randlett, 955 East Taylor St.
Hitch, Robert E., 602 Fenton Bldg
Hodgson, C. W., Rockland Ave., Park Hill, Yonkers, N. Y.
Hoff, Magda M., North High School, Minneapolis, Minn.
Hogan, Clarence, 591 Borthwick St.
Holman, F. C., 558 Lincoln Ave., Palo Alto, Cal.
Holt, C. R., 216 Failing Bldg
Holt, Walter H, 586 East Davis St.
Horn, C. L, The Wheeldon Annex.
Howard, Ernest E., 1012 Baltimore Ave, Kansas City, Mo.
Hudson, Wm. Maurice, 637-9 Pittock Block.
Ivanakeff, Pasha, 246 Clackamas St.
Jaeger, E. J., 135 6th St.
Jaeger, J. P., 135 6th St.
Jennings, E. C., Milwaukie, Ore.
Johnson, Carrie R. H, 971 Hillsdale Ave.
Johnson, H. G., 618 Nicolet St., Minneapolis, Minn.
Johnson, Mary Virginia, Box 296, Hood River, Ore.
Jones, Frank I., 307 Davis St.
Jones, Thos. R., 226 So. C. St., Arkansas City, Kans
Joyce, Alice V., 595 Lovejoy St.
Kach, F. G, 369 E. 7th St. N.
Karnopp, J L., Ry. Exchange Bldg.
Keith, E. Grace, 810 Corbett Bldg.
Kendall, Arthur C, Hotel Arthur.
Kerr, D. M. G, Canadian Bank of Commerce, Keremeos, B C.
Kerr, Dr. D. T., 556 Morgan Bldg.
Ketchum, Verne L, 705 Y. M C A.
Knapp, Mary L., 656 Flanders St.
Koerner, Bertha, 481 E 45th St., N.
Krebs, H. M, Washougal, Wash.
Kunkel, Harriet, 405 Larch St.
Kunkel, Katherine, 331½ Montgomery St.
Ladd, Henry A, Care of Ladd & Tilton Bank.
Ladd, W. M, Care of Ladd & Tilton Bank.
Langley, Manche Irene, Forest Grove, Ore.
Larson, Edward G, Buena Vista Apts.
Lawffer, G. A, 309 Stark St.
Lawson, Agnes G, 767 Montgomery Drive.
Leach, John R., 4719 72nd St.
Leach, Mrs. John R, 4719 72nd St.
Leadbetter, F. W., 795 Park Avenue.
Lebb, David, 502 Corbett Bldg.
Lee, John A., Concord Bldg
Leppich, Elsa, 611 Corbett Bldg.
Letz, Jacques, Care of Scandinavian-American Bank.
Lind, Arthur, Care of U. S. National Bank.
Loucks, Ethel Mae, 466 East 8th St.
Luesing, Theo. N, 595 E. Pine St.
Luetters, F. P, 131 E. 19th St.
Lund, Walter, 191 Grand Ave., N
Luther, C. V., E 34th & Belmont Sts.
McArthur, Lewis A., 561 Hawthorne Terrace.
McBride, Agnes, 1764 E. Yamhill St
McClelland, Elizabeth, 323 E. 12th St. N.
McClure, Olga, 407 E. 50th St. N.
McCollom, Dr. John W., 556 Morgan Bldg.
McCray, Elmer, 328 Pine St.
McCready, Sue, 512 U. S. National Bank Bldg, Vancouver, Wash.
McCulloch, Charles E, 1410 Yeon Bldg.
McDaniel, Adrienne, 874 Laura Ave.
McDaniel, Ida, 784 Laura Ave.
McDonald, Laura H, 354 E. 49th St
McIsaac, R. J., Parkdale, Ore.
McLaughlin, Jessie A., 648 Patton Road.
McLennan, Margaret, Box 324, Honolulu, T. H.
McLeod, R. L, 468 E. 37th St.
McNeil, Fred H., Care of The "Journal".
McNeil, Florence J., 607 Orange St.
Mackenzie, Wm. R, 1002 Wilcox Bldg.
Mallahan, Clo., Y. M C. A.

Manners, C H, Underwood, Wash.
Marble, W. B, 3147 Indiana Ave., Chicago, Ill.
Markham, B C, 343½ Washington St.
Marsh, J. Wheelock, Banks, Ore
Marshall, Bertha, Station A, Vancouver, Wash
Martin, J. C, 524 East 51st St. North
Martin, Mrs J. C, 524 East 51st St, N.
Mason, J. M, R. F. D., No 2, Milwaukie, Ore.
Mathis, C. J., 149 Sixth St
Mears, Henry T, 494 Northrup St.
Mears, S M, 721 Flanders St.
Meredith, John D, 329 Washington St.
Merriam, Dr C. Hart, 1919 16th St. N. W Wash, D C.
Merten, Chas J., 307 DaVis St.
Metcalf, Alice K, 531 E Couch St.
Metcalf, Edna, 531 E Couch St.
Metcalf, Gertrude, 680 East Madison St.
Mills, S, American Power & Light Co, 71 Broadway, N. Y.
Millais, Mrs Ada M Rice, 415 10th St.
Millais, James A, 415 10th St
Miller, Beulah F, 629 East Ash St.
Miller, Jesse, 726 E 20th St
Miller, Maude Ethlyn, 767–15, Eugene, Ore
Mills, Enos A, Longs Peak, Estes Park, Colo
Monroe, Harriett E, 1431 East Salmon St.
Montag, John W, 883 Commercial St.
Montague, Jack R, 1310 Yeon Bldg.
Montague, Richard W, 1310 Yeon Bldg.
Moore, Edith, 547½ Sixth St.
Morgan, Mrs. C N, Box 144, Palms, Cal.
Morkill, Alan Brooks, Canadian Bank of Commerce, Victoria, B C
Munger, A R, 28th & Washington Sts., Vancouver, Wash.
Myers, Earl, Sunnyside, Wash.
Nelson, L. A., 410 Beck Bldg
Newell, Ben W., Care of Ladd & Tilton Bank.
Newlin, Harold V., 202 Fenton Bldg.
Newton, Josephine, 1350 Pine St., Philadelphia, Pa.
Nichols, Dr. Herbert S, 802 Corbett Bldg.
Nickell, Anna, 304 College St.
Nilsson, Martha E, 320 E 11th St. N.
Nissen, Irene, 969 E. 23rd St. N
Nordeen, Edith, 361 Graham St.
Norman, M Oscar, 499 E 9th St. N.
Northup, H. H, 599 Elizabeth St.
Nottingham, Jessie Ray, 271 E. 16th St. N.
Nunan, Cinita, 489 W Park St.
O'bryan, Harvey, McKay Bldg.
O'neill, Mark, Worcester Bldg.
Oglesby, Etta M, Baron Apt. 14th & Columbia Sts
Ormandy, Harry W., 501 Weidler St.
Ormandy, James A, 501 Weidler St.
Otis, Madge I., 724 Cascade Ave, Hood River, Ore.
Parker, Alfred F., 374 E 51st St
Parsons, Mrs. E T., Mosswood Rd., Univ. Hill, Berkeley, Cal.
Pastoriza, Hugh, Technology Club of N. Y., No 17 Grammarcy Park.
Patterson, Neva, 876 Gantenbein Ave.
Pattullo, A S, 500 Concord Bldg.
Pauer, John, 485 East 20th St N.
Payton, Perlee, G, 3916–64th St. S. E.
Peaslee, W. D, 125 East 11th St.
Penwell, Esther, 72 East 82nd St. S.
Peterson, Arthur S, 780 Williams Ave.
Peterson, E. F., 780 Williams Ave.
Peterson, Laura, H, 309 College St
Pfaender, Freda, 171 E. 29th St. W.
Phillips, Mabel, 335 14th St.
Pilkington, Thos. J, R. D. 2, Sebastapol, Cal
Pittock, H. L, 812 Overton St, Imperial Heights.
Plummer, Agnes, 3rd & Madison Sts.
Porter, Fannie G, 1010 Jackson St., Oregon City, Ore.
Powell, Mary, E, 1330 E. Taylor St.
Powers, Paul B, 502 Spalding Bldg
Prevost, Florence, East Broadway Bet. 8th & 9th St
Pridemore, L. F., Government Camp, Rowe, Ore.
Rauch, G. L, 510 Yeon Bldg.
Rea, R W, 403 E 16th St
Reed, Rose Coursen, 208 Eilers Bldg
Reese, Lela, Hot Springs Hotel, Stevenson, Wash
Reid, Prof. Harry Fielding, Johns Hopkins University, Baltimore, Maryland.
Reist, Linn Landis, 600 Chamber of Commerce
Richardson, Jean, 888 E Washington St
Richardson, Nettie G, 888 E Washington St.
Rice, Edwin L, 1191 E. Yamhill St.
Riddell, Geo X, Hotel Frye, Seattle, Wash.
Riddell, H H., Yeon Bldg, City
Riddell, Morse, Care of H H Riddell, Yeon Bldg
Riker, Dr Leah S, 1020½ Belmont St.
Riley, Frank Branch, Chamber of Commerce Bldg.
Riley, Laura, Madison Park Apartments.
Riley, Lotte Brand, 61 Lucretia St.
Riseling, R F, 776 E Yamhill St
Roberts, Arthur L, 609 Selling Bldg.
Roberts, Ella Priscilla, 109 E. 48th St.
Robinson, Earl C., 658 Morgan Bldg.
Robinson, Lloyd, 683 E Burnside
Roblin, C. W, 40 No. 9th St
Roemchild, Mrs Otto R, 211 Miller St. Knoxville, Pittsburgh, Pa.
Rogers, Homer A., Pardkale, Ore
Roosevelt, Theodore, Oyster Bay, N. Y.
Root, Mrs E T, 405 Henry Bldg
Rosenkrans, F A, 335 E. 21st St. N.
Ross, Rhoda, 1516 E Oak St.
Ross, Willis W, 494 Yamhill St.
Routledge, Fred A, 159 67th St. N.
Royal, Osmon, 735 Y M C A.
Rucker, Willard, 681 E Ash St.
Rustin, Marie, Dwand Hospital, Pekin, China.
Sammons, E. C., 69 E. 18th St.
Sammons, Reta, 69 E. 18th St.
Sammons, Aldon, Century Club, 7 West 43rd, New York City.
Schlieske, Louise, 818 Clackamas St
Schmidt, Lucie C., 667 Everett.
Schneider, Marion, 260 Hamilton Ave.
Schuyler, James T, 655 Kearney St.
Scott, Elsie, 1565 Knowles.
Searcy, Robt. D, 1334 Northwestern Bank Bldg
Sharp, J. C, 785½ E. Main St.
Sharp, Mrs J. C, 785½ E Main St
Sheehan, John D, M D, 701 Dekum Bldg
Sheldon, Mrs E C, 822 Halsey St.
Shipley, J. W, Underwood, Wash.
Sholes, Charles H, Box 243 City
Sieberts, Conrad J., 623 E 16th St.
Sill, J. G, 539 Vancouver Ave.
Silver, Elsie, 100 6th St.
Skelton, Effie A, 1627 Peninsular Ave.
Smedley, Georgian E, 262 E. 16th St.
Smith, Dorsey B, 116 3rd St
Smith, Mrs. Dorsey, No 8 Trinity Place Apts.
Smith, Geo. Choate, 727 Chamber of Commerce Bldg
Smith, Kan, Ketchikan, Alaska
Smith, Mrs. Kan, Ketchikan, Alaska.
Smith, Leotta, Oswego, Ore, R F D.
Smith, Warren D, 941 E 19th St.
Smith, Wm L, 328 Schuyler St.
Snead, J. S. L., 505 Weidler.
Spaeth, J. Duncan, Princeton, N. J.
Sparks, J. C., 416 City Hall
Spurck, Nell I., The Campbell
Stanford, Mildred, 1811 Water St., Olympia, Wash.

STARR, NELLIE S, 6926 45th Ave, S. E.
STEARNS, LULU, 4148 65th Ave. S. E.
STONE, W. E., Purdue University, LaFayette, Ind.
STONE, Mrs, W. E., 146 N. Grant. St., LaFayette, Ind.
STRONG, ANNA LOUISE, 508 Garfield Ave., Seattle, Wash.
STUDER, GEORGE A., 1114 Williams Ave.
STURGES, DANIELA R., 648 Gerald Ave.
TAYLOR, RAYMOND E, 320 Morrison St.
TAYLOR, VERA E., 604 Spalding Bldg.
TENNESON, Alice M., Care of High School, North Yakima, Wash
THATCHER, GUY W., 302 Sacramento St
THAXTER, BENJ. AUGUSTUS, 994 Bryce St.
THORINGTON, DR. J. M, 2031 Chestnut St., Philadelphia, Pa
THORNE, H. J., 452 E. 10th St. N.
THORNTON, OLIVER C, 691 Locust.
TIFFT, ARTHUR P, 710 Chamber of Commerce.
TINDOLPH, A. G, Title & Trust Bldg.
TOWNE, EDWARD W., 416 City Hall.
TREICHEL, CHESTER H, 535 Mall St.
TREICHEL, GERTRUDE, 535 Mall St.
TRENHOLME, ARTHUR K, 333 E. 44th St.
TUNZAT, MARIE, 367 Wygant St.
UPLEGER, MARGARET, University of Ore., Eugene, Ore.
UPSHAW, F. B, 594 Ladd Ave
VAN BEBBER, L., 501 Fenton Bldg.
VAN ZANDT, DEAN, 849 Front St.
VEAZIE, A L., 695 Hoyt Street.
VERNON, HOWARD W., 22 Reade St., New York City.
VESSEY, ETHYLE, 1207 W. 18th St., Vancouver, Wash.
VIAL, LOUISE ONA, 580 E. Main
VOLKMAN, S A, Mult, A. A Club.
WADE, GERTRUDE, 390 Hall St.
WAGNON, COLOMA M, 603 Sixth St.
WALDORF, LOUIS W., Western, Neb.
WALKER, M L, Corvallis, Ore
WALTER, WILLIAM A S, 55 N 21st St.
WARNER, CHARLES E., Santa Barbara, Cal.
WATTERS, REV. DENNIS A, 321 East 8th, N.
WATTERSON, LOUELLA, Carlton Hotel.
WEER, J. H, Care of West Coast Groc. Co.
WEICHELT, O. H, 6015 Hillegan Ave, Oakland, Cal
WEIR, L. H., Albuquerque, New Mexico.
WEISTER, G. M, 653 E 15th N.
WENNER, B. F., Route No. 2, Milwaukie, Ore.
WHITE, WM JR, 1302 Commonwealth Trust Bldg., Philadelphia, Pa.
WILBURN, VESTA, Lincoln Apts.
WILDER, G W., 226 14th St.
WILLARD, CLARA, 112 W. 10th St.
WILLIAMS, A BOYD, 131 East 19th St.
WILLIAMS, GEO M, 713 "F" St., Centralia, Wash
WILLIAMS, JOHN H., 938½ Pacific Ave., Tacoma.
WILLIS, ARZA M, Touchet, Wash
WILSON, A, 1324 Cascade, Hood River.
WILSON, CHAS. W, Bellevue, Idaho.
WILSON, RONALD M, The Hill Hotel.
WINTERS, C. L, 240 East 32nd St.
WOLBERS, HARRY L, 208 16th St.
WOLD, HAROLD L, 423 Abington Bldg.
WOODWORTH, C. B, Ladd & Tilton Bank.
WORTMAN, CHAS H, 6034 46th Ave, S. E.
WYNN, FRANK B, DR, 421 Hume-Mansur Bldg, Indianapolis, Ind
YORAN, W C, 912 Lawrence St, Eugene.
YOUMANS, W. J., 687 E 9th St
YOUMANS, MRS W. J, 687 E. 9th St
YOUNG, BEATRICE, 228 Alberta St.
YOUNG, EMILY, 360 E 9th St, N.
YOUNG, KATE E., 360 E. 9th St., N.
YOUNGER, NELL, Hotel Ramapo.
ZIEGLER, MAE, 89 Mason St.
ZWEINER, THERESA, Blooming Prairie, Minn.

US
US
W. S. KIRK'S
ARMY and NAVY GOODS STORE
Corner Third and Stark Streets, Portland, Oregon
A Mecca for HIKERS, Hunters and Autoists
BLANKETS KNAPSACKS SWEATERS
SHIRTS SHOES KHAKI CLOTHING
HATS SOCKS SHELTER TENTS
LEGGINS PONCHOS MIDDIES
We also carry a very large line of U. S. Army war implements converted into useful and ornamental decorations for homes, dens, etc.
We invite your inspection of our stock
Catalog upon request

Set in Monotype and printed by James, Kerns & Abbott Company, Portland, Oregon

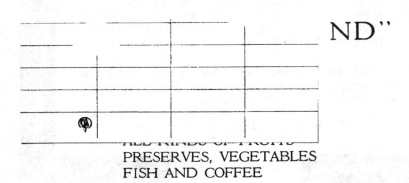

AZAMA

A Record of Mountaineering in the Pacific Northwest

NESIKA KLATAWA SAHALE

Published by THE MAZAMAS
213 NORTHWESTERN BANK BUILDING
PORTLAND, OREGON, U. S. A.

Fifty Cents

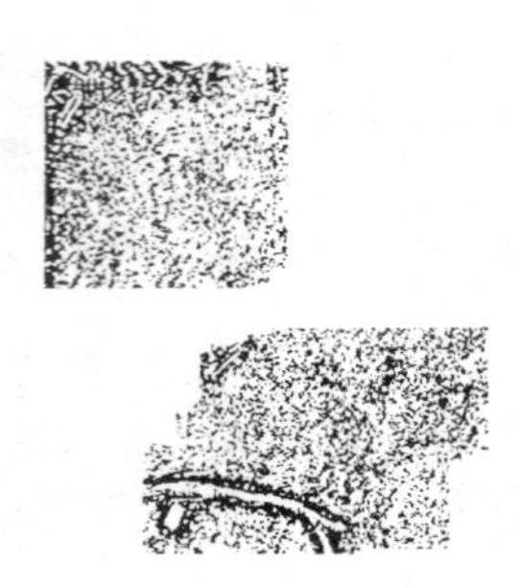

MAZAMA

A RECORD OF MOUNTAINEERING IN THE PACIFIC NORTHWEST

Publication Committee

BEATRICE YOUNG — ALFRED F. PARKER, *Editor* — PAULINE GEBALLE

VOLUME V — PORTLAND, OREGON, DECEMBER, 1917 — NUMBER 2

Contents

The feeding of the rivers and the purifying of the winds are the least of the services appointed to the hills. To fill the thirst of the human heart for the beauty of God's working—to startle its lethargy with the deep and pure agitation of astonishment—are their higher missons. They are as a great and noble architecture, first giving shelter, comfort and rest; and covered also with mighty sculpture and painted legend.

—RUSKIN

Mt. Jefferson from Mazama temporary camp in Jefferson Park

MAZAMA

VOLUME V | DECEMBER, 1917 | NUMBER 2

Mt. Jefferson Outing, 1917

By MARION SCHNEIDER

Just ten years following the last previous official Mazama outing to Mt. Jefferson, a party of over fifty left the Union station, Portland, at 8 p. m., on Saturday, August 4, 1917, for Detroit, Linn county, the railroad terminus. The Pullmans were sidetracked for several hours at Albany, but reached their destination, as scheduled, at 4:30 a. m. on Sunday. Everyone set out at once in the gray of the morning, as five of the twenty miles to our camp at Pamelia lake were to be covered before breakfast. For a mile or more the party trudged along the railroad track. At Hoover the path followed an old road grade. The underbrush had grown up in dense profusion on both sides and gave the impression of walking through an enormous corridor. Breakfast was served in a grove of large trees on the bank of the Santiam river. And such a breakfast—fried ham, eggs, coffee and biscuits, served as only Chef Weston can serve them.

After a short rest the party continued on their way, following for miles the shore of the wild and picturesque Santiam. Gradually the trail led up into the heavier timber, crossing and recrossing the ever-present mountain streams. At Woodpecker ranger station, a mile above the junction of Whitewater creek and the North Santiam river, a stop was made for lunch.

The remaining seven miles to the lake were a source of constant delight. The trail wound in and out among gigantic trees. Suddenly and most unexpectedly a big tent loomed up right in the heart of the woods. Because the feeling of being in the mountains was entirely absent, some doubted that this tent was the Mazama kitchen. However, the sight of Jack Benefiel, serving soup to the advance group, was most reassuring.

Just in front of the big tent were the usual out-of-door tables, but this time they were made in steplike sections because of the slope. Beyond was a beautiful lake, bordered with great trees and protected by the high, barren ridges above. Mt. Jefferson was not visible, but a trail led a short distance around the lake to a point where one could glimpse the rugged pinnacle. No snow-fields or glaciers were in view; just the bare, rocky crest and the tops of intervening foothills.

The first day in camp was spent in much the same manner as all previous first days—in getting settled. Along the lake shore were innumerable camp-sites from which to make a selection. There was a plentiful supply of boughs for beds and an abundance of wood for the camp-fires. A path was cleared from the women's quarters to the lake, in preparation for the early morning plunge that was so thoroughly enjoyed each day by nearly everyone in camp. Rafts were built, and even before the close of the first day enthusiastic fishermen were bringing back in abundance the delicious trout which formed such a welcome part of the camp menu.

On Tuesday morning the training for the final climb was begun. Mr. W. C. Yoran guided a party of thirty-five to Smith's Cove. The trail led through heavy timber, with a gradual increase in elevation, although at no time was there a broad, sweeping view of the country. In the cove was an attractive mountain meadow, with the ever-present little lake and snow, and flowers in wild profusion.

While the untried were being initiated into the delights of mountain climbing, eight hardy Mazamas made the trip to the summit in order to scout the best route for the official climb. They found the climb difficult and tedious, owing to the character of the mountain. Although undoubtedly of volcanic origin, the typical cone shape of the volcanic mountain was no longer in evidence. The main body of the ridge was found to be composed of ash and loose pieces of lava, thus making alternating sections of hard and soft material. The summit, at an elevation of 10,523 feet, was formed by an abrupt, sheer mass of jagged, crumbling rock. Ropes fastened at the crest and at various places where secure rock could be found, were left hanging against the face of the pinnacle to serve as moral support, and physical if needed, to the less experienced

Wednesday was a most unusual day for Mazamas on an annual outing, for nearly everyone stayed in camp. The air at this elevation, 3500 feet, was so mild and pleasant, the camp was so comfortable and warm, that incentives for action were lacking. There was plenty of activity, however, in the afternoon; for the entire camp enjoyed an hilarious swimming frolic.

On Thursday the desire to become better acquainted with the mountain region returned more vigorously than ever, One party of four men started on a two-day trip around the mountain; another party climbed up Milk creek to Lighthouse rock, experiencing the joy of contact with the mountain itself; a third party journeyed to Hunt's Cove; a fourth climbed the ridge behind the lake to a tract of meadows

View in Hunt's Cove, showing the summit of Mt. Jefferson in the background. *Photograph by R. J. Davidson.*

Mt. Jefferson from the east end of Hunt's Cove. *Photograph by F. I. Jones*

Showing structure of rock on pinnacle *Photograph by F. I. Jones*

Looking south from the snow-fie ds o Mt. Jefferson, showing the Three Sisters, Mt. Washington, and Three-Fingered Jack (left to right)

Photograph by Winter Photo Company

known as "Grizzly Flats" from which a far-reaching view could be enjoyed.

Friday was the day selected for the beginning of the official climb. At four o'clock in the afternoon two parties, forty persons in all, set out for bivouac camp, at the timber-line. One group climbed the rough and precipitous bed of rock and gravel known as "Dry Gulch." The other, led by Dr. W. E. Stone, followed an easier grade through the timber and over the ridges. At seven, everyone had reached the destination. The location was ideal from every standpoint,—plenty of sheltered sandy spots, an abundance of boughs for beds, and great masses of snow available for the water supply. The party was loath to settle down to rest. The air was pleasantly warm, the ridges and meadows were visible for miles, and the horizon was radiant with the gorgeous coloring of sunset. But as there was a strenuous day ahead, by nine o'clock the peace and quiet of the mountain had settled down upon the little gathering.

At 4 a. m. everyone was up and ready for the final climb. The route selected led along the top of the ridges to the base of the pinnacle. Very little snow was encountered, the walking being mostly on crumbling rock and gravel. With every step, a little avalanche of dirt and shale went sliding down the mountain.

At 10:30 the first company, in charge of Mr. J. R. Penland, had reached the base of the pinnacle. The sight of that vast mass of crumbling rock, looming up abruptly for some four hundred feet, was awe-inspiring. Alpenstocks were useless for such work. The intrepid mountaineers climbed up the dizzy height, inch by inch, until after an hour's exertion the entire party of ten were assembled on the summit. The air was perfectly clear. Mt. Hood, fifty miles to the north, loomed up as if but a stone's throw away. Many other peaks of the Cascade range, from Rainier to Shasta, towered up to repay the effort of the climb.

As there was so little space on the crest, and as the danger from falling rock was ever present, the second company had to wait at the foot until the first had made the thrilling and hazardous descent. It was not until late in the afternoon that the last of the twenty-eight successful climbers returned to the base of the pinnacle. Because of the long wait at this point, it was 9 p. m. before the entire party had arrived at permanent camp.

Sunday was indeed a day of rest. At 11:30 services in charge of Dr. A. J. Montgomery were held in the grove. Dr. Henry Marcotte, whose genial and kindly personality had made him one of the most popular members of the camp, preached a sermon.

On Monday a party of sixteen, led by Dr. Stone, departed for bivouac camp. The party included Mr. J. E. Bronaugh and his son, who had come to camp to participate in the second official climb. On Tuesday morning Dr. and Mrs. Stone left the main group in order to try another route to the summit. Dr. Montgomery guided the others in a most efficient and successful manner. Both parties reached the top about the same time, and rested for an hour or more. The day was hazy and the distant view unsatisfactory, but the glaciers and parks about the mountain itself afforded a prospect which was thoroughly enjoyed. While the group was at the top, a gigantic roar was heard and an enormous avalanche of rock from the very pinnacle itself went thundering down over the glacier. No time was lost in climbing down to safety. The homeward journey was shortened by following a new route over the snow-fields. At 6:30 p. m. the entire party, under Dr. Montgomery's able leadership, marched up to the cook tent, ready to do justice to another of Chef Weston's delicious dinners.

An account of the outing would not be complete without a few words concerning the gatherings at the camp-fire. Mr. Yoran directed the singing and furnished much entertainment during his short stay in camp. Dr. Montgomery and Dr. Marcotte graciously accepted the responsibility of leadership during the last week. State Biologist William L. Finley gave entertaining and instructing talks on birds. Mr. H. T. Bohlman told about the birds in the immediate vicinity. Dr. Stone gave informal talks on mountaineering. Mr. Hall, Forest Supervisor of the Santiam Reserve, told about the work of the U. S. Forest Service. Mr. Ira A. Williams, of the Oregon Agricultural College, gave a talk on the geological formation of the Mt. Jefferson country. Intermingled with the serious discussions were merry, light-hearted songs and jests.

On Thursday thirty left camp for an outing in Jefferson Park, on the north side of the mountain. The main group followed the ridges across the very face of Jefferson and dropped down into "Hanging Valley," a veritable paradise of mountain parks.

The park itself extends about two miles from east to west and a mile from north to south. It includes hundreds of lakes and pools. In many places the snow-fields extend down to the edge of the water. At intervals, little ridges covered with fir and hemlock rise between the meadows. The mountain itself, with its massive rocky ridges, its wonderful glaciers and long snow-fields, dominates the view. The meadows, one above the other in a shelf-like arrangement, are a never-ending source of joy. The little streams meander through hedges of brilliant flowers, which seem to grow in masses of the same color;

here a bed of gorgeous paintbrush, there of purple lupine; yonder, against the snow, a mass of blue gentian; then a patch of yellow; mountain flowers of all kinds in wild and glorious profusion.

Every minute of Friday was filled with joyous and enthusiastic explorations of the park. Regret was constantly expressed that but one day could be spent here.

On Saturday one group left with a pack train, bound for Detroit by way of Wild Cheat Meadows; the other by way of Breitenbush hot springs. The first group followed Whitewater creek to the lower end of the valley. From here a well defined trail led up the ridge, across the edge of a twelve-hundred-acre burn, down to Wild Cheat ranger station. The party camped here for the night.

Early on Sunday morning the three groups, one from Pamelia lake, another from Wild Cheat Meadows, and the third from Breitenbush, started down the trail. There was a merry reunion at Detroit, where the parties boarded the specials for Portland. Another successful chapter in mountaineering was closed.

LIST OF PERSONS WHO CLIMBED PINNACLE OF MT. JEFFERSON ON MAZAMA OUTING OF 1917

ALLEN, ENID C., Minneapolis.
AYER, ROY W., Portland.
BENEDICT, LEE, Portland.
BENEDICT, MAE, Portland.
BLAKNEY, C. E., Milwaukie, Oregon.
BRONAUGH, GEORGE E., Portland.
BRONAUGH, JERRY E., Portland.
BOHLMAN, HERMAN T., Portland.
CHAMBERS, MARY E., Eugene, Oregon.
DAVIDSON, R. J., Portland.
DAVIDSON, MRS. R. J., Portland.
EVERSON, F. L., Portland.
FAGSTAD, THOR, Cathlamet, Wash.
FORD, G. L., Portland.
GIRSBERGER, MABEL R., Portland.
GOLDAPP, MARTHA OLGA, Portland.
GRIFFITH, B. W., Los Angeles, Calif.
HATCH, LAURA, Northampton, Mass.
HARDESTY, W. P., Portland.
HENTHORNE, MARY C., Portland.
HOGAN, CLARENCE, Portland.
IVANAKEFF, PASHO, Portland.
JONES, F. I., Portland.
KNAPP, MARY L., Portland.
LOUCKS, ETHEL M., Portland.
LUETTERS, F. P., Portland.
MARCOTTE, HENRY, Kansas City, Mo.
MARSH, J. W., Underwood, Wash.
MILLER, BEULAH, Portland.
MONTGOMERY, ANDREW J., Portland.
MONTGOMERY, BERNARD, Portland.
NILSSON, MARTHA E., Portland.
NORDEEN, EDITH, Portland.
PENLAND, JOHN R., Albany, Oregon.
PETERSON, ARTHUR S., Portland.
PETERSON, E. F., Portland.
ROYAL, OSMON, Portland.
SCHNEIDER, MARION, Portland.
SHERMAN, MINET, Portland.
SMITH, LEOTTA, Oswego, Oregon.
STONE, DR. W. E., Lafayette, Indiana.
STONE, MRS. W. E., Lafayette, Indiana.
TOWEY, WM. J., Portland.
WOLBERS, H. L., Portland.
ZIMMERMAN, DARL, Eugene, Oregon.
ZIMMERMAN, T. J., Portland.

Mt. Jefferson

By Ira A. Williams

Oregon Bureau of Mines and Geology

If I were asked which of all the varied features of the Cascade range in Oregon offers the strongest appeal to the real lover of the out-of-doors, I should find no alternative but to reply, without hesitation or tremor of voice, the Mt. Jefferson region. This is so not because there are no other sections of the range dominated by prominent snow and ice-clad peaks, deep canyons and innumerable lakes. For there are, between the California border at the south and the Columbia river, an intermittent serial chain of them. Does Mt. Jefferson and its scenic environs stand out, then, because of its position, as the master jewel in this two hundred and fifty-mile string of most splendent gems? It is true that the mountain itself can be seen from far beyond Mt. Hood at one end, from Mt. McLoughlin in the Crater Lake region, and even from Mt. Shasta beyond the California line at the other. It occupies a medial location and, at a distance, an apparently commanding one. But it is no more pre-eminent a mountain peak than are the Three Sisters forty miles to the south of it, and even less so than Mt. Hood, which overlooks the Columbia far to the north.

The cause for its acknowledged appeal must therefore be sought at close range, and will be found only by those who are able to visit the mountain, who study its features, and particularly its surroundings. If we would recognize the best in any human friend, we cultivate acquaintanceship. So, too, will the lure of mountain and forest grow with close association; and just so will the individuality of Mt. Jefferson be revealed only to the sympathetic person who seeks it in a spirit of friendship and appreciation.

This great mountain rises from the very summit of the Cascade range, in truth one of nature's brilliant gems. It is the sculptured form of a once lofty volcano whose belchings forth in the time of greatest activity spread volcanic materials from its crater over a large range of surrounding country. That man-cut gems may properly display their beauty, and thus evidence their intrinsic value as articles of adornment, suitable and costly settings are provided whereby native scintillations are vivified, elusive sheen or play of color enhanced, the jewel itself made a thing of abounding life. No less does nature do, and more effectively does she proceed often to surround and to support her masterpieces with a setting or environment from which they, as nuclei, shine forth in all their fascinating splendor.

The spreading lower slopes of Mt. Jefferson are aproned with forest. Alpine types have crept persistently upward until brought to a dwindling and protesting halt by the downcoming mantle of ice and snow. About its feet and for miles along the Cascades summit to north and south, as uncounted as they are indescribably beautiful, lakes, some of them snow-encircled the year through, are sprinkled with methodical indiscrimination amongst the clumps of firs and hemlock, beneath impending cliff, at the brink of precipice. The interstices among the trees and lakelets, which are frequently of such proportions as to be appropriately termed meadows, are occupied with grassy sod, often boulder-besprinkled, and invariably adorned with the more brilliant, not to forget the modest hues as well, of springtime flowers. As a rule, these small lakes are unconnected, except during a portion of the year when the accumulated snows of winter reach from one to another, and designedly protect sward, flower patches, lakelets and all, until each summer's sun brings them forth to life again. In places, however, the waters drain into the deep canyons that quickly drop away from the mountain to east and west, and they do so by way of meandering streamlets from lake to lake that sooner or later plunge over rock cliffs, some of magnificent height, in repeated series of splendid falls, foaming rapids and cascades.

We who have visited the Mt. Jefferson region have grown into the habit of calling these beauty places about the base of the mountain coves;" where they are of considerable expanse and fairly even surface the term "flat" is a much used though widely inaccurate designation. More comprehensive still, and quite appropriately, is the name "park" applied to suggest and to describe the larger, better defined lake- and pond-dotted, rock-studded areas where attractively spaced tree clumps, shrub clusters and interspersed meadow lend character and a comforting charm. Thus we know and revel in our knowledge of Smith and Hunt coves, for example, and Grizzly flat to the south of the mountain; while Jefferson Park at the north base has never failed to inspire even the most casual visitor with the enchantment of its beauty, and the wildness of its surroundings, which are as yet largely unexplored.

In a setting such as this, there is abundant reason for the superior attractiveness of the great mountain and for the irresistible charm of the Mt. Jefferson region. It is the mounting of the jewel, therefore, as much as it is the gem itself, that makes of it an object of distinction. It must be remembered, however, that the ignoble gem, as the man of sterling worth, is rarely lifted into the distinguished class by surroundings alone. An unquestioned inherent integrity, a perfection of modeling, the gleam of quality, must exist, or commonplaceness remains in the presence of even the most elaborate of environment. Mt. Jefferson has been made a wonderful mountain through the cooperative influence of its surroundings, although, alone, it would stand out as a thing of inspiring beauty and nobility.

Let us briefly observe some of its particular characteristics. At once we recognize its volcanic character. Every piece of rock of which it is composed has come forth from within the earth by volcanic cruption. Its vast height and bulk are due to the heaping up of materials that issued from its crater. But today there is no evidence of a crater in its top, nor is its form that of the finished volcano. Something—many things—have happened since it ceased to grow.

Again we may revert to the simile of the gem. This one of nature's mountain children was brought into being by the forces of volcanism. It was doubtless originally immensely larger, more symmetric, and we know not how much loftier than now. But it would seem that the task, if such it may be regarded, was not completed by mere accumulation of a great heap of rock materials. The evidence we have for this assumption is the fact that forces, nature's own agencies, went to work promptly to shape, to alter and to remodel the new-born mountain, just as does the lapidary with the native stone. Indeed, we have no gem at all until its innate qualities are revealed from beneath the rough.

Mt. Jefferson's sides were first etched by the wind and rains. Streams of water trickled, then cut and gouged their way into the slopes of the former mountain until they became coursed with a series of gullies, gulches, canyons. Its great height made of it a vast theatre for the reception of long centuries of snowfall. At intervals untold quantities of the snow must have moved down its slopes in enormous avalanches, then melted away and thus added to the eroding strength of the ever-growing streams. As the higher slopes became more and more roughened, and on account of variations in the hardness of the rocks encountered, its configuration finally became sufficiently irregular that, instead of moving downward suddenly as avalanches, the snows accumulated, not for merely a year or so, but for many years, and in various positions upon the mountain's sides. So long did the snow pile up that by its own weight much of it became compacted into solid ice. After the quantity of ice and snow had become very large, so large that their inclined position was an unstable one, they responded to the pull of gravity and began to move downward. Thus were the glaciers born. And at once, when they came into being, another relentless molding force was brought into action in the remodeling of our mountain. Immense thickness and weight gave to each ice stream a power to bite deeply into the mountain, to excavate and to carry away unmeasured amounts of its substance.

And so we liken unto the polisher of gems these powers of nature, wind, water, and ice, in their busy transfiguration of a noble mountain. The deep niches they have chiseled into it are the facets of the cut gem, and the rock ribs and projecting promontories upon its sides the dividing angles and edges between contiguous refracting faces.

True, nature has dug deeply into the mountain, so deeply that even portions of its bony framework appear to be exposed. And what a dull, comparatively lifeless object to look upon had the process been one only of digging and removal! The lapidary makes first cuts with rough tools, grinds away first with coarse abrasives; but follows always with painstaking polishing processes that uncover and bring out innate quality to best advantage. May we believe that the great glaciers and their attendant snowfields on Mt. Jefferson are but the concrete expression of a method pursued to show forth her primitive beauty and proportions; that the gleam of the great draping cloak of white by which she is seen and recognized from near and far is a master and finishing stroke of the sculptor whereby this mountain gem has been made splendid and immortal in the eyes of man? It is an enticing thought indeed to entertain. And we may carry the analogy further. Infinitely beyond the artificial limitations of the polisher of gems, nature varies with the seasons the extent to which she enlivens the dullness of the mountain by the gleam of snow and ice. In winter, when the clouds hang low, as though to pierce the fog with its glare of whiteness, she covers it over, nearly or quite entirely, with a mantle of fresh snow. In summer the dark rocks show through, the escalloped borders of the enshrouding frigid skirt contract, as though again to more effectively display, by contrast between snow and rock, the mountain and the forest, the bulk, the majesty and the grandeur of one of Oregon's most splendid mountain peaks—Mt. Jefferson, the Matterhorn of the Pacific Northwest.

The Glaciers of Mt. Jefferson

By LAURA HATCH

It has been known for some time that there are glaciers of fair size on Mt. Jefferson,* but as few details had been gathered, and practically nothing written about them, they offered an attractive field for investigation by the writer. So few people, indeed, had visited this section of the Cascade range of Oregon, that the three most important glaciers on Mt. Jefferson, those on the north and east slopes, were not named until 1915, when Mr. Ira A. Williams of the Oregon Bureau of Mines did so in his delightful description of the scenic features of the neighboring region.†

Although the time at the disposal of the writer was very limited, a number of interesting facts have been gathered concerning the glaciers, and with these in hand for comparison it will be easier at some future time to make observations of greater value.

The glaciers of Mt. Jefferson are practically confined to the north, east and southeast slopes of the mountain. On the south and west the snowfields are small and seem barely able to cling to the very steep slopes of the mountain (Fig. 1), and no glaciers are seen from the distance. On closer examination, however, a thin ribbon of stagnant ice is discovered at the head of Milk creek, which drains the western slope of the mountain, and which occupies a very narrow and deep canyon near its source. As this ice is undoubtedly the remnant of a former active glacier, it will be called for convenience the Milk Creek glacier. It is not now connected with any large snow-field and during August, 1917, water from the small patches of snow above was seen falling over the cliffs above the glacier before dropping beneath it. (Fig. 2.) The roar of the river beneath the ice could be distinctly heard throughout its length of about half a mile.

The Milk Creek glacier in 1917, beside being stagnant and covered to a certain extent with debris from the neighboring walls, was really only a shell or crust over the water that flowed beneath. This condition was best seen at the lower end of the glacier where the combined attack of the sun's rays above and the river below, has thinned the ice so that sections of it have fallen in, leaving caves and ice bridges. (Fig. 3.) For 300 feet or more farther down the valley, ice can be

*See brief notes by I. C. Russell in Journal of Geology, 1904, p. 261, and also in Bulletin 252 U. S. Geological Survey 1905, p. 124.

†Bulletin 1, volume 2, Oregon Bureau of Mines and Geology, 1915, pages 43-44.

Mt Jefferson from Jefferson Park

Photograph by Winter Photo Company

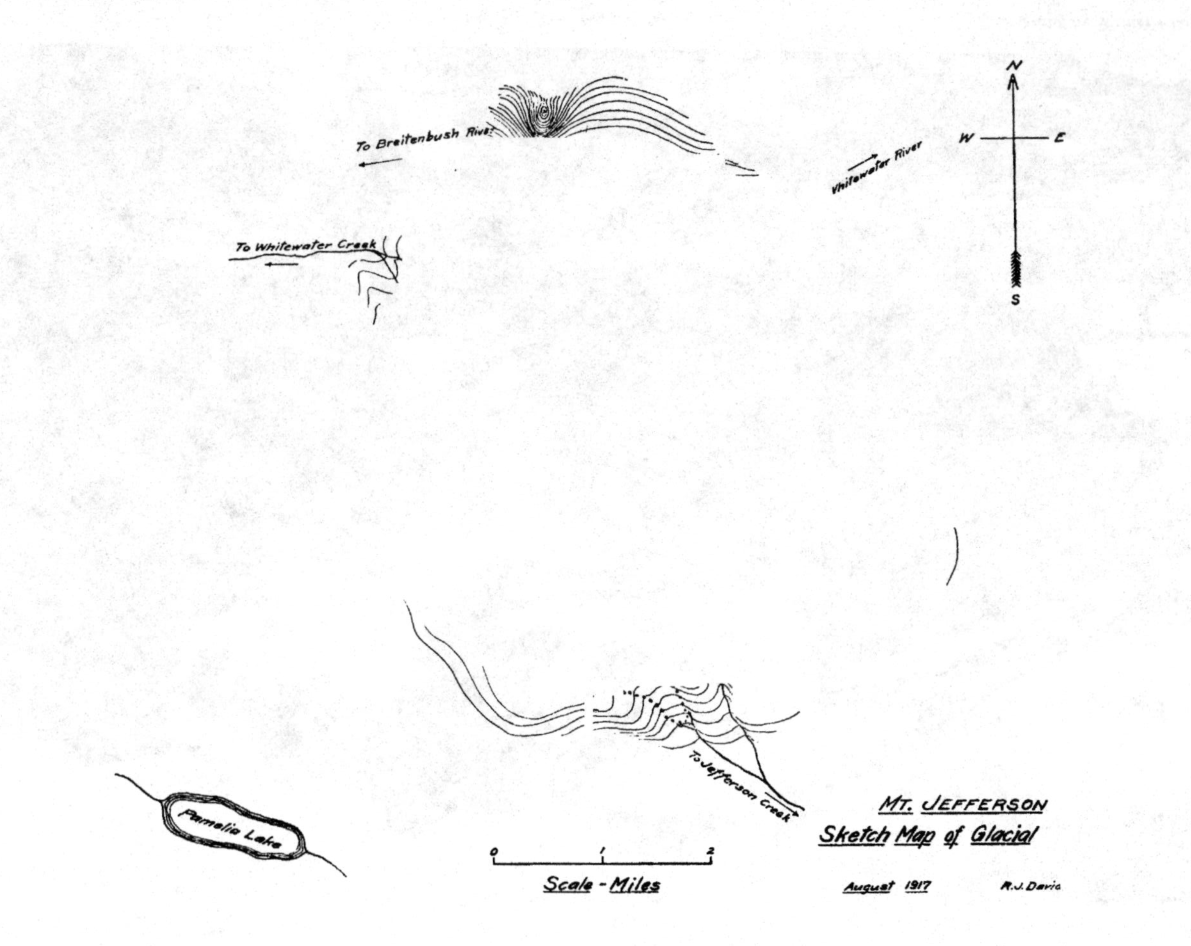
N
W
E
S
To Breitenbush River
Whitewater River
To Whitewater Creek
To Jefferson Creek
Pamelia Lake
0
1
2
Scale - Miles
MT. JEFFERSON
Sketch Map of Glacial
August 1917
R.J. Davis

found in the valley wall, showing a very recent and rapid retreat of the glacier.

As this ice tongue is so narrow (from twenty to fifty feet wide) and so thin, one might be willing to consider it just the result of the accumulation of snow of the preceding winter, if it were not for the abundant evidence of strong glacial action on all sides. Below the glacier the stream has intrenched itself in a sheet of coarse glacial drift, only here and there uncovering a ledge of lava over which it develops falls. As the valley widens to the west this sheet of morainic material is spread out over a fan-shaped area and terminates about a mile from the ice. No distinct terminal ridge was noted, showing that the glacier did not remain long in this position. On this coarse moraine only bushes and very small trees were found, proving it to be of very recent date. The writer was fortunate enough to meet Dr. A. J. Montgomery and his son, who had seen the glacier in this position in 1907. This date corresponds with that which had been worked out from the size of the vegetation, the largest trees being only about ten years old.

The size of the glacier in 1907 was great, of course, compared to its present extent, but it was slight compared to what it must have been when it built the high terminal moraine another half mile down the valley. (Fig. 4.) Here it must have been two or three hundred feet thick at least near its terminus and probably much thicker in the narrower part of the valley above.

The lack of connection of the glacier with the present snow-fields, and their very small size (at least in the late summer) make it hard to see what could have brought about these former glacial advances. If the head of the valley is studied it will be seen that the snow from practically all the western slope of the mountain would be coneentrated at the head of Milk Creek canyon. As this valley is so very narrow and deep a small change in the amount of annual snowfall might bring about a great change in the amount of ice in the canyon. The advances might have followed a series of years with a somewhat greater snowfall than usual or with cooler summers, or with a combination of climatic factors bringing about the greater accumulation of snow in the valley. The exact causes and conditions for the advance could of course be worked out only after careful records had been kept for a number of years and these compared to records from neighboring regions. Unfortunately no data has so far been obtained concerning the position of the ice in the few years preceding 1907, nor of the snowfall or other climatic factors, but as Mazamas and others have visited the region before that time, it is hoped that some information may be obtained.*

On the north and east slopes of the mountain, however, are glaciers that compare favorably in size with other glaciers in the United States. The largest one is that which covers the whole eastern face of the mountain and is fully five miles in greatest width, and a mile or two in length. It is more or less fan-shaped, radiating from the base of the pinnacle on the top of the mountain. From here it spreads out into five separate lobes to the edge and partly below the shoulder between the main mountain mass and the general plateau level. This glacier, or at least, the northern lobe of it which can be seen from Jefferson Park, Mr. Williams† has called the Whitewater glacier, because it drains eastward into the Whitewater and Deschutes rivers. The name is rather unfortunate, as there is a Whitewater creek on the northwest side of the mountain and as a glacier also drains into that, the two may well become confused. A view of the three middle lobes of the Whitewater glacier as seen from the summit of Mt. Jefferson is given in figure 5.

To the southeast is Waldo glacier which although in line with the lobes of the Whitewater glacier occupies a separate basin. In the summer of 1917 it was particularly characterized by three great crevasses which crossed it near its head. (Fig. 6.)

On the north slope of the mountain, overhanging Jefferson Park are two glaciers occupying distinct hollows at somewhat different elevations. The one farthest east, Mr. Williams has named the Russell glacier,‡ and the other one the Jefferson Park glacier. Excellent views of these glaciers can be obtained from the ridge at the head of Jefferson Park. (Fig. 7.) Little detailed work could be done on these glaciers for lack of time. Pictures, however, were taken from advantageous points of their termini, and it is hoped that they may be used at some future time to help determine the extent and rate of glacier movement.

The most interesting thing observed was distinct proof of a very rapid recent decrease or recession of the ice. The best evidence was found at the lower end of the Whitewater glacier where a double-crested moraine fringed the ice. In the inner ridge, glacier ice is found up to twenty and twenty-five feet above the present surface of the glacier. (Figs. 8 and 9.) The double-crested moraine is probably to be

*Anyone able to give information on these points is earnestly asked to communicate with the writer as soon as possible through the corresponding secretary of the Mazamas.

†Loc. cit.

‡Loc. cit. As there are already a glacier and a fiord named after Russell, in Alaska, it would be wise to rename this glacier too, if possible.

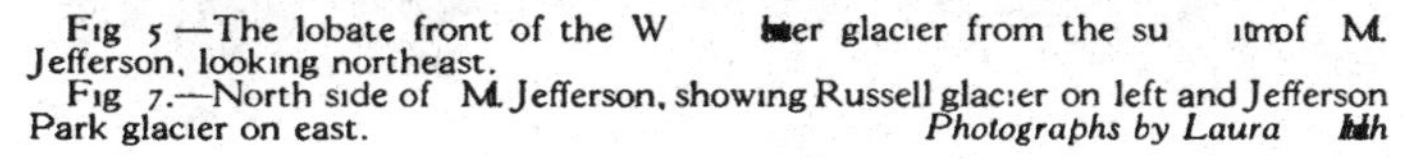

Fig 5.—The lobate front of the W hiter glacier from the su itm of M. Jefferson, looking northeast.

Fig 7.—North side of M. Jefferson, showing Russell glacier on left and Jefferson Park glacier on east.

Photographs by Laura hh

Fig 6.—Waldo glacier as seen from ar the su itm of Mt Jefferson and the irregular patches of snow on the plu to the south

Fig 8 - Ed moraine of the Wer glacier, showing evidence of ice hn the i er moraine up to twenty feet above the pent glacier surface

Fig 1.—Southwest slopes o M. Jefferson as en from Gly Ed. uth of Pamelia lake.

Fig 2—Head of Milk Creek glacier, August, 19, showing water from snow-fields above falling over rock cliffs before dropping h ae o gl.

Fig. 3 Ice caves ad bridges developed nrl er end of Milk Ck glacier.

Fig 4.—High terminal moraine ridge but oe ad one-half mls below terminus of Milk Creek glacier.

Fig. 9.—Near rew showing glacier ae in the inner moraine.

Photographs by lia hh

interpreted as due to the melting down of the surface of the glacier, rather than to a second advance after the first had been deposited. The two moraines are very nearly parallel and the inner one usually stands at about the same relative height compared to the outer higher ridge. The idea seems to be that the ice by melting has lowered its surface so much that the inner side of the moraine has gradually sunk down, making a separate ridge. As the glacier ice is still found so high in this inner moraine, it shows that the lowering of the surface has been very recent and very rapid. If this twenty-five feet is added to the twenty to forty feet that the inner ridge must have slumped down if it was ever continuous with the outer moraine, it makes the total thinning of the glacier a considerable amount.

Whether this recession of the glaciers in the Mt. jefferson region points to a gradual amelioration of the climate or is a local or temporary phenomenon, can only be determined after observations have been continued over an extended period of time, and compared with evidence from other parts of the world.

Other interesting things were noted but can be only mentioned at this time, for instance, the crevassing of the glaciers. All of them (except the Milk Creek glacier) had some cracks, showing that the ice was moving, but the most interesting ones were seen in the second lobe of the Whitewater glacier. From the top of the mountain the eye is caught by the unusual grouping of the crevasses into radiating curved lines. Nearer the surface of the glacier it is observed that the ice is really very much shattered, and impassable below a certain level on the mountain side. A very sharp hill evidently rises from the floor over which the glacier passes, for a mass of rock is exposed, on the upper side of which the layers of ice are upturned and deeply cracked. The structure of the ice is well shown by the layers of debris. The ice rides over the summit of the hill but breaks off just beyond and reforms below, making a tumbled mass of dirt-covered ice blocks which at a distance give the appearance of a medial moraine. In all the glaciers the deep crevasses near the heads (*bergschrund*) are most noticeable, but as these are clearly defined and the ice or snow between them is relatively smooth, they do not offer serious obstacles to the crossing of the glaciers, as do the crevasses farther down. The author hopes at some future time to go on with the study of these little-known glaciers, and in the meantime to gather data concerning the former positions of the ice and conditions of snowfall from any who have visited the region and are willing to furnish it.

An Unofficial Ascent of Mt. Jefferson

By W. E. Stone

On the southwestern face of Mt. Jefferson are to be seen two deep channels or gullies, which, commencing near the summit pinnacle, descend in nearly parallel courses and finally converge to form Milk Creek canyon, which then makes a sharp turn to the westward below the timber-line. These gullies are filled with snow and serve as chutes for a stream of falling stones, which start at the pinnacle and hurtle downward with incredible speed, coming to rest a mile below on the snow which fills the bottom of the canyon. The sharp ridge lying between these gullies is conspicuous in any view of this face of the mountain, and at first glance seems to offer a direct and interesting route to the summit. The official route for the ascent was however chosen upon the long curved ridge lying to the south or right of these gullies, as being less difficult, although of greater length. (See photograph on opposite page.)

On the second official ascent (August 14) the writer, accompanied by Mrs. Stone, left the main party on the ridge just above the bivouac camp, with the purpose of climbing by this central ridge, which so far as could be learned, was an untried route. To reach the foot of the ridge, it was necessary to cross the long scree slope descending into Milk Creek canyon from the south. Keeping our altitude from the point of departure we swung around the side of the great ampitheater and in about an hour reached the snow at the foot of the right hand gully or couloir. We crossed this without difficulty, there being few falling stones in evidence at this early hour, and by means of the second narrow tongue of snow reached the summit of the ridge. The preliminaries thus accomplished we now were face to face with the problems of the ascent. Seen at close range the ridge appears to be broken into benches, with perpendicular faces, alternating with steep slopes of broken rock and volcanic cinders, and interspersed with pinnacles or gendarmes. The lower third of the ridge is composed of a reddish yellow formation abruptly ending in a horizontal stratum of columnar basalt which afforded an interesting bit of climbing. Above this there was a long stretch of tedious work over loose sliding rock, lying at its natural angle of repose. We conjectured that this repose had not been disturbed during some ages and received the full benefit of the long accumulated resistance to gravity. Never have I encountered so much loose material resting on a hair trigger. Every projecting rock was the precarious support of an overflowing mass of

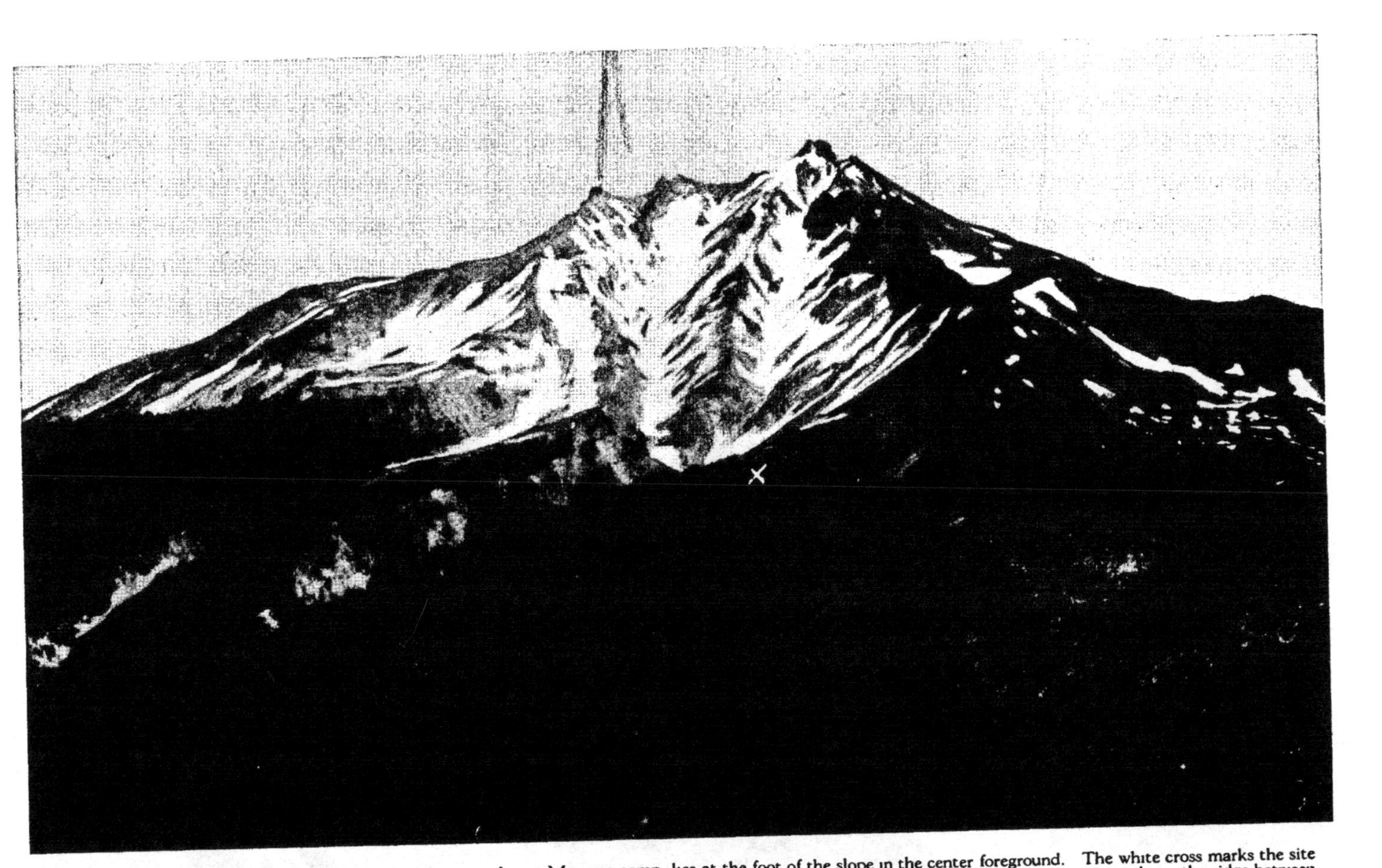

Mt. Jefferson from the southwest Pamelia lake, site of 1917 Mazama camp, lies at the foot of the slope in the center foreground. The white cross marks the site of the bivouac from which the official climb started. The official route was on the dark ridge to the right of the snow-fields. One ascent was made up the ridge between the snow gullies.

Photograph by W. E. Stone

Hanging snow-field at the head of Milk Creek glacier, Mt Jefferson
Photograph by Ira A. Williams

debris which at the least touch would come pouring down upon us. Evidently this would be an impracticable route for a large party of climbers. With only two of us, the utmost care was necessary to avoid an accident from falling stones.

Having accomplished about two-thirds of the distance, we came, just below the long snow field which extends up to the base of the final pinnacle, to a mass of conglomerate with a perpendicular face standing right across the ridge and threatening to block our ascent. On either hand its sides fell away so sharply into the gullies and discharged such a continual rain of falling stones that a flanking movement appeared to be impossible. There seemed no way but to go straight over the top, but the steepness of the face and the treacherous nature of the rock which offered no holds beyond the projecting constituents of the rotten pudding stone, soon showed this to be out of the question. A survey of the right hand slope gave no better results, and we turned to the left with little hope of success. A bold shoulder cut us off from a nearer view into the gully, but when by dint of hacking at the soft stone with the ice-axe, a precarious foothold was gained, the sight was far from comforting. Against the face of the rock pressed the solid ice walls of the gully resting at an angle of fifty to sixty degrees. Looking up one could see some fifty feet away a safe position whence the ridge resumed its normal slope. Looking down, the fearful ice slope disappeared behind a turn of the couloir suggesting unknown depths below. We hesitated to accept the challenge of this forbidding ice slope. Without ropes and with but two in the party, to venture upon such an enterprise involves serious risks. The only alternative was to turn back, which we were in no mood to consider, so after a careful study of the situation, I strengthened my foothold on the rock and then began to hack out the first step on the ice slope. Finally it was completed, of ample width and depth for both feet. Standing thus with the chasm below and the steep slope of blue ice rising before me, one firm step after another was slowly cut until the score reached fifty. Close behind came my companion, to whom I could give no assistance other than words of encouragement. Slowly we reached the summit of the ridge—the outstretched foot felt the firm rock and with an overwhelming feeling of relief we had surmounted what will always be remembered as a "nasty bit of climbing."

We were now clear of the rocks and standing at the foot of the conspicuous snow-field lying below and to the west of the summit pinnacle. Fortunately it was still firm from the night's freezing and not so steep as to require further step cutting. At twelve o'clock, seven hours from the bivouac, we were resting and lunching at the foot of the imposing rock pile which caps the mountain.

The main party had been in view on the sky line of the southwest ridge from time to time during the morning, but for the last two or three hours had not been seen. Our plan now was to skirt the base of the pinnacle around to the south and join them in the "red saddle" but they were nowhere to be seen and the intervening gully, bombarded by a constant stream of falling stones, prevented further progress in that direction. Returning to the northwest foot of the pinnacle we selected a route to the summit which proved not at all difficult and arrived a little before the main party, whom we discovered in the last chimney of the official route, awed to silence by the fear of falling stones.

The route of our ascent is practicable for small parties without difficulties, save in the one place below the last snow-field, where the ascent of the rotten rock face or of the ice slope in the left hand gully ought not to be undertaken without ropes, and then only by careful climbers.

△△△

Pamelia Lake

By G. W. WILDER

High up in the Cascades, in a charming defile at the foot of Mt. Jefferson, lies beautiful Lake Pamelia. An oval-shaped body of transparent, crystal water about a mile in length, fringed with massive timber, it is the dream of woodsmen, fishermen and nature lovers alike. Placid and serene, it rests between the steep and towering sides of Grizzly Flats and the more gradual basic slopes of America's Matterhorn. Fed by the eternal snows of the high Cascades, its pure and sparkling waters find egress in the wild primeval canyon leading down to the north fork of the Santiam river. Well within the boundaries of the Santiam National Forest, far from the usual haunts of man, surrounded by the virgin forests, its solitude bespeaks the untarnished handiwork of nature.

To reach this peaceful scene of rest and contentment one leaves the train at Detroit, the eastern terminus of the Corvallis & Eastern Railway, and follows the trail on the old unfinished grade some ten miles along the north bank of the Santiam river, whose cold blue waters come pouring down the rocky canyon. For another ten miles the trail leads through timber, up hill, down hill, Indian fashion, across two creeks to Kingfisher cabin, and then up through the forest of big firs till, unexpectedly, one bursts out upon the shores of the wonder lake. Thanks to the Forest Service, the trail is well defined and easy to follow. Sign-boards mark the turnings and give the directions and distances. Trees are blazed along the way and the grades are easy for both man and beast. A number of small creeks and springs furnish abundant opportunities for rest and refreshment.

Camping among the big trees along the shore is a rare treat. The gently sloping forest floor and the absence of thick underbrush combine to make the choice of a camp-site almost an embarrassment, for suitable spots are everywhere. Plenty of wood and water, the two chief requisites of camp life, are in abundance; while the making of a soft bed of fir boughs is a matter of only a few minutes. Hanging moss in huge quantities affords luxurious pillows. In the late summer months the days are warm, but not uncomfortably so, and the nights are delightfully cool and quiet. The absence of winds in this sheltered spot leaves the air laden with forest perfumes that induce long, refreshing dreamless sleep that makes one want to retire early and sleep late. There is no need for tents or other shelter unless it rains, and even then, at this high altitude, the moisture evaporates quickly, leaving everything dry again.

The south shores of the lake rise precipitously a thousand feet or more to Grizzly Flats. The slopes are partially covered with fallen timber and an ascent is necessarily slow work. Along this side of the lake, a trail zigzags over the fallen trunks, leading to the upper region of Hunt's Cove, over the divide and away to Marion lake, some twenty miles beyond. On the north side, beyond the camp-site, the shores rise gradually for several hundred yards and then more abruptly to the upper flats. This region is so thickly covered with fallen timber and underbrush as to be almost impenetrable. No trail extends beyond the dry water-course up which the Red Rag trail leads towards the summit of Mt. jefferson. Several attempts to go around the lake were given up after trying to get through this maze of logs and brush. Some of the fallen trees are real giants and tempt one to walk along on them in the vain hope of making headway through the endless tangle of undergrowth. They lie so thick that one may walk in any direction without touching the ground.

Many of the fallen trees have found their way into the lake, and the surface is dotted with them. Sometimes this mass of debris floats to one end of the lake or collects in some cove, and the remainder of the surface is free; at other times it is scattered over the entire surface of the lake. Drifting silently here and there, forming new combinations only to break them, it seems to be on a never-ending search for a final resting place. Many of the trunks extend down into the water, their stumps resting on the bottom far below; and, as the water is clear, it is interesting to trace the giant forms in the uncanny depths. These snags often bar the progress of the floating timber, and fishermen who use rafts are never sure of their routes because of the constantly changing barriers thrown in their way.

Although the water in this lake is usually cold, there are days when the shallower parts near the shore become warm enough for very comfortable bathing. The big logs are used in this sport; and when they are too thickly crowded into a favorite spot, it is no small task to clear a space large enough for a swimming pool. Diving is not popular, as one strikes cold water a few feet beneath the surface. Riding the logs or running from one to another, playing tag, seems to be far more interesting. Any visitor who has not, on a trig little raft, provided with a paddle and a box, explored a greater portion of the lake, has missed a great pleasure. Many wrecks of such craft along the shore testify to the efforts of those who have previously visited this charming place.

In addition to the usual surface outlet, Pamelia lake has two underground passages leading from central portions of the lake to the canyon about half a mile below. These subterranean outlets carry a considerable volume of water and appear in the canyon as huge bubbling springs. During the long summer months, the snows in the higher altitudes recede and the incoming waters decrease in volume to such an extent that the underground outlets carry off more than flows in. This causes the water of the lake to lower and the surface outlet to become dry. It is said that in a few instances the lake has almost disappeared, or at least has been reduced to the size of a small pond. Evidence of this may be seen in the clear water of the shallow portions of the lake, where one can see great ditches in the bottom, cut through the alluvial soil by the creeks after the bed of the lake has been exposed. During autumn the rains and early snows quickly restore the lake to its normal level, where it remains until another season. The moss on the trees shows that during winter this region is covered with snow to a depth of from eight to ten feet.

The flora and fauna do not differ materially from those of other well-known districts in the Cascades. The casual observer cannot fail to notice, however, the humming-birds and the "camp-robbers" (Oregon jays), and the few but well organized "yellow-jackets," the latter two always being in evidence at meal-times. An occasional fight be-

Mt. Jefferson from the west side of Pamelia lake This photograph was taken in 1916, and shows more snow on the summit than in 1917

Photograph by Winter Photo Company

The famous pinnacle of Mt. Jefferson viewed from the south summit
Note the climbers on the first slope

Photograph by Winter Photo Company

tween a water-snake and a "water-dog" always attracts attention, and hours may be spent in watching an army of ants working around an old stump. "Spanish fly-traps" grow along the edge of the creek, ready for unwary insects, and hosts of various little flowers, peculiar to high altitudes, abound on the broad acres of Grizzly Flats.

Pamelia lake is a place for rest and retrospection. Here nature has slumbered for untold centuries. Mt. Jefferson is old, as mountains go, and there is no evidence of recent upheavals as in other portions of the Cascades. The big fir stumps show an age of three hundred years, which, although almost a mere instant from a geological standpoint, shows that for some time, as measured by man's life-time, this region has been quiet and peaceful. Here one may rest and contemplate the beauties of nature without hardship; one's wants are few and easily supplied. A delightful summer camping place,—quiet, secluded and restful—memories of it awaken a longing for the gentle murmur of the sighing boughs, the cool shade, the mirrored waters and the perfume-laden air. To commune with nature in this lovely spot is to have lived, and life is better for it. To receive a new inspiration that puts determination into the soul is well worth a sojourn amid the beautiful surroundings of Pamelia lake.

△△△

A Trip to Three-Fingered Jack

By JOHN R. PENLAND

Of the many delightful side trips of the 1917 Mazama outing, the pack-sack jaunt to Marion lake and Three-Fingered Jack will be remembered by those participating as a most eventful, pleasing and noteworthy expedition.

On Saturday evening, August 11, upon our return from the first official climb to the summit of Mt. Jefferson, Ed and Art Peterson and the writer decided to make an expedition southward to the Marion lake and Three–Fingered Jack country. Accordingly, on the following morning we made up our packs, taking enough provisions for a three days' jaunt; and while the rest of the Mazamas were preparing for Sunday services we set out on our journey. We followed an old trail southeastward as far as Hunt's Cove. At this point we left the trail, swung up the mountain, crossed the east end of Grizzly Flats and then went over another ridge into Bingham Meadows. Here we were fortunate in meeting with a sheep-herder who was familiar with the country and who pointed out to us an old Indian trail which led in the direction of our objective point.

The day was hot and our packs by this time were getting heavy. The perspiration welled from every pore. But we kept persistently on the hike except for a few minutes when, on the bank of a little stream, we stopped for lunch and relaxed our muscles. At 5 p. m. we reached Marion lake, a magnificent body of water about one and a half by two miles in extent. Here, stretched upon the ground beside three small knapsacks we found three pilgrims of the trail, who had arrived from Detroit about five minutes ahead of us. Apparently, they were exhausted from heat and the long journey; but Art declared it was worry from lack of appropriate words in their already voluminous vocabularies to describe the horrors of the trail. Knowing that a plunge in the lake would be invigorating as well as cooling, and being of a helpful disposition, we invited our weary friends to take a dip with us but their only reply was groans, accompanied by the words: "What fools even here."

We spent a couple of hours surveying the scenery about the lake, and then darkness drove us to camp. We crouched around the campfire, talked and wondered about our Mazama friends at Camp Hardesty on Pamelia and our families at home, exchanged a few reminiscences, and snuggled down in our sleeping bags without even a thought of what the morrow might be.

Morning soon appeared, and as we sat in camp on the lake shore partaking of a real breakfast prepared by three real chefs we watched the morning sun break over the mountain-tops and pour out its gold upon the crystalline lake. Above the fringe of green and far in the distance, clothed in a thin film of haze, stood Three-Fingered Jack, stubborn and bold, seeming at first to beckon us, then to dare us, and upon the next look to say: "You cowards!" Whereupon we determined to attack and conquer this giant. Notwithstanding the fact that we had just circled Mt. Jefferson, crossed her dangerous glaciers and climbed her lofty and perilous summit, we realized that this, without doubt, would prove to be the most prodigious task of all. Accordingly, at 6 a. m. we set out, unhampered by pack-sacks and determined to prove or disprove the questionable statement of man's ability to conquer this monster.

The trail led us along the north and east border of the lake, across numerous cataracts and waterfalls, through dense forests, clouds of mosquitoes and out into an old burn. We were gaining elevation constantly. The country was becoming more open and the scenery more splendid. The trail was on the west side of the mountain and about a mile from its base. We could see from a distance that ascent from the north or west would be impossible, so we left the trail on the west and followed a southeasterly course. Then, unexpectedly, we discovered that we were in a natural park; prairies, meadows, clumps of trees, lakes and myriads of flowers which, with their various and innumerable hues, made a gorgeous picture in the sunlight—a feast for artist or botanist.

We next crossed a snow-field, and at this point the real climb began; first, a long, steep stretch of loose, sliding rock, which not only tested the muscles but also the soundness of the lungs and heart.

We were now on the main divide of the Cascades separating western from central Oregon. This divide consists of domes, spires and knife-like ridges of hard lava, volcanic ash and pumice-stone. The climbing here was not extremely difficult, although at times somewhat slow and dangerous, the ridge being in some places so narrow and sharp that it was necessary to resort to crawling. At length we came to a pinnacle which seemed impassable on either side or over the top. Here we found a small can in which were recorded two names, with the date, "July, 1910." A few feet away another name was found.

All hope of going further seemed futile. Realizing that we now had more need for prayer-books than for alpenstocks, we stacked arms. Determined that no possible passage-way should escape us, we crept along a narrow ledge under the hanging wall on the east, and from here we made a slide for a narrow tilted ledge some distance below us. We then scaled some seventy-five feet to the top of an almost perpendicular wall 2500 or 3000 feet in height. A few minutes later we were at the base of the topmost spires; the highest towering about eighty feet above us, the next highest about forty feet. In a few seconds we had agreed that the tall spire was "unclimbable" without the aid of tools and ropes with which we were unprovided. We climbed to the top of the next highest, took two or three photographs, left a small box containing a record of our ascent, and returned by the route by which we ascended.

I live not in myself, but I become
Portion of that around me; and to me
High mountains are a feeling, but the hum
Of human cities torture.

—*Byron*

A War-time Ascent of Mont Blanc

By J. Monroe Thorington

American Ambulance Corps, 1917

It is with some hesitation that one writes of climbing adventures already described in the early periods of Alpine history; the days of crowding tourists, with mountaineering equipment consisting so largely of guide-books and basket lunches, do not form an attractive background for mountain adventure, however thrilling. Nowadays, the true Alpinist, peculiar creature, often prefers to remain by his own fireside, dreaming of old battles with great peaks and of days full of joy in the solitude and grandeur of nature, when high heels at high altitudes were things unknown.

But now, with the period of the great world war, there develops a new page in mountaineering history—a page not so enthralling as earlier ones, perhaps, but resembling them. The ever present "tripper" has vanished from the land of high peaks; many a one, God knows, gone to give up his life for ideals higher and nobler than any which have ever inspired the mind of a mere mountaineer.

And yet it is with a sense of relief that one strolls over an Alpine meadow in the absence of luncheon debris and without having sublime thoughts disturbed by the weird noises from a Teutonic mouth-organ.

Gone for the present is the tourist and closed are the expensive mountain inns which are the curse of an economical Alpinist. The towering mountains are again in silence and are beckoning; history is in a measure repeating itself; and for those whose memory reflects old days amongst great peaks and silent unfrequented valleys, this little sketch is written.

In tourist days, the ascent of Mont Blanc brings forth little comment. In its way, it is recognized as one of the "stunts" which must be attempted by any visitor to Chamonix who is either slightly out of his mind or who openly boasts of his powers of pedestrianism. The average tourist, however, generally finds that the trip to the Grand Mulets is quite sufficient for all purposes, and returns thence leaving the upper snows untrodden and with vastly changed ideas concerning the difficulties of mountaineering in general.

In war time, however, things are a bit different; the Grand Mulets is closed and the scarcity of climbers makes six or seven ascents a record season for Mont Blanc. A party setting out for the mountain is a novelty; climbers are watched through telescopes from the valley, and on returning are looked upon with admiration and awe by the few summer visitors.

When we started in August, 1917, for a week of climbing in the Chamonix district, it was with very little hope of making an attempt on Mont Blanc; the weather had been unusually bad and the fall of new snow was very large. However, my guide, Favret, was optimistic, and came around one evening when the weather looked very uncertain, insisting that we make ready to start on the morrow. It seemed absurd to me, but he won the argument, and the next day was clear.

I was introduced to the second guide, Claret-Tournier—the wildest looking person imaginable, short and smiling, with red hair and a pair of black and white checked pantaloons which would have made an English sporting tailor turn green with envy. Tournier, however, proved to be a most capable and companionable fellow and later succeeded in bringing four liters of wine intact over the crevasses of the Bossons glacier to the Grand Mulets, a feat for which I have only profound admiration and respect.

We started out of Chamonix with heavy packs and with feet which, for all our hobnails, seemed very light. The trail leads through a beautiful forest, carpeted with moss and pine needles, toward the cascade du Dard and the Bossons Glacier, the tip of the latter seen ahead of us as it comes far down in the valley and encroaches on the nearby village. Mounting rapidly in zigzags, the trail leads into the most unexpected places; into dark mossy nooks wet with the spray of silvery waterfalls, through the forest, and out again over luxuriant mountain meadows with old ruined chalets and a view of the valley.

The morning sun made us remember that our packs were loaded, and our progress was therefore most leisurely. And then, too, the crevassed icefall of the Bossons Glacier, the corniced snow arête of the Aiguille du Midi above us and the jagged soaring pinnacles of the great aiguilles further eastward, are natural beauties too wonderful to pass by in haste.

We turned the corner at the Pierre Pointue inn, closed as was the Montanvert, and continued up the trail, which now became narrower and steeper. The rocks of the Grand Mulets, hitherto hidden by the Bossons icefall, became visible as we ascended, but the peak of Mont Blanc had apparently receded and was quite out of sight, the peak of the Dôme du Gouter dominating the view ahead.

We stopped for lunch at the last grass slope, beside a sparkling stream rushing down from the Aiguille du Midi, and paradoxically lightened our burdens by the transfer of food from our packs into our own interiors. This being accomplished in the thorough manner known only to mountaineers, we proceeded, crossing several turbid glacial brooks and much toilsome morainal debris on our way to the edge of the glacier.

The first stretch of ice is almost level, dirt-covered and with no crevasse of any importance. There is a complete system of surface drainage, with little cascades pouring down the seracs and ending in crystal-clear rills and rivulets which cut channels in the surface of the blue ice and finally disappear with a rush into some underground chasm to join the subglacial streams below.

It was quite a walk over the ice before a wide crevasse required the use of a rope, and from that time on the real pleasures of the day began. We cut steps down the sides of a huge crevasse to a narrow slanting ridge which gave us a fragile path to the snow on the other side, up which we climbed, giving careful attention to the rope and with fingers numb from attempts to find handholds in the ice. This was our first difficulty, but we met no other in crossing the little snow-filled basin which leads in a series of miniature plateaus to the junction of the Bossons with the glacier de Taconnaz.

Here, however, we were soon in the midst of great seracs and towering ice pinnacles of the most fantastic shape, some apparently needing only a breath of wind to send them crashing down upon us, We passed through an icy corridor and stopped to admire a huge arch of snow at least fifty feet high, like a great window through which we looked out at the blue sky and a part of the valley beneath.

The crevasses became enormous, but a zigzag route brought us through without difficulty. The slope was now quite steep, and our sky-line ahead was broken by jutting rocks, on a corner of which we could make out the Grand Mulets cabin.

We arrived at the great crevasse bridged by a rickety wooden ladder, and crossed on all fours, gazing down about two hundred feet into blue depths lined with glistening icicles which seemed ready to impale one in case of a slip. Then came some easy slopes leading upward by a winding path through crevasses which had seemed impassable from below, and bringing us to a particularly nasty-looking place which for a moment seemed as if it would necessitate a wide detour in search of a new route. A great crevasse lay before us, bridged by an insecure ribbon of snow which ended blindly in an overhanging snow wall on the far side.

Favret at last solved the difficulty by cautiously crossing the bridge, cutting steps in the snow wall and traversing sideways for about twenty feet to a notch through which he wriggled out of sight. He anchored, and we followed one by one, making good use of our ice axes to secure ourselves. The notch in the little snow cornice gave us trouble for the moment, as the snow was soft and our hands were almost numb with cold. Once on top, however, we found easy snow slopes leading up to the Grand Mulets.

We had the cabin all to ourselves, and in fact had to smash the lock of the door to get in. Everything appeared to be in good order and there were no signs of recent visitors. We built a roaring fire in the kitchen stove and soon had our wet clothing steaming above, while we sat comfortably near and toasted our toes. Then Tournier brought out the provisions and wine, quite unhurt, and we heaved a sigh of relief and contentment.

Evening came on before we realized it, deepening the blue coloring of the crevasses in the glacier around us, while the gleaming white snows of the Dôme and the peak of Mont Blanc gradually shaded into a delicate pink which turned to fiery red as the sun set over the shadowed valley westward. The coloring extended to the rocks of the Aiguille du Midi, Mont Blanc du Tacul and to other peaks nearby, gradually changing to purple, blue and finally to a leaden grey.

It grew cold rapidly and we were glad to retreat to the fire, where Favret was busily at work over a huge kettle of steaming soup. We crawled into our bunks very early in the evening, both because we were tired and because it is necessary to start for Mont Blanc at an hour of the morning which, to say the least, is not respectable.

It is frankly admitted that no one of us, even the guides, slept very much that night. First one and then another would prowl out to have a look at the weather, in spite of the fact that the sky was cloudless above the dull white snows of the Dôme, while far below us in the valley the lights of the villages gleamed and twinkled like reflections of the stars above us.

The alarm clock went off with a bang at one a. m., finding us quite ready to get started. Favret already had the fire crackling in the stove, and hot tea and toast were soon before us, momentarily demanding our entire attention.

We made use of the trench trick of covering our socks with paper before putting on our shoes, and later in the day confirmed the opinion of others that it is a great protection against cold. The guides lit two small lanterns and roped us together before leaving the cabin; and then, with everything in order, we started.

The snow was frozen hard and we made rapid progress upward, the flickering lanterns casting weird and gigantic shadows among the seracs and crevasses about us, while far below we could see the lights of towns in the valley. It is remarkable how steep a perfectly easy gradient will appear when one's only source of illumination is an indirect light coming from between the legs of the man ahead.

We climbed for nearly an hour through crevasses and up steep snow slopes toward the Dôme, until a sudden gust of wind extinguished our only sources of illumination. Stumbling along in the darkness, we reached a more sheltered spot where it was possible to relight the lanterns, but the guides then admitted that we must have wandered considerably from the route. Not being willing to retrace their steps, they tackled a steep wall of snow ahead, cutting nearly two hundred steps to the top and bringing us out, to the surprise of all, at the slope which leads directly to the Petit Plateau. The short cut had saved us nearly an hour's hard climbing.

It grew light as we entered upon the plateau and the lanterns were extinguished. We passed on up the long and weary slopes to the Grand Plateau, resting a few minutes before circling the base of the Dôme du Gouter toward the Col du Dôme. The snowy peak of Mont Blanc, rising across the Grand Plateau, was red in the morning light and the sun reached us as we cut steps up a little icy slope into the pass and climbed up toward the Vallot refuge.

We had a second breakfast in the rocks below, warming ourselves in the sunlight and kicking our half-frozen feet together to restore circulation. The Vallot cabin was snowed up solidly; and being unable to obtain entrance, we continued on toward the snow ridge of the Grand Bosse du Dromadaire. This, in the early morning, had a very evil look, and the frozen arête required a great deal of laborious step-cutting. In spite of the energy of Tournier, the icy wind prevented rapid progress, and we spent nearly half an hour in getting to the top.

Then, over flat snow slopes and the easy ridge of the Petit Bosse, the route leads to a knife-edge of snow, known as the Mauvais Arête. This is undoubtedly the most dangerous portion of the ascent, requiring a hundred-foot traverse on a frozen snow slope so steep that a slip could not possibly be checked and would probably send the entire party slithering down four thousand feet to the Italian glacier du Miage. It was a pleasure to see Tournier for once cutting steps which would hold one's entire foot.

The distances on the last portion of the ascent are most deceptive, the peak of Mont Blanc seeming constantly to recede, but our worst difficulty was soon behind us and we made our way over the slopes to the long final arête and were soon on the summit.

We had a cloudless day, and the view could not have been more perfect. Northward, over Mont Blanc du Tacul, we could see a corner of Lake Geneva and the triple peaks of the Dent du Midi. Far over the Mer de Glace and the Aiguille Verte, we could distinguish the Jungfrau and the Finsteraarhorn among the Oberland peaks. More to the eastward, snowy Monte Rosa, the rocky tooth of the Matterhorn, and many other Zermatt giants appeared. Southward, we looked across the Italian glaciers to the peaks of the Graian and Maritime Alps, while to the southwest lay the grim serrated wall of the Dauphine peaks, with the Meije, Ecrins and Monte Pelvoux easily singled out. The western view has the summit ridge in the foreground, with the Jura and the plains of France stretching to the horizon.

Everything seemed far below us, and it was hard to believe that the low rocky ridge to the north-east was the Aiguille-massif which soars skyward above the valley of Chamonix. A cold wind sweeping up across the Brenva glacier cut short our stay on the summit and we started down with a rush, reaching the Col du Dôme in a little over thirty-five minutes. The difficulties during descent, contrary to expectation, were much less apparent than in the reverse direction.

It was warm and sunny in the pass and we rested for half an hour in the glistening snow-saddle before descending. Then, taking the steepest slopes of snow, we glissaded rapidly down into the tremendous basin of the Grand Plateau, passing onward to the Petit Plateau, where we stopped to pick up our lanterns, which we had stowed away on a convenient serac earlier in the day. Then on downward, knee-deep through the snow, soft and wet in the afternoon sun, a shower of spray following us as we glissaded toward the Grand Mulets.

Arriving, we gathered our extra belongings together and demolished what little still remained of our provisions. Leaving the cabin shipshape, another hour brought us to the trail at the edge of the ice and in two hours more we were walking through the woods close to Chamonix. We had conquered Mont Blanc, but we had been traveling many hours and, in spite of ourselves, thoughts of the inner man tended to crowd out more sublime ideas.

A tremendous amount of food and a good bed brought us forth in the morning with a better balanced sense of proportion. Mont Blanc rose cloudless in the sunlight, and through a telescope we could pick out our tracks as high as the Bosses du Dromadaire. Our respect for the wonderful mountain had increased immensely.

But our climbing was over. Our leave of absence had nearly expired, and we were soon to be back in our places. It is a far cry from the peaceful quiet of the mountains to a hospital full of wounded soldiers. We tried for the moment not to think of wounded men.

Before leaving, we gave our guides a real party in honor of our successful week, and we waved goodbye as our train pulled out of the little station. As we rumbled down the valley, the Chamonix Aiguilles rose behind us in a serrated massif sharply outlined against the sky, while the gleaming snowy heights of Mont Blanc seemed to bid us a farewell and to invite our return. But, somehow, I cannot but think that our next visit to the Alps will be just a little bit different from our "war-time vacation." C'est la Guerre!

△△△

Mont Blanc is the monarch of mountains;
They crowned him long ago
On a throne of rocks, in a robe of clouds,
With a diadem of snow.
—*Byron*

△△△

With the Prairie Club in Glacier Park

By Laura H. Peterson

After a two days' journey in a special car over the Great Northern, there arrived at Glacier Park, Montana, on the evening of August 4, 1917, a happy party of thirty men and women. They were members of the Prairie Club of Chicago on their annual western outing. They had chosen for this year the newest park under federal supervision, and had decided to begin the journey from the eastern entrance, which is far more attractive than that by way of Belton. The approach is along the banks of the Flathead river, and from the train can be seen glimpses of the mountains; at first mere hills, which seem, however, to grow gradually higher until one is in the very heart of the Rockies. At the station were seen several of the Blackfeet Indian guides with their glistening black braids and gay-colored bandannas. Sometimes may of them come from the neighboring reservation to sing and dance dimes out of the pockets of the tourists.

The prospectus of this trip had announced that the personnel of the party would include members of the Prairie Club, non-members who were vouched for by them, and persons affiliated with other outing organizations. The writer was glad to be welcomed as a member of a sister club, the Mazamas, and to be given all the privileges accorded one of their own number.

The evening of our arrival was the only opportunity we had to inspect the attractive Glacier Park hotel, situated but a stone's throw from the station. It is completely finished with huge fir logs and deco-

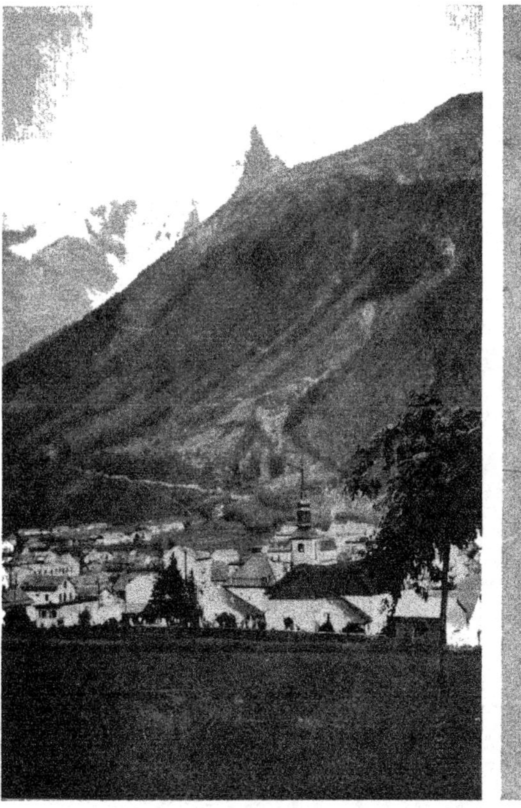

Left—Village of Chamonix, with the Col des Grands Montets and the Aiguille du Grand Dru
Center—Crossing a crevasse on the Bossons glacier below its junction with the glacier de Taconnaz
Right—Grand Mulets with the Dome du Gouter, Mont B anc

Photographs by J. Monroe Thorington

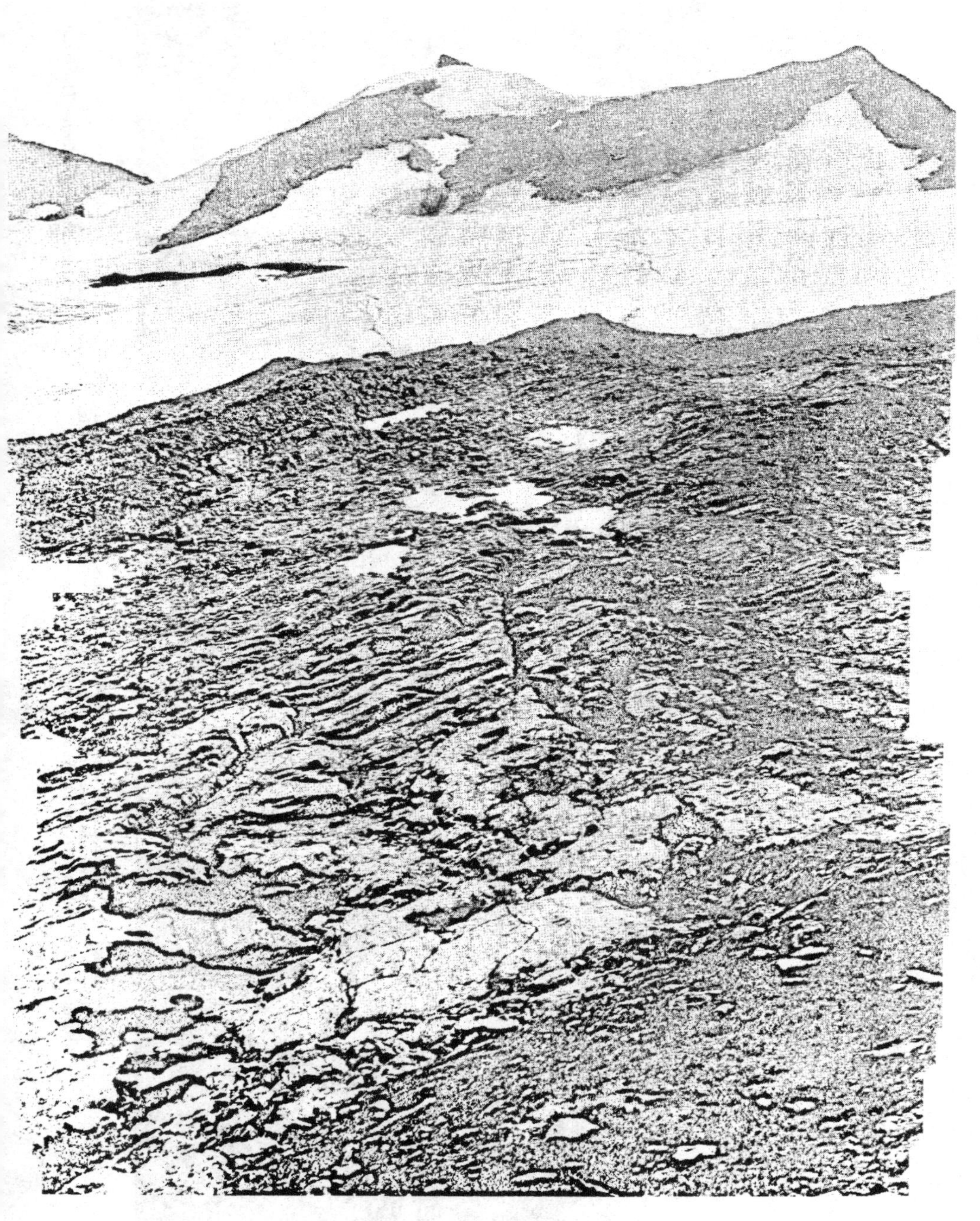

Sperry glacier. Glacier National Park

rated with buffalo and bear skins, while from the ceiling hang electrically lit Japanese lanterns. Grouped about are potted pines and in the center a cheerful indoor camp-fire. We were loath to leave this spot, but our schedule called for an early start on August 5, so we laid aside all thoughts of business cares and responsibilities and in khaki and boots plentifully hobbed began our tramp through the park.

Yes, there are automobiles and horses too, so that each traveler may have his choice of modes of travel. The autoist, however, will see but little of the real beauty spots, for the roads skirt the region and allow stops at but few places. Nothing but a trail was good enough for Prairie Club people and there were surely plenty of trails; rocky, to be sure, but wide, safe and plain, so that there was no danger of becoming lost. Sometimes these paths wind up a valley beside a splashing stream; then they zig-zag up a mountain side. Sometimes they cross snowfields, and again they wander deep into the woods and through flowery meadows that rival the parks about Rainier.

Our trail the first day led us over Mt. Henry. The party was lined up in true Mazama style with an official leader and a rear-guard; each with a whistle to signal for occasional rests as we trudged up the flank of the mountain. Although it was only 8,875 feet in elevation, many of us at the end of the two weeks' tramping, looked back to this as the hardest thing we had done. It was this walk which tested our endurance, the distance we could cover in a day, and the pace we must keep up in order to accomplish our daily schedule.

Our start was never earlier than eight o'clock, and it was necessary to reach the days' destination by dinner time. For the first two weeks we slept and took our meals at the various hotels and chalets situated at convenient intervals through the park. Within a few days everyone was in condition and ready for the average walk of eighteen or twenty miles a day. Trails in Glacier Park are excellent places to develop knee troubles, and very few of our party escaped. So general was the affliction that, at the suggestion of one of our number, we formed a branch of the main club and gave it a real Greek name, "Nu Kappa Psi." Nearly every mountain trip furnished us with new members.

One day was spent in and about Two Medicine camp, with a side trip of fourteen miles to Dawson pass. A snowstorm was encountered there; the only bad weather of the entire trip. In spite of the fact that it took all afternoon and evening to dry out, we felt thoroughly repaid; for during this excursion we saw our first bear, tracking through the new-fallen snow.

The following days' trip was all the more beautiful because of this snowfall. Every flower held its chalice of ice, and the bleak sides of Rising Wolf mountain kept their soft white blanket. Access is gained from this valley to the next by Cut Bank pass, a rocky ridge only thirty feet in width and justifying the name "Backbone of the World," given to all this region by the Indians. Down on each side, 3,000 feet below, is a marvelously beautiful blue lake. It was indeed with the greatest reluctance that we left this ledge and began our descent into the neighboring valley.

Our attention was here called to the geology of this section. All through the park there are innumerable lakes, blue, clear and cold. Over 250 of them hide among the rocky hills; some of them forming long chains through the valleys as far as the eye can reach; others filled with floating ice, as is Iceberg lake; while many of them are situated just as this one near Cut Bank pass. They lie in the perfectly formed cirques made by the action of glaciers which, though small now, must at one time have been much larger and have cut into the sides of the mountain and then, as they advanced, left behind these symmetrical circular basins. It is here, also, that one sees for the first time examples of the so-called Lewis overthrust. Geologists tell us that it belongs peculiarly to Glacier National Park; that the western rocks, pushed up and over those to the east, and then the pushing of the mountain rocks over those of the plains produced a wrinkling or overthrust fault. The more recent movements were very strong, and the rocks were forced up a great distance. Then the intense stress caused the crust to begin to move, and the strata broke in several places. Later, the streams wore away the softer rocks and left the precipitous walls of limestone as we see them now. This uplifted mass has been carved by the action of the water, and gradually the huge valleys are formed, with their chains of lakes and streams. One of the curious results of this action of the water is seen in the double cascade of Trick falls. The water disappears in the ground some distance above the waterfall, reappearing again through a slit in the face of the rock, while on occasions when the volume of water is great, it comes tumbling over the top of the cliff at the same time, thus forming a double cascade which justifies its name.

Of course we had a row (please pronounce the '*o*' long) on St. Mary's lake, perhaps one of the best known spots of the park. We enjoyed our trips on the various lakes, particularly McDermott and MacDonald. Several of the glaciers were visited, though we were unable to study Blackfoot, the largest one. In the spring of this year a great avalanche had swept down, destroying the Gunsight chalet, which was the stopping place at that point. This made it necessary for us to make in one day a trip which originally was planned to cover parts of two days, with some time on the glacier. Sperry glacier, however, affords one a taste of the real thing. Whether you climb the bold scarpment by the iron ladder fastened to the rocks with huge bolts, or take the goat trail over the snow-field, you will have some respect for the guide, with his ice-axe and life rope, as he leads you gingerly over the blind crevasses and carefully around the open ones.

Our trip was not all strenuous tramping. No matter how hard or long the trail had been, some of the party might always be found after dinner in the ballroom of the hotel or chalet; and neither tired knees, bedroom slippers nor khaki trousers barred us from the floor. No restrictions as to hours were imposed, provided we were ready in time for the next day's start.

The only baggage taken was stowed in small duffle bags which were transported part of the way by wagon and part by pack-horse. Due to the careful management of Mr. Henry Leissler, chairman of the outing committee, no dunnage bags were lost or missing when their owners daily rushed to find them just before the dinner-bell rang. A move every day means unpacking and packing the bags every twenty-four hours, and any who have had such an experience for three weeks will not wonder that we soon called them "devil" bags.

There were many real mazamas in the park, but they all traveled too high and too fast for ordinary hikers. Most of the time we could see them picking their way cautiously along lofty, rocky ridges in search of tufts of grass and herbs which persisted to the very tops of the almost bare rocky mountains. At Granite Park, however we met a Mazama of another kind at close range. This was our friend Dr. Barck, of St. Louis. He shared our camp-fire one evening and gave us information as to accessible points of interest in the vicinity.

Granite Park is a poor place to suggest for any very strenuous mountain work; it is such a wonderful spot that it holds you there. The little Swiss chalet nestles almost at the "ridge-pole of the roof of America," just at the foot of the famous Garden Wall. All about are magnificent, snow-crowned peaks, and down below is virgin forest. It is without doubt the finest place in the entire region. Sunrise over the Garden Wall gives a feeling never to be forgotten, and sunset over Heaven's Peak makes you want to linger there forever.

Over every pass we went until we quite appreciated Mary Roberts Rinehart's definition of a pass as "a place where the impossible becomes barely possible." At the end of two weeks the party divided, some of them starting homeward and others going into the western and northern parts of the park on a real camping trip where trails are few and difficult. There they surely were at the top of the continent; traveling on or near the irregular crest line, close neighbors of the continental divide, which we had already crossed several times during the first two weeks. The last climb of the final two weeks, up and over two passes in one day, Gunsight and Lincoln, skirting the edge of the highest lake in the park, Lake Ellen Wilson, to Sperry chalet and then down to Lake MacDonald and out of the park, brought to a close one of the most pleasant outings ever participated in by the writer; and every member of the party felt that there is truly a feeling of fellowship in the swing of a shoulder-pack and the clatter of hob-nailed boots.

△△△

A Trip to Crater Lake on Skis

By R. L. Glisan

Crater lake has always proved a powerful magnet in drawing me there at different seasons, and I have made my pilgrimages in various ways—by wagon, horseback, mule-team, auto and snow-shoes. I decided last March to attempt the trip on skis. I knew that only a camera-crazed enthusiast would venture there in the face of a threatened snowstorm, and I found such an individual in Frank I. Jones.

We left Portland Saturday evening, March 10, 1917, registering at the White Pelican hotel, Klamath Falls, the following evening. We left there Monday morning on the branch railroad. It was a cold, clear day. We followed the shore of Upper Klamath lake, Mt. Shasta and Mt. McLoughlin, better known as Mt. Pitt, appearing across the broad white expanse, for the lake was a solid sheet of snow-covered ice. Autos had crossed repeatedly; ski tracks were visible, closely paralleling the shore, and bands of cattle were seen traveling along the ice, drinking at water holes cut in the thick ice. Ducks, herons and an occasional pelican rose from the larger water spaces, where hot springs kept the ice open.

At Chiloquin we bundled into a straw-filled sleigh; thirteen persons occupied the seats, with a big red rooster in a crate as rear guard and superstition chaser. The snow was well packed, the grade easy, and the horses hardly slackened speed as we glided through the pine forest to the edge of Klamath marsh, on past the Indian agency, and the picturesque site of the old fort, arriving at the small town of Fort

Klamath in such brief time that we regretted the thirteen miles were not doubled.

On the way we saw large bands of cattle, patiently standing in long runways cleared in the snow, where the rapidly lessening hay was sparingly doled out in the meagre hope of the scanty supply lasting longer than the snow. Calves, just born, staggering on bent, unsteady legs, made the sight more pathetic.

The snow had gradually deepened to over four feet as we neared Fort Klamath. Mr. Kirkpatrick welcomed us at the hotel bearing his name. The temperature that morning was five degrees below zero; on the first of the month it had been twenty degrees below.

The outlook Tuesday morning was not promising. Over a foot of snow had fallen during the night. It was still snowing, and the heavy gray sky gave no assurance of any immediate change for the better. We took the sleigh four miles to the Copeland place at the end of the beaten road.

Breaking road was no sinecure. Nine men and twenty horses had averaged only two miles a day on another road they were opening for travel. At 8:45 we continued our way on skis. Our packs averaged over thirty-five pounds each, Jones' camera equipment for color photography being more than twenty pounds. I also had my camera and tripod. In addition to the provisions and personal effects, we had snow-shoes strapped on our packs, for emergency use.

At the Scott place we stopped to chat with five young trappers, and inspected their winter catch of fox and marten skins. We then pushed on through the soft snow, taking turns breaking trail through the pine forest. Another snowstorm about mid-afternoon shut out the sun and we looked for mile posts or signs. Cheered by the sight of a blue enameled sign on a near-by pine, we turned aside to investigate. After poking the snow away I unearthed, or rather unsnowed, an ice-cream sign. For the first time it failed to awaken a responsive chord. About five o'clock a peaked snow mound, rising slightly above the level, announced our destination. A shovel thrust in the snow under the peak gave us the means to clear an entrance, and we soon ferreted below and entered the cabin of the Wildcat ranger station at the park entrance. A fire soon warmed the cabin, which being deeply encased in snow retained the heat all night.

Wednesday morning promised fair, sunshine and blue sky following a starlit night. We left our snow-shoes behind as useless luggage and started up the road, tall, high-crested yellow pines casting long shadows on a spotless floor of white. Soon we neared the rim of Anna Creek canyon, frequently enticed to the very edge for the enchanting view of the stream, a green twisting ribbon far below. White slopes alternated with sheer walls of colored rock, columns and spires upthrusting here and there.

The sun, at first most welcome, became too effusive, and the softened snow stuck in wads on our skis, requiring abusive whacks from our poles to dislodge it; our energy being finally equally divided between the forward push of the ski and the sideblow of the pole.

Resting at high noon, we tramped wells for our feet, sat on our skis, nibbled lunch and cussed and discussed our situation. In despair, turning the skis over we rubbed off the snow and rubbed on some prepared floor wax, and, to our great relief, found they gave the snow the slip.

After seven hours of continuous plodding, having failed to make Bridge creek, which we understood was half way, we cached our camera equipment and provisions, hanging them from a high branch out of reach of the pine martens, whose tracks had evidenced their roving disposition. Another hour, and we reached the deep-set curve where a timber-cribbed opening under a deep floor of snow showed us Bridge creek, the only bridge on the road. We found out later that this was five and three-quarter miles from Wild Cat.

Passing the Garden of the Gods, made doubly impressive by the long afternoon shadows pointing clear across the canyon and up towards the sentinel peaks, which guard Crater lake rim, inspired by the view we pushed on. It was after sunset when we reached Headquarters, where we were most cordially welcomed by H. E. Momyer, our coming having been announced by the reverberation of the telephone wire which we struck as we stooped to unfasten our skis.

Fourteen feet of snow on the level necessitated going down a snow stairway to the front door. Mr. Momyer, as acting superintendent of the park, was in charge, monarch of all he surveyed, his only companions, bluejays, feathered camp robbers, and a pine marten, all so tame that a robber ate from his hand, and the marten overcame all caution in his eagerness to secure scraps of the fresh meat we had brought. His dark lithe body appeared like a shifting silhouette against the snow stairway.

Thursday morning registered seven degrees above zero—clear, cold and snappy. We left after breakfast to reclaim our cache. Gliding easily down the tracks of yesterday without any appreciable effort, we passed the Garden of the Gods, in and out rounding every down curve, by Bridge creek, and all too soon checked our speed as we saw our cache hanging safe above the marten tracks. Our cameras secured, we photographed the trees loaded with yesterday's snow, lunched on our return

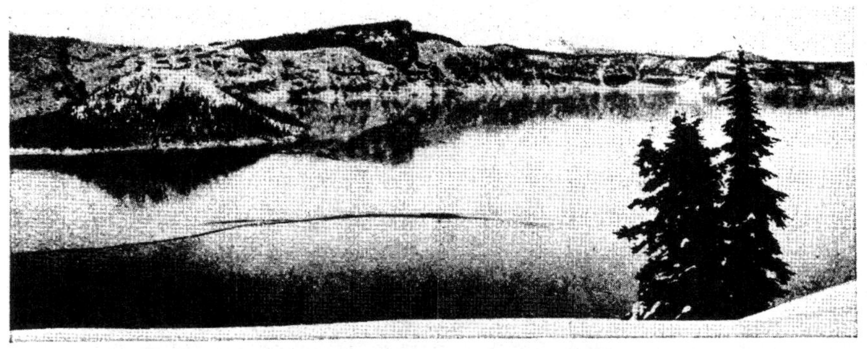

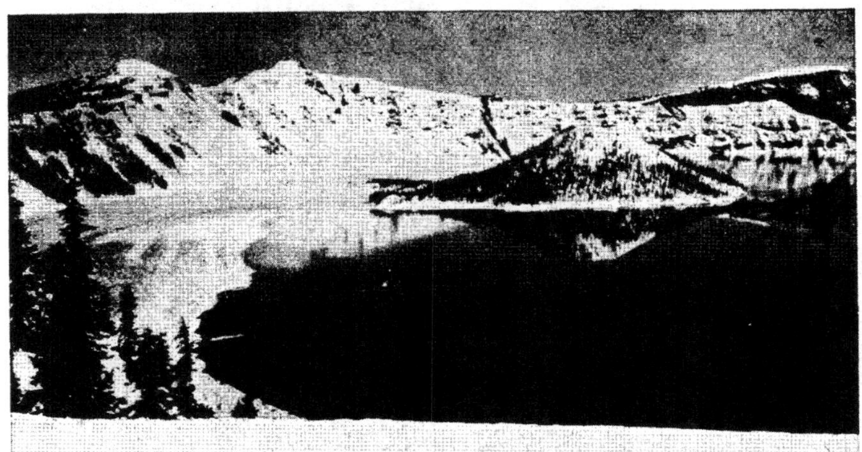

Views on ski trip to Crater lake   Upper—Llao and Mt Thielsen.   Middle—The Watchman, Llao and Wizard island.   Lower—Crater Lake Lodge.

Photographs by R. L Glisan

Mt Jefferson from the south s ope o Mt Hood, fifty miles distant, with the Three Sisters appearing at the left

Photograph by M. H. Barnes

at Bridge creek, lowering a flask by a long cord to draw water from the south or California side, which we called the wet side, as we failed to fill it on the Oregon side. The day proved as easy as the previous day had been strenuous, and the sun and blue sky were a marked contrast to the clouds with which we started. It seemed incredible there could be such a difference.

Friday proved another perfect day. We leisurely made the engineer's camp just below the river and stopped at the shelter cabin courteously left available for winter visitors. The peak of the high-pitched roof rose a few feet above the snow, resembling a fair-sized dog kennel. Again we ferreted down, opening a small trap-door under the peak. On the upper floor was a bed with comforters. Descending a wall ladder into the dark abyss below, we lighted a lantern, and rejoiced at the sight of a snug pile of firewood. The stove started, the cabin was heated in short order and retained the heat like a fireless cooker. In the afternoon, we went up the low gap where the old road meandered to the rim, and came out on the lake at the base of Castle Crest. The sun was setting, giving a warm glow to the snow in the light, and cold grey to the snow in shadow. In the shadow below lay Wizard island, a white cone; The Watchman, Glacier and Llao rose on the western rim, kindled by the last rays, which in turn brought out Thielsen in sharp relief to the north, with blue sky above. The scene was sublime, one feature only missing—the marvelous blue of the lake. To our great surprise, the lake was frozen, fully three-quarters of its surface being ice-covered. We had been told that the lake never froze, and could not freeze, because of its phenomenal depth, constant temperature and surface-ruffling winds. It was a great disappointment to Jones, with his color plates. To me the unexpected was the more appreciated.

A biting wind came with sunset. My tripod froze and refused to close. Our gloves stiffened with the cold. Hastily we slipped our parkas over our shoulders and drew the monk hoods over our heads. Thus enveloped, we were soon in a glow again. While Jones waited for the sun to reappear under a low banked cloud, I returned to the shelter cabin to prepare supper. Following down the ravine, along the sides of which the old road coursed, having to check my speed with the poles as brakes, I soon came out at the head of the valley, picked up our ski tracks, reached the cabin, and had supper ready by the time Jones returned. He had more difficulty in the uncertain light and stiffer snow crust.

Saturday morning, an overcast sky prevented sunrise views. We returned to the rim after an early breakfast, and spent the morning along the rim photographing the lake below and Crater lake lodge with its snow banks reaching up to the front dormer windows. Shasta and Union peaks appeared to the south, from different points on the road leading westward along the rim towards the Watchman. The sun frequently burst through the white clouds to reward our patience. Returning, we skied down the slope traversed by the new road, to the shelter cabin, for a late lunch. We then returned to Headquarters.

Sunday found a slight snow falling, giving a cushion for the skis and smoothing out irregularities. Reluctantly bidding our host farewell, we started down to Fort Klamath.

Our skis needed no urging and no guiding. Down the broad road and around the broad curves on a gently descending grade they kept the deep grooves, so arduously made on the ascent, and nothing could ever be more wonderfully enchanting and exhilarating. Spending over an hour at Wild Cat, another hour at the trapper's cabin, after crossing the broad Klamath marsh with the range of peaks beyond the broad white sheet, we lingered twenty minutes at the Sisemore ranch, photographing the band of elk left there by the state warden. We reached Fort Klamath in ample time for the Sunday chicken dinner, our keen-edged appetite prompting us to render ample justice. Another enjoyable sleigh ride on Monday morning brought us into Chiloquin for the train, returning to Klamath Falls for the night and taking the train Tuesday for Portland.

△△△

From Hood to Jefferson in April

By Chester H. Treichel

On April 2, 1917, Dean Van Zandt, C. E. Blakney and I left Portland on the interurban train for Bull Run, to begin a trip from Mt. Hood to Mt. Jefferson. We wore water-proof mackinaws, khaki trousers, and high-top boots. We carried sixty-pound packs, besides our alpenstocks, ice-axes, snow-shoes and skis; we also had one small toboggan, which we were to take turns in carrying.

After a night at Aschoff's, we set out the next morning for Casa Monte. We arrived here about six in the evening, after a difficult day. The roads were muddy and tedious. The following morning, April 4, we struck off through the soft snow, about a foot deep, near Rhododendron. We loaded all our packs on the toboggan, taking turns alternately pulling and pushing it; but we decided at the end of the day that carrying our packs on our backs was far less tiring, and accordingly we abandoned the toboggan at Cold Springs, before going on to Government Camp, where we spent the ensuing night.

Our hosts here took us out for several hours' skiing the next day. We were very awkward and had any number of spills and tumbles in the soft snow, but we refused to be discouraged.

Another morning found us striking out at eight a. m. for the summit of Mt. Hood, which rose bold and clear before us. We wore snow-shoes and dragged our skis, reaching timber-line in less than two hours. Here we left the skis. Traveling became much slower, owing to increased steepness. Wonderful views stretched away from us in all directions. Mt. Jefferson loomed in the distance among surrounding hills, all covered with uninterrupted snowfall of many days.

When we were about two-thirds of the way up, the air suddenly became bitterly cold and storm clouds enveloped us, sweeping down from the summit of the mountain. We could not see farther than a few feet in any direction. We kept on climbing, nevertheless, and at two p. m. reached the base of Crater rock. Having made no preparations for spending the night on the summit, we decided that it would be folly to attempt to go farther under such conditions. Accordingly we retraced our steps back toward camp. The storm increased in fury. Only by exercising careful judgment in keeping just to the right of White River glacier and in constantly relying upon our compasses, were we able to reach the place where we had left our skis. We put these on, and continued our downward journey in a series of sudden starts and plunges into snow banks. Finally, however, after following Zigzag canyon for a couple of miles, we became fairly proficient in the art of making graceful curves on skis, and we reached Government Camp in time for supper.

Mr. Fox and Mr. Pridemore started out with us the next day, but left us at the Summit House, whence we plodded on alone, through the white forests and twelve or fifteen feet of snow. We made good time in reaching Big Meadows, keeping up a slow but steady march. We made our camp under a clump of large firs which gave good protection from storm. We had fir-bough mattresses and slept warm and snug in our sleeping bags. We rose next morning to realize that it was Easter Sunday. Several more inches of snow had fallen during the night.

We broke camp about eight o'clock, after a good breakfast. Conditions for skiing were not at their best. It was necessary to climb a grade of several miles to the summit. We encountered much difficulty in getting over logs and crossing a number of ravines. We had planned to reach Clackamas Lake ranger station, sixteen miles away; but we had not realized what a hard task we had set ourselves for

Easter Sunday. Sometimes we were able to follow a telephone line, but most of the way this was covered with snow and we were obliged to rely solely upon our compasses for direction. After stopping a short time for lunch at Clear lake about noon, we kept steadily on until about five p. m. when we saw a ranger trail sign projecting a bit out of the snow, which gave the distance from there to the ranger station as three miles. We were tired, but we kept on. Every so often the leader would drop behind, the second man taking his place and breaking trail. It was about seven o'clock when we distinguished the outlines of a cabin surrounded by a number of trees in a large open meadow. We knew at once that it was our long-looked-for destination, and nothing could have pleased us more than the sight of it.

Inside, we hurriedly built a fire, and after resting, drying and dining, we crawled into our sleeping bags, thoroughly tired out.

Monday we rested. We rose late and spent the whole day about camp. The day was fairly clear, but toward nightfall it began to storm as usual.

Warm Springs was set for our destination the next day. A heavy crust on the snow made skiing very difficult in the morning, and more snow kept coming down all day. By this time, however, we were more skillful in the use of skis, and we made fairly good speed, reaching Warm Springs at four o'clock in the afternoon.

Here we found a little cabin about seven feet long by six feet wide, completely covered with snow except for a little opening at one end. We built a fire, chopping away the snow to make a draft and let out the smoke; but our scheme failed to work, and our little home under the snow was nothing but a smoke house. So we crawled into our sleeping bags early.

Breaking camp at nine the following morning, we soon left the smoky cabin far behind. About five inches of snow had fallen in the night. Much of this stuck to our skis, making progress very tiresome and slow. At two-thirty we reached Lemiti ranger station, a distance of about seven miles from Warm Springs. Stopping there about an hour for lunch, and to wax our skis, we decided to go as far as Olallie Meadows ranger station that afternoon.

Six o'clock brought us to this spot just as the last rays of sun shone over the hills. Mt. Jefferson, a little to the right of Olallie Butte, made a wonderful picture, like a huge sparkling diamond.

By this time, cooking our meals under difficulties was an old story to us. We took turns, a day each, in acting as cook. We spent the time after dinner in drying ourselves and our equipment, many things being quite wet from the storm, together with the snow that kept falling from the trees as we passed under them. At nine o'clock we went to bed, but we had little sleep that night, as a heavy storm soon began to rage, continuing until late the following day. It was very cold.

We passed a second night in this cabin, and arising at six o'clock on Friday, the 13th, we soon set out for Jefferson Park. We crossed several lakes, Olallie being the largest. All were frozen and covered deep with snow. Shortly after two p. m. a fierce storm closed in about us, making it necessary again to use our compasses for direction. By five the storm was raging with such fury that we were compelled to descend quite a distance into a valley and seek a spot for camp.

We made a lean-to shelter out of boughs and a canvas. The wind played havoc with our fire, for which we had no dry wood. We were bitterly cold until we got into our sleeping bags. The storm raged all night, driving snow through every crevice of our shelter.

In the morning, the thermometer registered a little above zero. Our clothing was frozen. It was nearly nine o'clock when we set out, to climb some 1500 feet to the summit of the ridge. The storm was still blustering. We had to cross many steep snow slopes. I remember one in particular, where each step had to be tested before any weight was placed upon it; a slip would have meant a slide of several thousand feet. We were barely able to distinguish each other through the storm, although we traveled only a ski-length apart. By afternoon the storm became a regular blizzard. Coming to the end of the ridge, we started down into the valley, thinking it might lead into Jefferson park. The descent was too steep for snowshoes or skis. We plunged through snow over hip deep, sometimes sliding and falling. On reaching the floor of the valley we realized that we were not in Jefferson Park after all, but in a sort of box canyon with sheer cliffs on all sides.

We camped under a large fir, and were pleased to find, the next morning, that no snow was falling and the wind had ceased. Studying our maps, we concluded that we were in Gorge Creek* canyon, and that by following the creek we should reach Breitenbush hot springs.

Skiing was good this morning, the down grade being gradual, although we had to keep dodging in and out among the trees. At nightfall we could find no place suitable for a camp, we so had to make the best of very bad conditions.

We were glad to get up at break of day. We had a disagreeable time with breakfast, and made little progress during the morning. We changed from skis to snowshoes shortly before noon, finding we could make much better progress with the latter. We began to notice blazes

*Incorrectly named on Oregon National Forest Map. Should be South Fork of Breitenbush.

along the way, and an occasional forest fire warning sign. It was about four o'clock when we reached Breitenbush hot springs.

We picked out the most comfortable cabin and made ourselves at home. Cooking on a stove and sleeping on springs were novel sports to us, and we enjoyed the hot water of the springs. We spent two days resting and bathing, mending and drying our clothes.

When we set out again the weather was fine. The snow was soft in the lower valleys and we had to snowshoe all the way to Detroit, a distance of twelve miles. The crossing of numerous streams made it necessary to take off and put on our snowshoes many times; we had great difficulty in following the trail around cliffs. The hot baths of the preceding two days seemed to have weakened us, and we had to rest often.

At length, rounding a curve in the trail, we saw to our delight the town of Detroit some distance below us. It was Thursday, April 19, at four o'clock, when we reached Detroit and saw the first human beings, besides ourselves, since leaving Government Camp.

Had the weather been a little milder, our trip would have been a very pleasant one; but as it was, it often taxed us nearly to the limit of our endurance. We learned a number of important things, however, about taking care of ourselves in the worst of weather and conditions; making comfortable camps in snow, and cooking over little fires in raging storms. And best of all, we accomplished something that, so far as we could learn, had never before been attempted at that season of the year.

△△△

Phases of Vulcanism as Shown in the Cascades

(*An Abstract*)

By **Pauline Geballe**

"It seems probable that the Cascades are formed to a large extent of tilted blocks of basalt which were originally horizontal; and belong to the same series as the Columbia river lavas farther east, which have been, in comparison, only moderately disturbed. The Cascades, at least in Washington, do not seem to have been formed mainly by the piling up of erupted material, as has been suggested in explanation of their origin farther south, but are due to the uplifting and tilting of previously consolidated lava sheets as well as granite and coal bearing strata which occur high up on each flank of the mountain and even form portions of the main divide. The great volcanoes which appear so prominent along the general trend of the range are secondary to the main mountain building."*

In the northern Cascades, "the structure is highly complex, and is by no means in a single great north—south anticline or simple monoclinal block sculptured by erosion. The Cascades as we know them seem to have been carved from an upraised peneplain."* Mr. George O. Smith says Prof. Russell's hypothesis rests upon the presence of truncated folds, stream erosion having reduced the whole region to nearly sea-level. "During the later portion of the time of base-leveling, the widely spread sheets of the Columbia river lava were poured out. The date of the period of planation is shown approximately by the fact that the folded beds of Eocene were truncated. The broad peneplain must have reached the greatest degree of perfection in the late Tertiary time, probably extending into the Pleistocene."*

The lava flows as seen in canyon walls show parallel layers. During the period of eruption there were no general movements such as would cause unconformities within the Columbia river series. This lava is believed to have been poured out in a "series of vast inundations over a deeply eroded land surface." It should be 4000 feet thick, because it is believed that no erosion took place, but it is in some places only 3000 to 3500 feet thick, so Prof. Russell argues that there has been subsidence in the eastern portion because there are no marine deposits imbedded in the lava sheets.

†The Cascades have a north–south trend. From Oregon to Mt. Rainier, the range is made up of basaltic and andesitic lavas of the Tertiary age. Farther north the rocks are older and the topography more varied. These are called the Northen Cascades. The oldest rocks known in the Northern Cascades are of Paleozoic age. They are largely metamorphic. Some traces of the original show they were of both sedimentary and igneous origin and their character indicates that the conditions of sedimentation and volcanism were quite similar to those of the same period in the Sierra Nevada mountains and British Columbia. The rocks themselves are strikingly similar to the rocks found in the Blue mountains and Okanogan valley.

During the Mesozoic time sandstones and other sediments were laid down in portions of the Northern Cascades, though the Cretaceous rocks lying just south of the Canadian border show that a Cretaceous

*Prof. Russell.

†Information obtained from the Mt. Stuart and Ellensburg portfolios.

sea extended southward from British Columbia. Similar rocks are found in the John Day valley and Blue mountains of Oregon, showing that Central Washington at that time was probably land with Cretaceous seas to the north and south of it, or else any marine Cretaceous material which was deposited has been eroded away.

Then these Cretaceous and older rocks were folded and uplifted, accompanied by intrusions of molten magma. The chief of these in the Northern Cascades was the intrusion of the massive granitic rock which formed the great granodiorite batholith of Mt. Stuart. Then came a period of erosion and the mass was carved into bold relief. At the beginning of the Tertiary age, the Northern Cascades was a comparatively rugged country, though not necessarily greatly elevated.

During the early Eocene period sedimentation occurred in the arm of the sea where Puget sound now is and in the bodies of water, both fresh and salt, that existed elsewhere. The Swank formation was laid down in this period. Then came slight elevation and erosion and the first eruptions of basaltic lava, the forerunner of those great lava flows of the Miocene period. Later there was sedimentation and it was at this time that the large lake beds in central Washington were filled with the sediment and organic matter which forms the Roslyn and Swank coal beds.

That great erosion took place at this time is shown by the stratigraphic break between the Eocene and Miocene. Where the older rocks are exposed near Ellensburg, it is found that the first flows of Yakima basalt covered a surface of considerable relief, but that the lava was so thick that it filled the valleys and covered the hill-tops, making it a monotonous waste of black rock. The lava poured out of great fissures in quiet streams flowing for long distances and finally consolidated into sheets from twenty-five to one hundred feet in thickness. Dikes can be traced to the older rocks below showing the channels through which the lava issued. In Yakima canyon ten or more separate flows can be counted. That these flows occurred after considerable intervals of time is shown by petrified trees and stumps, also thin strata of sedimentation. The maximum thickness according to Prof. Saunders is 5000 feet. Sediment from the old Cascade mountains was washed in by the streams and buried near the margin of the basaltic plain.

In the latter part of the Miocene, the eruptions ceased and there was a depression of this area toward which the streams flowed and made deposits. The sediment consisted of light-colored andesitic pebbles, sand and boulders which had been emptied farther west. These andesitic eruptions in the west began even before the basalt flows had ceased, because andesitic lavas and pumice are found in the Ellensburg sandstone beneath the latest basalt flow. The coarseness of the material and the stream bedding show that the streams were of such volume as to transport huge boulders. The material was of the explosive type and could be easily carried. The deposit spread out in wide alluvial fans over the comparatively level surface of the basalt.

In the early Pliocene period, there was folding and uplift; then erosion by streams which reduced it all, sandstone and basalt alike, to one level plain—a peneplain. The peneplain was then subjected to orogenic forces, the level surface was raised in the form of a dome along a long axis and this was the birth of the present Cascades. It was an extremely slow process, but of considerable elevation, so that the larger rivers, such as the Columbia and Yakima, kept their meandering courses across the ridges and were entrenched. The valleys between the ridges furnished natural routes for surface waters and so many tributaries of the Yakima river have courses consequent upon warped hollows of the peneplain. This was the closing event of the Pliocene period.†

"The present Cascades are called Cascade or Pliocene Plateau and the present valleys and ridges are the result of recent erosion since the last uplift. During the uplift or at its climax, volcanic cones, such as Hood, Adams, St. Helens and Rainier, were built upon the old block surface. Other peaks of the Cascades are remnants of erosion left as monadnocks above the peneplain."‡

Thus the present Cascades consist of an old comparatively level platform of granite upon which rest these newer volcanic cones, built up chiefly of material of the explosive type. Their eruptions have ceased so recently that many of the older Indians still relate tales of the fires that issued from the bowels of the earth; though, if we except Mt. Lassen, there has been no recorded volcanic activity within the memory of the white man.

†Information obtained from the Mt. Stuart and Ellensburg Portfolios.

‡E. J. Saunders, "Geological History of Washington."

O Nature! a' thy shews an' forms
To feeling, pensive hearts hae charms!
Whether the summer kindly warms,
Wi' life an' light;
Or winter howls, in gusty storms,
The lang, dark night!

—*Burns*

Some Birds of the Higher Cascades

By WILLIAM L. FINLEY

Oregon State Biologist

Many people think that wild birds are more abundant in the remote sections of the state less frequented by man. As a rule, it is just the opposite. The ordinary bird takes to living about a settled community. While we may find an occasional robin in the higher mountains, and in the less settled portions of the state, yet as a rule, the robin likes a lawn and he prefers to live near where people live. The meadow-lark, for instance, may be found in the semi-desert or in the sagebrush country, but he is much more at home near a plowed field.

Among the typical bird residents of our Cascade forests, one may find some of the woodpeckers, chickadees, creepers, nuthatches and kinglets, feeding continually on the insects that infest our forest trees. The woodpeckers are especially equipped by nature to bore into the tree-trunks for grubs and other insects. The foot of the woodpecker is different from that of the robin. Instead of three toes in front and one behind, the woodpecker has two toes in front and two behind, so that it can more easily cling to the upright trunk. The tail has strong sharp, pointed feathers that catch in the bark and act as a prop for the bird's body. The beak is long and chisel-pointed, making a very effective instrument for cutting wood. The tongue is elastic, with a sharp point and barbed on the sides so that it can be thrust deep into the burrows of wood-boring insects.

One of the most striking birds that a mountain climber may discover in the high Cascades is the Clark's crow (*Nucifraga columbiana*) or *nutcracker*. This bird was first found by Capt. William Clark near the site of Salmon City, Idaho, August 22, 1805. While it is a crow in action, yet it is quite different in dress. Its whole body is white, but the wings and tail are black, with the exception of a white patch on the lower part of the wings and the outer feathers of the tail. It is a striking creature, typical of the high western mountainous country where the Alpine hemlock and jackpine thrive.

The best chances I have had to study these birds were at Cloud Cap Inn (on the north slope of Mt. Hood) and at Crater Lake Lodge. Like the Oregon jays, the crows have learned to stay about these hotels, where they get free meals at all hours of the day. At times, they become so tame that they will take a nut or piece of meat from the hand. In traveling through the higher Cascades, I had often seen one of these conspicuous birds launch out from the tree-tops, sometimes with a long swoop, opening its wings and curving up before the next drop. The continuous, harsh, rattling call, sounding like "Char-r! Char-r!," is a familiar sound, typical of pine timber and rugged mountains.

The Oregon jay (*Perisoreus obscurus*) is particularly a camp bird. He loves the outdoor abode of a man in the woods. Careless campers almost universally call him "camp robber." He has the good trait of paying frequent visits about camp and picking up odd scraps that drop from the table or are thrown away, a habit which no one can question. If a camper is careless about leaving his meat out in the open or the butter uncovered, the natural conclusion reached by the birds is that this means: "Help hourself, so it won't go to waste."

At Camp Hardesty, on the border of Pamelia lake, during the past summer, food was served on a row of thin shakes that were meant for tables, but which might not have been recognized as such by the birds. When the butter eas set out in a pan at six a. m. on these forest lunch counters and no one came to eat, I saw an old jay drop down, poke in his bill and pull it back with a pleased expression and a whistle that said: "Come on, children; breakfast is now ready." And the children came. I could see no cause for two late-sleeping Mazamas raising a hue and cry and calling these birds "camp robbers." Anyone might think that food set out in the open woods was to be eaten and not to melt and spoil in the sun.

The Oregon jay can easily be recognized by his fluffy dress, rather long tail and general gray color with no sign of blue. The top of the head and upper part of the back are blackish. The rest of the upper part of the body is brownish-gray, while the under part of the bird is white. The young birds that follow their parents about during the summer look like a different species. They are of a dull sooty brown, darkest on the head and a little lighter below, but lacking the white under parts of the old birds.

The Steller's jay (*Cyanocitta stelleri*) is the common blue jay of the Oregon forests. His dark blue body and high crest easily distinguish him from the Oregon jay. While the Oregon jay is often bold enough to take food from the hand, the Steller's jay is more wary and afraid of man.

One might not expect to find humming-birds in our higher mountains, yet if the ordinary tramper has his eyes open, he will see them, especially late in the summer. The rufous humming-bird (*Selasphorus rufus*) is the common species living on the west slopes of the Cascade range. The humming-birds that live and breed in the lower hills and canyons seem to follow the flowers and work up into the higher mountains during the late summer. The young humming-birds are ready to leave the nest and fly about in June and July. By this time, the early flowers are past blooming in the lower altitudes, but up nearer to the snow line there are myriads of blossoms.

Along almost every wild mountain stream is a typical bird of our western country, the water ouzel or dipper (*Cinclus mexicanus*). This bird is sometimes called the "teeter-tail," because of his continuous bobbing motion. It is slate-gray in color, with a short wren-like tail. While it is not a web-footed bird, yet it dives and swims in the swiftest water, picking up water insects and larvae at the bottom. As a rule, it makes an oven-shaped moss nest with a door in the side. It is placed on a shelf of rock, often behind a waterfall, where the bird may have to fly through the spray or a thin wall of water to reach its nest.

The western evening grosbeak (*Cocothraustes vespertinus montanus*) and the varied thrush (*Ixoreus naevius*) are both birds of the higher mountain regions, which come down into the valleys to spend the winter. The grosbeak is a strange-looking bird, often taken for an imported songster rather than an Oregonian. It has a heavy light-colored beak, with black on the top of the head, wings and tail. The wings have large white patches which show distinctly when the bird is in flight. The main part of the body is olive and yellowish-green. Flocks of grosbeaks are often seen during the winter and spring season about the valleys, picking up seeds under the maple, locust and cherry trees.

The varied thrush is often called the Alaska or Oregon robin. It nests in the wilder mountains, where the timber is dense. The bird has a weird and mysterious note, a sort of a monotone song that can be imitated by using a combination whistle and voice note. When he is driven down from the high mountains by the snows of winter, the later fruits are still hanging on vine and tree. He seems to be ravenously hungry for the sweet-tasting fruit that has been planted by man. His taste sometimes turns to grapes and apples to such an extent that some farmers think him a nuisance.

The varied thrush has the size and actions of an ordinary robin. but has a very different dress. He can easily be recognized and distinguished from the ordinary robin because, instead of a brick-red breast, his breast is yellowish-brown, crossed at the throat by a black band.

Among the birds of prey, one may find hawks, owls and eagle in the higher mountains. The little sharp-shinned hawk (*Accipiter velox*) is the most destructive among our smaller birds. While camped at Pamelia lake last summer, Mr. Herman T. Bohlman was watching a flock of tree swallows darting about and catching insects over the

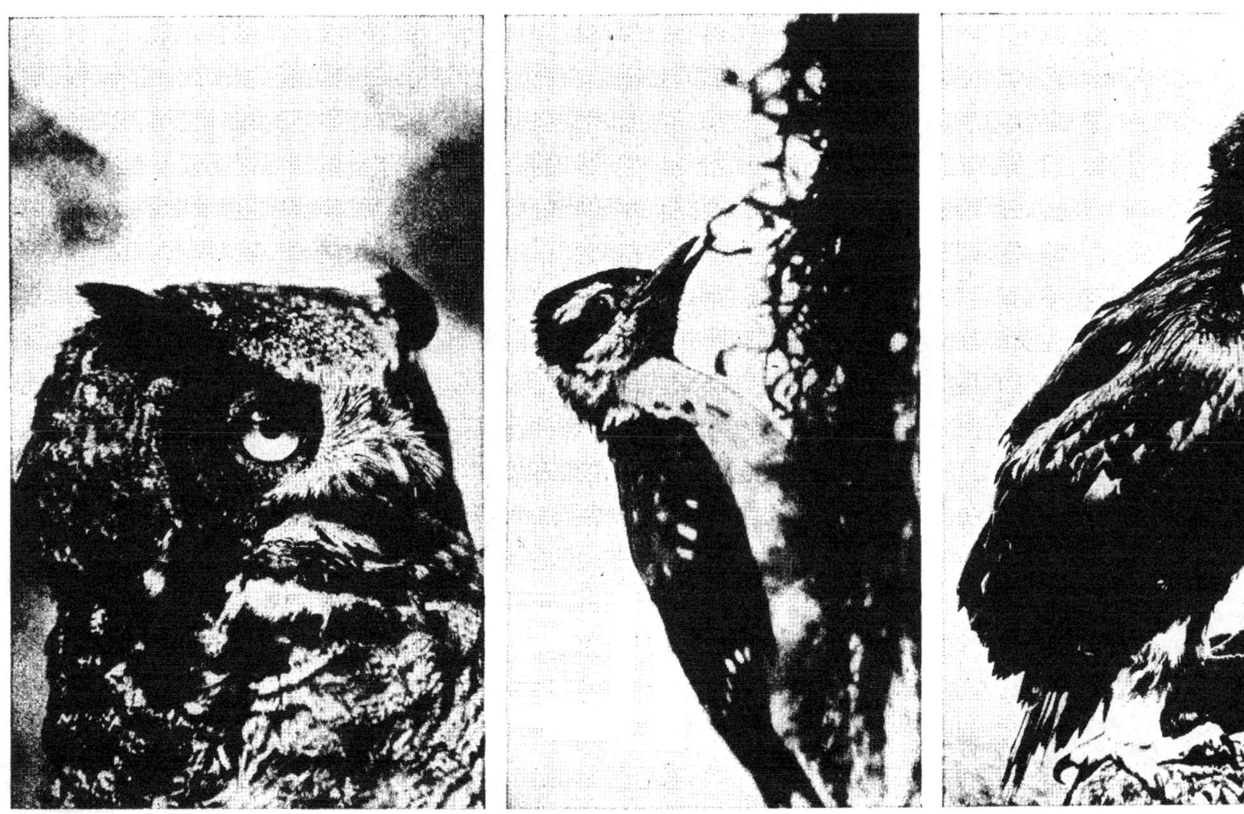

Left—Dusky horned owl. Center—Harris' woodpecker hunting for grubs. Right—The golden eagle, one of the most powerful birds of prey. The eye of the eagle is the most perfect organ of sight in existence

Photographs by W. L. Finley and H. T. Bohlman.

The Prouty memorial tablet.
Photograph by James L. Quinn

Carving crypt
Photograph by Mary C Henthorne

Location of tablet
Photograph by Mary C. Henthorne

lake one afternoon. Suddenly, one of these little hawks came like a shot out of the sky and caught a swallow in mid-air, carrying him away for a meal.

The dusky horned owl (*Bubo virginianus saturatus*) is the largest of the owls that inhabit the Cascade mountains. It is a powerful bird, living largely on game birds, rabbits and other small animals.

If one has his eyes open, he may frequently see an eagle sailing above the Cascades. In Oregon, we have two, the bald eagle, (*Haliaeetus leucocephalus*) and the golden eagle (*Aquila chrysaetos*). The term "bald" originated from the white head which is an unmistakable identification of the full-grown bird and at a distance gives the impression of baldness. The bald eagle has much the same general coloring as the golden eagle until it is three years old. In the time of Audubon, these young bald eagles were considered a separate species. At the age of three years, the bald eagle attains maturity and his white feathers appear on the head and neck. To distinguish one species from the other, look at the lowest joint of the leg. If this is covered with feathers down to the toes, it is a golden eagle; if the leg is naked, it is a bald eagle.

△△△

Two Voices are there; one is of the Sea,
One of the Mountains; each a mighty Voice.
—*Wordsworth*

△△△

The Harley H. Prouty Memorial

By John A. Lee

"To him who in the love of Nature holds
Communion with her visible forms, she speaks
A various language; for his gayer hours
She has a voice of gladness, and a smile
And eloquence of beauty, and she glides
Into his darker musings with a mild
And healing sympathy that steals away
Their sharpness, ere he is aware.

Many have read and pondered these majestic lines of the poet. To few did they have a deeper meaning than to Harley H. Prouty, ex-president of the Mazamas and to them affectionately known as "Uncle Jed," who departed this life on September 11, 1916.

To those nearest to him he had frequently expressed the wish that "when the summons comes to take his chamber in the silent halls of death," his remains should repose amid the mountains that he loved so well. Jerry E. Bronaugh and James L. Quinn, his executors, the former himself an honored ex-president of the Mazamas, determined that this wish of their departed friend should be observed. In the actual carrying out of the wish, what would be more fitting, as it seemed to them, than to have the last sad rites performed by the loving hands and hearts of the Mazamas, with whom the deceased had been so closely identified; whose life indeed, for a number of years, had in large measure been his life? Accordingly, in response to the request of the executors, a committee of the Mazamas was appointed, to be known as the Prouty Memorial Committee. This committee consisted of Rodney L. Glisan (chairman), Mary C. Henthorne, Anne C. Dillinger, Jerry E. Bronaugh, Edgar E. Coursen and the writer, all of Portland; W. C. Yoran, of Eugene, Oregon; and Dr. W. E. Stone, President of Purdue University, Indiana.

The remains had been cremated, and the committee decided that the urn containing the ashes, together with a bronze tablet suitably inscribed, should be deposited somewhere in the heart of the Cascades, amid the haunts that he had so often courted in life. Just what point to select was a matter of some debate, for many had been the mountain fastnesses that his tireless feet had explored. The committee finally decided that no more suitable place could be chosen than some point in the region of the Three Sisters, where, in 1910, he was first ushered into the Mazama brotherhood; where, on the occasion of the outing of that year, he performed his most notable feat in mountaineering; and where, in 1916, he bade what was to be a last fond farewell to his beloved mountains, with their glaciers and rugged cliffs and slopes, their singing streams and heathery meads.

To review and perhaps further perpetuate the memory of the exploit just referred to we trust will not be amiss in this connection.

As has been stated, it was in 1910 that Harley H. Prouty first became associated with the Mazamas, and it was then that the writer first met him. We were receiving the registrations that year for the summer's outing, and Mr. Prouty came in to have his name enrolled. Brief as was the interview, we were deeply impressed with the personality of our visitor and concluded then that somehow, in some manner, this personality would impress itself upon the outing. The members of the party were not long in coming to appreciate the splendid qualities of heart and mind of our departed brother, but it was not until the day of the official climb that he performed the feat of daring, nerve, endurance, and all-round skill in mountain climbing that will make his name remembered in Mazama circles for all time.

The official climbing party of the Mazamas had completed the ascent of the Middle Sister and were assembled on its summit. Across a sharp and declivitous canyon to the north, through which poured the ice stream of Collier glacier, reared the North Sister, a jagged cone of rapidly disintegrating lava. Up to that time the southerly and highest pinnacle on the summit of this peak had, so far as known, never been scaled. Because of the steepness of its walls and the loose and friable nature of the rock composing it, the ascent of this pinnacle had been deemed scarcely feasible. Aware of this, but nothing daunted, Mr. Prouty left the others of the party on the summit of the Middle Sister and set out alone to ascend this second unclimbed peak. Balked in his direct attack, he descended for a distance, executed a flank movement to the west, and ere nightfall had accomplished his purpose and was back in camp. A few days later he showed his skill as a guide by conducting a small party to the summit of this peak; and he duplicated the feat again in 1916 on the occasion of the Mazama outing of that year. Of the twelve names in the Mazama registration box on the summit of the North Sister, that of Harley H. Prouty appears three times. The name Prouty Peak has been given to this highest pinnacle of the summit of the North Sister in recognition of his achievement.

In selecting the memorial tablet, the committee decided that it should be modest in size and simply inscribed, in keeping with the reserve and quiet dignity of the deceased. Its dimensions are 18 by 24 inches, and it bears the inscription:

HARLEY H. PROUTY
A Lover of the Mountains
1857--1916

On the evening of Friday, August 31, 1917, the two executors and the Mazama committee set out from Portland to perform the final part of their mission. For various reasons, three of the committee, Rodney L. Glisan (chairman), Edgar E. Coursen and Dr. W. E. Stone were unable to accompany the expedition, much to their own regret and that of the other members. The party was joined in Eugene by W. C. Yoran, of the committee, James McAlpin, an experienced stonemason, and Edward Parks, the driver of the conveyance that was to carry the expedition up the McKenzie river to Frog Camp, where the trail leading to the Three Sisters begins. Arthur Belknap, packer, was to be in waiting at Frog Camp with pack-horses in sufficient number to transport the outfit of the party to the park region at the base of the Three Sisters.

All went well, and by nightfall of Saturday the party was comfortably established in camp at the precise spot where the Mazamas had encamped in 1910. The tables and the fireplace prepared by the party of that year were still intact and were utilized by us.

The site of this camp is pleasing. It nestles in one of the most charming dells of a region bountifully supplied with alpine parks and meadows. Its elevation is about 6400 feet and a clear and lusty tributary of White Branch flows through it. Across the stream to the south, Obsidian Cliff rises with some abruptness, and to the east the summits of both the Middle and North Sisters are plainly visible. The outing party of 1916 was not able to utilize this, camp as it was still buried deep in snow. The camp site of that year is beyond Obsidian Cliff, a mile or so to the southwest.

We were up and stirring early on Sunday morning, as the time allotted for the accomplishment of our mission was none too long. As we were busy about the camp, the morning sun rose above the summit of the North Sister, the scene of Mr. Prouty's exploit of seven years before, and its rays were observed to light up a jutting promontory of Obsidian Cliff, directly opposite the camp and not more than 500 feet away. The projection stood out conspicuously and appeared to be composed of light gray stone, firm in texture, and its face smoothly eroded by glacial action. Some of us mentally decided, then and there, that this was to be the location of the crypt and tablet, and thus it proved. Other sites were examined, only to be rejected in favor of the one first considered, and the choice was unanimous—spontaneous, it might be said.

Little time was lost in getting to work. Willing hands soon suspended from the rock a platform upon which Mr. McAlpin could stand while wielding mallet and chisel, and almost before the evening sun had left the rock in shadow a crypt had been excavated, the urn containing the ashes set therein, and the tablet, so placed as to act as a covering for the crypt, bolted and cemented to the face of the rock.

After everything had been placed in readiness for Mr. McAlpin and he had begun his work, a little group, consisting of Messrs. Quinn and Parks, Miss Dillinger and the writer, climbed to the summit of the Middle Sister and wrote into the Mazama registration book a simple record of the expedition, giving the location of the crypt.

Monday morning, just before starting on our homeward journey, we all gathered about the crypt and a brief and simple service was held. This consisted of a few short talks and the reading of Thanatopsis. Just above the tablet there chanced to be a circular pocket in the rock. Slight additional excavation fashioned this depression into a natural vase. In this were placed small branches of alpine fir, mountain hemlock, and white-bark pine, the three trees (and the only three) found at this elevation; clusters of asters, lupines, and Indian paintbrush, the flowers that perhaps most of all render bright and cheerful our alpine meadows; and some sprigs of heather, both the white and the purple.

As the contents of the vase were being arranged, "Jimmie" Quinn whose boyhood had been spent among the heath-clad hills of bonnie Scotland, let fall a remark that impressed his companions deeply: "The white heather! It is known in Scotland as the badge of eternal friendship."

By the morning of Tuesday we were all back at our homes, feeling that an appropriate act of fraternal affection and esteem had been well performed.

The crypt is on the most natural line of travel for those visiting the Three Sisters region, and will not fail to be noted by any observant person passing that way.

△△△

Yon mountain's side is black with night,
While, broad-orbed, o'er its gleaming crown
The moon, slow-rounding into sight,
On the hushed inland sea looks down.
—*Whittier*

△△△

The 1917 Mazama Outing to Mt. Hood

By JEAN RICHARDSON

One of the most important activities of the Mazamas is their annual expedition to the summit of Mt. Hood. Many people of Portland and its vicinity whose vacations are short and whose time for recreation is limited, hail with pleasure this opportunity to form a closer acquaintance with their nearest and best-loved mountain peak. Daily, with interest and awe, these busy residents look out across the intervening hills and valleys at its white, cloud-surrounded heights, silent, majestic, ever on guard. They feel its fascination, and when announcement is made of the Mazama annual outing, they respond to the call.

The 1917 excursion to Mt. Hood was planned to take place at the close of the N. E. A. convention, July 14 and 15, in order to give the visitors, as well as the people of Portland, an opportunity to participate. Registration headquarters were opened in the Mazama club-rooms, with Miss Harriett E. Monroe in charge.

After a week of careful preparation, about a hundred persons gathered at the Union station on the morning of July 14 and boarded the Mazama special train for Hood River. The rail-autos carried them on to Parkdale. From here they were conveyed in motor trucks to the beginning of the Sand Canyon trail, two miles from the chosen campsite on the north side of the mountain. At about sunset, they began winding their way up a ridge along one side of the beautiful Sand canyon.

The committee had decided to make the ascent from a point on this ridge about one and one-half miles southeast of Cloud Cap Inn, and here the party encamped for the night. The route up the mountain from this point is a particularly advantageous one, as there is no loss of elevation at any point.

On July 15, about 4 a. m., Mr. Roy W. Ayer, now president of the club, chief guide on many previous Mazama outings, summoned this band of would-be Alpinists from their fir-bough beds. Breakfast was eaten in the semi-darkness, and at about 5 o'clock the company assembled to listen to a brief lecture by Mr. Robert E. Hitch on the evils of eating snow, starting rolling rocks and spoiling steps. These final instructions over, Mr. Ayer took charge of the party and the ascent began. Out across the snow, under the firs and hemlocks, they marched to timber-line, where a halt was made to study the weather conditions. The sky had become overcast and rain was falling in a disheartening manner. Most fortunately, the storm was of short duration and half an hour later the party was well on its way to Cooper's Spur. A very direct route up the side of the mountain was taken and by one o'clock eighty-three successful climbers were on the summit, enjoying the hospitality of Elijah Coalman, the forest ranger stationed at the fire lookout and learning, many of them for the first time, how this old world looks from away up in the clouds. A party of about forty climbed the south side of the mountain on the same day. After spending an hour enjoying the view, exploring the summit, and registering names in the Mazama record, the company began the descent.

The return trip was made without accident, and at 4:30 on Monday morning the excursionists reached Portland, sunburnt and tired but triumphant and happy.

This outing, on the whole, was most successful. The party was made up of persons from all parts of the United States. They were unanimous in voting it a wonderful experience and departed full of enthusiasm over the trip and admiration for our Oregon scenery.

NAMES OF THOSE WHO REACHED THE SUMMIT OF MT. HOOD, JULY 15, 1917

Roy W. Ayer
Sarah H. Bacon
J. H. Balmanno
G. G. Battey
N. W. Battey
Geo. W. Bissell
W. Boyshuk
Gus Brockman
Emma Brown
Geo. T. Brown
W. F. Buse
Charlotte Callwell
David Campbell
Grace Campbell
Harriet Campbell
W. H. Campbell
Cornell Carlson
Arthur Crawfort
F. R. Cook
C. B. Compton
Anne Dillinger
Robert Duniway
Marie Evans
Elaine Ewell
G. L. Ford
Herbert Foster
R. H. Fox
H. Geiseike
Lucie George
W. A. Gilmour
Harry C. Grey
Margaret Griffin
J. T. Hazard
Mrs. J. T. Hazard
J. E. Holden
Mrs. J. E. Holden
George I. Howe
Sophie Huff
Myrtle Huff
Herbert Huff
L. W. Hughett
J. Jackle
L. Jeannin
Alice M. Johnson
Jas. C. Kendricks
F. M. Kiger
Mary L. Knapp
Ernest E. Knight
Margaret Kreiner
Johanna Kruse
Agnes Lawson
Ethel Loucks
Lucia Macklin
Ella Sabina Mason
R. P. Mercer
Geo. Meredith
W. McDougal
C. T. McGilvra
Mrs. C. T. McGilvra
Jean McKercher
Max Neitsch
E. J. Newcomer
Mabel Newcomer
B. W. Newell
D. G. Onthank
Geo. A. Patterson
Edward Patzelt
T. D. Phillips
R. W. Rea
Edwin L. Rice
Hulda Scheel
Emma D. Scholes
Nora Self
J. Duncan Spaeth
O. B. Sperlin
Vera E. Taylor
Violetta Wentworth
Mrs. Margaret B. West
Margaret L. West
J. C. Windham
Mrs. J. C. Windham
L. L. Woodward
Harry L. Wolbers

THOSE WHO MADE SIDE TRIPS TO ELIOT GLACIER AND COOPER'S SPUR

W. P. Hardesty, Leader
R. H. Atkinson
Mrs. R. H. Atkinson
Ralph Duniway
Harold Deming
Robert E. Hitch
Mrs. L. W. Hughett
Josephine Kelly
Iva Kendricks
J. C. Martin
Harriett E. Monroe
Alda McCready
Mrs. McCready
Marguerite Overhuls
Jean Richardson
Lucy J. Smoot

Mt. Shasta in History and Legend*

By Allen H. Bent

Mt. Shasta has had a most interesting history in spite of its lateness in getting on the map. It was not discovered until late in 1826, when Peter Skene Ogden, "humorous, honest, eccentric, law-defying Peter Ogden, the terror of Indians and the delight of all gay fellows, making an unusually long journey for beaver skins from the Hudson Bay Company's post at Fort Vancouver on the Columbia, caught sight of it. This was thrity-four years after Vancouver had discovered and named Mts. Hood, St. Helens, Baker and Rainier, eighty-five years after Mt. St. Elias had been discovered, nearly three centuries after the Sierra Nevada, the southern Rocky mountains and the Appalachian chain had been seen by Spanish explorers, and more than three hundred years after the first ascent of Popocatepetl, Mexico's great smoking mountain. Notwithstanding all this, Shasta was the first of the higher peaks on the Pacific slope to be climbed and for several years was supposed to be the highest mountain in the United States. It was seen in 1841 by Lieutenant Emmons of the Wilkes Exploring Expedition, in 1846 by John C. Fremont, on his third western journey, and in 1851 by Robert S. Williamson, of the U. S. Topographical Engineers, during his search for a railroad route from the Columbia river to the Sacramento valley. All three referred to it in their official reports, Wilkes calling it Mt. Shaste, Fremont the Shastl Peak and Williamson Shasta Butte. The Wilkes expedition included several distinguished scientists, including James D. Dana, the geologist. Fremont's guide was Kit Carson. Wilkes and Williamson in their reports, the first printed in 1845, the second in 1855, each show a picture of the peak. Williamson also has lithographs of Lassen's Butte, as he calls Mt. Lassen, and Mts. Hood, Jefferson and Three Sisters. The first ascent to which I find any reference was made by Capt. Henry Prince, who in 1852 planted the American flag on its summit. Prince was a native of Eastport, Maine, a graduate of West Point, and later a soldier of distinction in the Civil War, at the close of which he was brevetted brigadier-general. The year before this ascent gold was discovered at Yreka—Shasta Butte City it was originally called—some thirty-five miles northwest of the mountain, and the first settlers came into the region. The village at the foot of the mountain, at first called Strawberry Valley, later contracted to Berryvale, takes its present name of

*Parts of this article have appeared in Appalachia, Volume 13, Numbers 1, 2, 3, 1913-1915.

Sisson from J. H. Sisson, who settled there about 1857 or 1858. He took care of the few travelers that came, hunted and trapped and later guided occasional climbers to the top of the mountain.

The second ascent of which I have found any record was made by Israel S. Diehl of Yreka in October, 1855. He seemed proud that he had no equipped and noted travelers, officers, literati, or blooming lively belles" for companions, but that he was solitary and alone on this "stupendous and unknown undertaking."* After two days' journey on horseback he made camp at tree line after dark. To keep off the grizzlies and mountain lions he made a big fire which burned everything around him. The grizzlies did not get him and he had a fine day on the mountain.

The next ascent, the following April, was early in the year for a mountain of this magnitude, and it is not surprising to learn that the thermometer carried to the top registered twelve degrees below zero. Anton Roman, who made this ascent in 1856, was a native of Germany, had come to America at an early age, crossed the plains in '49' mined at Yreka and later opened a book-shop in San Francisco, where Thomas Starr King, Mark Twain, Charles Warren Stoddard, Bret Harte, and the other writers of the day were wont to gather.

Thomas Starr King, known to every mountain lover in New England by his enthusiastic descriptions of the White mountains and to Californians as the man whose eloquence saved their state to the Union during the Civil War, had designs upon Shasta soon after he went to the Pacific coast. In January, 1861, he wrote to his friend, President Ballou of Tufts College: "Last week I went to Marysville. From the church tower I saw the Sierra in saintly whiteness along a horizon of two hundred miles. They were a hundred miles away but seemed not over thirty, and far on the north two hundred and thirty miles air-line, the pyramid of mighty Shasta peeped over the dim plain—a knob of steady flaming gold. Do come out here and go with me to see it and Oregon. We'll go to the summit of Shasta and laugh at Mont Blanc. I mean to."

In the meantime, probably about 1854, Joaquin Miller appeared on the scene. In the story of his life—"*My Own Story*" is the title of the book—he says:

"I ran away from school in Oregon at the age of thirteen to the gold mines of California. It was late in the fall—I was alone, a frail,

*The account of Diehl's climb, with a reference to Prince's ascent, will be found in "Scenes of Wonder and Curiosity in California: a Tourist's Guide to the Yosemite Valley." By J. M. Hutchings. New York and San Francisco, A. Roman & Co. 1870. There is a small engraving of Shasta.

sensitive, girl-looking boy, almost destitute. As I descended the stupendous and steep mountain that fronted the matchless and magnificent glory of Mt. Shasta, I fell in with an old mountaineer, by the name of Mountain Joe, one of Fremont's former guides, who was on his way to the Rio Grande to get a band of Mexican horses." Miller accompanied him, after being wounded in an Indian fight at Castle Rocks, and spent the winter in Arizona and Mexico. Upon his return the following year, he settled down among the Indians. Of this period he wrote, "Here were the tawny people with whom I was to mingle. There loomed Mt. Shasta with which my name, if remembered at all, will be remembered." As the winter approached he left the Indians and made his way to Yreka, and "The Forks," where he remained with James Vaughn Thomas until the following September, when they moved nearer Shasta and mined quietly for a year. Then his companion left. Miller married an Indian girl and went to live with her people near the mountain. Listen to his description of the coming of spring in those happy days: "At last the baffled winter abandoned even the wall that lay between us and the outer world, and drew off all his forces to Mt. Shasta. He retreated above the timber line, but he retreated not an inch beyond. There he sat down with all his strength. He planted his white and snowy tent upon this everlasting fortress, and laughed at the world below him."

He tells of Paquita, the Indian girl he married, making an excellent charcoal picture of Mt. Shasta on the door of their cabin. In an earlier volume, "*Life Amongst the Modocs,*" dedicated to the Red Men of America, is an account of one of his ascents, which he says was his last. It must have been about 1858. He acted as guide for a couple of missionaries. They got to the top, but made hard work of it, scattered some of their tracts and immediately started down. This to a boy in his teens, who had been living a wild free life for several years, coupled with the fact that they paid for his services in prayers and printed sermons, was too much for him, and the missionaries were given a sound lashing with his pen. At the time he simply stabbed the tracts with his bowie knife.

Josiah D. Whitney, chief of the Geological Survey of California, and his assistant William H. Brewer, who climbed and measured the mountain in September, 1862, were the first scientists to reach the top. They had been told that it could not be climbed, but when they got there they found, as Brewer expressed it, "A mixture of tin cans and broken bottles, a newspaper, a Methodist hymn-book, a pack of cards, an empty bottle and various other evidences of a bygone civilization."

As a result of the description of their ascent which Whitney and Brewer sent to their alma mater, Yale College, Clarence King, who later became the first director of the United States Geological Survey, but at that time was only a student of twenty-one, started for California, where his first ascent, in September, 1863, with Brewer, was up the now famous Lassen peak. Less than a year later he discovered and named Mt. Whitney, the highest mountain in the United States, but his interest here lies in the fact that he was one of the first to spend a night on Shasta. With him were Samuel F. Emmons, F. A. Clark, and A. B. Clark, all connected with the Government Survey of the Fortieth Parallel. This was in September, 1870. In his "Mountaineering in the Sierra Nevada," one of the earliest and at the same time most fascinating of books upon American mountaineering, he tells the story. After a night on the rim of Shastina and a perfect day on the summit, the view from which he describes in detail,—"what volumes of geographical history lay in view," he exclaims—the little party went down to the hot springs and prepared for the night. "We built of lava blocks a square pen about two and a half feet high and banked it with sand. I have seen other brownstone fronts more imposing than our Shasta home, but I have rarely felt more grateful to four walls than to that little six by six pen. The zephyr, as we courteously called it, had a fashion of dropping out of the sky upon the fire and leaving a clean hearth. This tempest descended to so many absurd personal tricks altogether beneath the dignity of a reputable hurricane, that at last it seemed to us a sort of furious burlesque. Not so the cold; that commanded entire respect, whether carefully abstracting our animal heat through the bed of gravel on which we lay, or brooding over us hungry for those pleasant little waves of motion, which, taking Tyndall for granted, radiated all night long in spite of wildcat bags from our unwilling particles." All this was in the interest of science, for King and Emmons were studying the geology of the mountain and the Clarks carried their surveying instruments. A few years later, on the last day of April, 1875, John Muir spent an unwilling night with a single companion near the same spot. An account with illustrations appeared in Harper's Monthly for September, 1877. The ascent, made with Jerome Fay, "a hardy and competent mountaineer," was for the purpose of making barometrical observations on the summit, while Capt. A. F. Rodgers of the United States Coast Survey made similar observations at the base. Until early afternoon the weather was fine; at one o'clock the thermometer stood at fifty degrees and a vigorous bumble-bee zigzagged around their heads. Soon afterward thin clouds began blowing over the summit, and just after three a storm broke in terrific fury, hailstones resembling small mushrooms falling heavily. "The thermometer soon sank below zero, hail gave place to snow and darkness came on like night. The wind, rising to the highest pitch of violence, boomed and surged like breakers on a rocky coast. The lightnings flashed amid the desolate crags in terrible accord, their tremendous muffled detonations unrelieved by a single echo." Near the hot springs Jerome concluded that it was impossible to proceed (though Muir thought they had a good chance to get down), so at the fumarole they stayed. Sometime in the early part of the night, after two feet of snow had fallen, the storm vanished into thin air as quickly as it had come. They dared not go to sleep, even if their sufferings would allow them, for fear the volcanic gases would collect in sufficient quantities to suffocate them. The dawn came, but the early morning sun seemed to have no heat. "At length," says Muir, "about eight o'clock on this rare first of May, we rose to our feet, some seventeen hours after lying down, and began to struggle homeward. At ten o'clock we reached camp and were safe. We were soon mounted and on our way down to thick sunshine; violets appeared along the edges of the trail, and the chaparral was coming into bloom, with young lilies and larkspurs in rich profusion. How beautiful seemed the golden sunbeams streaming through the woods, and warming the brown furrowed boles of cedar and pine. The birds observed us as we passed, and we felt like speaking to every flower."

In the summer of 1878, B. A. Colonna of the United States Coast and Geodetic Survey spent nine days on the mountain, the last four alone sleeping at the hot springs 300 or 400 feet from the top. He lost fifteen pounds in weight during that time. His pulse in repose was between 100 and 105, while the slightest exertion sent it up to 120. On the first of August he succeeded in accomplishing what he went up for, heliographing to Mt. Helena, 192 miles to the south. The same year Capt. A. F. Rodgers of the same Survey placed on the summit the metal signal tower that stood until 1905. Colonna succeeded in getting twenty Indians to carry his equipment up, but they did not remain, in fact did not even go to the top, although within a few hundred feet of it.

For the story of Mt. Shasta before its discovery by white men we must look to the Indian legends and to the mountain itself. The geological history of the mountain has been most interestingly told by J. S. Diller: "It has long been the field whereon was fought the battle between the elements within the earth and those above it. In the early days the forces beneath were victorious and built up the mountain in face of wind and weather, but gradually the volcanic energy reached its climax, declined, and passed away. Fiery lava has been succeeded by arctic cold."*

"The Indians have a very different idea of its creation, but it is equally interesting. The Shasta tribe, from whom it takes its name, is practically extinct—the proximity of a mining camp is not conducive to length of days in either red or white men—but fortunately their story of the mountain has been preserved for us by Joaquin Miller.

"The Great Spirit created this mountain first of all. He pushed down snow and ice from the skies through a hole which he made by turning a stone round and round, then he stepped out of the clouds on to the mountain top, and descended and planted the trees all around by putting his finger on the ground. The sun melted the snow and the water ran down and nurtured the trees and made the rivers. After that he made the fish for the rivers out of the small end of his staff. He made the birds by blowing some leaves which he took up from the ground among the trees. After that he made the beasts out of the remainder of his stick, the grizzly bear out of the big end, and made him master over all the others. He made the grizzly so strong that he feared him himself, and had to go up on the top of the mountain to sleep at night. Afterward, when the Great Spirit wished to remain on earth, and make the sea and some more land, he converted Mt. Shasta by a great deal of labor into a wigwam, and built a fire in the center of it and made it a pleasant home. After that his family came down from the Heavens and they all have lived in the mountain ever since. Before the white man came they could see the fire ascending from the mountain by night and the smoke by day.

'One late and severe springtime many thousand snows ago there was a great storm about the summit and the Great Spirit sent his youngest and fairest daughter up to the hole in the top, bidding her speak to the storm that came up from the sea, and tell it to be more gentle or it would blow the mountain over. He bade her do this hastily, and not put her head out, lest the wind would catch her in the hair and blow her away.

"The child hastened to the top, but having never yet seen the ocean (where the wind was born), when it was white with the storm she put her head out to look that way, when lo! the storm caught her long red hair, and blew her out on to the mountain side. Here she could not fix her feet in the hard smooth ice and snow, and so slid on and on down to the dark belt of firs below the snow.

---

*Quoted from Mr. Diller's "Mt. Shasta, a Typical Volcano," in the Physiography of the United States. National Geographic Society Monographs, 1895.

'Now the grizzly bears possessed all the wood and all the land even down to the sea at that time, and were very numerous and powerful. They were not exactly beasts then, although they were covered with hair, lived in caves, and had sharp claws; but they walked on two legs, and talked and used clubs to fight with, instead of their teeth and claws as they do now. At this time there was a family of grizzlies living close up to the snow and the father found this little child, red like fire, hid under a fir bush, and took her to the old mother, who said she would bring her up with the other children. When their eldest son was grown up he married her and many children were born to them. But being part of the Great Spirit and part of the grizzly bear, these children did not resemble either of their parents, but partook somewhat of the nature and likeness of both. Thus was the red man created, for these children were the first Indians.'

△△△

Mt. Hood in Autumn

By MARGARET A. GRIFFIN

Across the miles of wooded ranges we have seen Mt. Hood, serene and calm, and seemingly changeless, and have felt that if we could make one pilgrimage to its summit, we would be satisfied. But our friend Elijah Coalman, of Mt. Hood fame, who has climbed the mountain more than 365 times, tells us that no two of these climbs were alike; and this has also been the experience of many of us who are less familiar with the mountain.

Knee-deep in soft snow, with all available wearing apparel pressed into service, we have climbed a blustery mountain and looked down on a sea of billowy clouds hiding the valley beneath. On crusted snow, yet under the bluest of skies, against which the great crags stood out in sharp relief, we have climbed a sunny slope and been rewarded with a far-reaching view from the summit. Then, again, in late September, when Mt. Hood had yielded its harvest of snow, we have climbed through banks of fog to quiet heights which seemed apart from all the world. Great walls of solid blue ice revealed a beauty that we had not dreamed was there; and where before our feet had trod in safety on firm snow-fields, deep chasms stretched across the way. Steel Cliff seemed many times its usual size and its worn and wrinkled sides were colored in dull purple, yellow, red and brown. Then, even while we rested on the summit, the scene changed. A mighty wind arose, filled the air with flying snow and sent us hurrying from our shelter, down the icy steeps. Faces gleamed red under the biting sleet, and hair and clothing gathered the frost, until we looked like strange inhabitants of another world.

So we have learned that Mt. Hood holds a never-ending variety for us in changing mood and season; and, never weary, we go again, and long to go again and yet again.

And, remembering that most wonderful climb of all, the autumn trip, we went again in September of this year. As there were forest fires near Government Camp, we had some doubt that we should be able to get through. We were prepared, however, with axe and spade, but had no occasion to use them, and in fact had no real difficulty, although the fires were close in some places and it was necessary to remove a few burning snags from the road. It was a wonderful sight, though a sad one, to see the smoldering trees gleaming red through the dusk like lighted windows of a hillside city.

After a night at the timber-line, we climbed the mountain leisurely, taking all the time we could wish for enjoying to the full old Mt. Hood in autumn. The snow-fields, as on the previous autumn climb, were melted to such an extent that all the cliffs towered above us, and "south-side Hood," as it is known to scores of summer climbers, was so changed that they would hardly have recognized it. We made our way carefully over the snow-fields, where crevasse after crevasse was skirted, or crossed where snow bridges were safe. Peering down into these crevasses, we saw the curving halls of blue ice, leading down and away, tempting the imagination with visions of beauty just out of sight. About a quarter of a mile below Crater rock we came upon some ice caves on White River glacier where entry was possible. We picked our way carefully as far in as it seemed wise to go. We looked down into the dark blue of the inaccessible depths, and across into chambers where the penetrating light seemed to bring in the blue sky itself to be held captive by icy fingers.

We reached the summit and Elijah Coalman's shelter, and were soon enjoying a good dinner, the material evidence of his hospitality, and we tried to show that we appreciated the fact that hot biscuits on the summit of Hood are something out of the usual. When darkness had fallen over that little band of friends on the top of the world, the full moon sent her light to show us the mountain and her slopes as we had never before seen them. We saw the big fire in Iron Creek canyon and smaller smoldering fires around the valley, but we could not see far on account of the smoke. That was a new experience—the fulfilment of our wish to pass the night "up there close to the stars."

Our descent the following day was made as leisurely as our climb had been. Crater rock at this time stood perhaps three hundred feet above the snow at the end of the hog-back, and we climbed to its summit. We found the rock to be somewhat loose and crumbling, and although there were no difficult places, the loose rocks made it necessary to exercise care. From the top we watched the rocks, great and small, come tumbling and thundering down the sides of Steel Cliff to the glacier beneath.

Several later parties have been to the mountain in the open autumn weeks that followed. A party of our young men spent a day on the glaciers, and brought home to the less adventurous a harvest of beautiful photographs. Two of the party, T. R. Conway and H. L. Wolbers, scaled Illumination rock, which at that time afforded a climb of about five hundred feet, roughly estimated. This point lies between Zigzag and Reid glaciers and consists of two pinnacles joined by a saddle, the whole following the general slope of the mountain. The climbers crossed the head of Zigzag glacier, and, as the hour was late, made no attempt to choose the most feasible route, but climbed from a point below the higher pinnacle, on the upper side of the rock. Unlike Crater rock, Illumination rock was found to be of hard smooth stone of columnar formation. The parts separated easily and made the climbing somewhat difficult; and the smoothness of the rock and the absence of holds made a demand upon the physical strength. A large sloping rock, with an actual level of two or three feet in diameter, forms the top of the higher pinnacle.

The United States Forest Service has recently erected a shelter at timber-line, which will serve alike the summer and autumn climbers. It is the intention, also, to erect a shelter on the summit, in addition to the one now used as a lookout station for the purpose of locating forest fires. This second shelter will be for the use of climbers, and may be occupied by those wishing to remain on the summit overnight. Just what accommodation will be afforded is not known at this time, but undoubtedly the number of climbers will be increased by the existence of these cabins. Doubtful weather will become a less serious consideration. Many who have feared the strain of a hurried trip will be able to make the climb leisurely by remaining on the summit over night, and will have time and strength to spare for side trips to the glaciers and for seeing what the region has to offer.

A climber making his initial trip to the summit of Mt. Hood might prefer to go in the summer months, when the treacherous crevasses are covered with tons of snow which make traveling much more safe, when the big crevasse is scarcely more than a huge crack, and sufficient

Views showing condition of the glaciers of Mt. Hood in October, 1917. Upper—Ice caves in White River glacier Middle—Great seracs in White River glacier. Lower—Seracs in Zigzag glacier.

Photographs by F. I. Jones

Views from Silver Star mountain, June 17, 1917
Upper—Mt St Helens   Middle—Mt. Adams
Lower—Mt Hood

Photographs by F. I. Jones

snow clings to the last thousand feet to weigh down the loose rock and make the climbing easier. But the true student of nature will find the autumn mountain much more interesting; he will read its story more clearly when the snow has gone and the rocky outlines are exposed to his view. The adventurous man will find more to attract him. The climber will find a mountain new to him, and usually much better climbing weather. The camera man will find an unlimited field. And lovers of the mountain itself for its own sake will find a new satisfaction in its different phases.

△△△

The Silver Star Trip

By MARION SCHNEIDER

Silver Star mountain, in Skamania county, Washington, about twenty-five miles northeast of Portland, has long been considered by the Mazamas as a desirable objective point for a local outing; but it was not until this year that we finally succeeded in making the trip, after postponing it twice on account of weather conditions.

On June 16, 1917, a party of over ninety Mazamas, under the leadership of Alfred Parker and J. G. Edwards, left the Union Depot at 2 p. m. on a special train for Moulton, Washington. Upon our arrival there at 4 p. m., we set out at once on the eight-mile walk to camp.

The first part of the journey was over an old wood road which wound about with such a gradual ascent that many of us were sorely tempted to loiter in the evening sunshine and enjoy the occasional glimpses of the Lewis river below us. Presently we were actually climbing up and up, through a wide region of burned timber. There was something pathetic in the charred trunks of great trees, which seemed to stand as sentinels guarding the desolate hills about. Here and there we would catch a glimpse of a lonely little farmhouse in a distant clearing, but the cabins we saw were mostly deserted and almost hidden by young firs and maples.

After a climb of several hours we left the road and followed a path along the ridge. Dusk was just beginning to close in about the ranges opposite, and as we tramped along the hills gradually became an indistinct blur in the twilight. Then suddenly we descended into the light of a cheery camp-fire, near an old deserted cabin, and were welcomed by several Mazamas who had gone up earlier in the day. After a jolly little social gathering, the peace and quiet of the big out-of-doors settled over the sleeping camp.

Just as the light was breaking over the distant mountains, we rolled up our dunnage and started to climb once more. For the first few miles the trail led through a veritable jungle of underbrush. Then we came to the real mountain, with its rocks, its flowers, its snow and the scrubby trees twisted by the fierce winds. Up and up we climbed into a meadow of brilliant mountain flowers; up across the ridges into the deep snow. Our progress was slow, for the view began to unfold below us and each of us wanted to linger and enjoy every step. Those of the party who had climbed Mt. Adams were constantly marveling at the familiar scenes brought vividly before them.

At noon we reached the summit, climbing up out of the timber and snow to the very crest of Silver Star mountain. At this vantage point of rock, 4358 feet above sea level, we stood thrilled by one of the finest views ever given a mountaineer to enjoy. It was a day of perfect sunshine; the horizon cloudless; snow peaks as far as the eye could see; Adams, Helens, Hood, Rainier and Jefferson all looming up before us, and ridge upon ridge below us, with a wonderful valley spread out at our feet.

At about one o'clock we left the summit. Some of the party spent the next hour in glissading on the long snow-fields; others explored the ridges and enjoyed the far-reaching view into the canyons beyond. At seven o'clock the entire party was on board the special bound for home, and about two hours later we were in Portland.

All who were so fortunate as to be able to take this trip will ever remember it as one of the most noteworthy of our local outings.

△△△

The Camera in High Places

By R. L. Glisan

The first time I took a camera on a mountain trip, my pleasure increased tenfold. On previous occasions I had selected views from the collections of others, thinking in this way I could get a better set at less trouble than if I took any myself. I soon discovered that the views I wanted most, no one seemed to take. With my own camera, the interest in the trip increased. Always having in mind the possibility of securing good views, I took more notice of what I saw, could recall the scenery more vividly, and the photographs I secured brought back a flood of memories which a purchased photograph never awakened.

When I took my first photographs, I could not puzzle out the failures, and the dread of repeating failures was discouraging. I began making notes of the subject, stop, speed and light, using a small note-book for that purpose. When my album of a particular outing was ready, I noted the time and stop on the lower corner of the page and, if I contemplated a similar trip another season, would look over the album and refresh my memory as to the exposures. Each album, page and film is numbered and films filed in sets on a card index system.

I found the camera requires a more careful handling on the higher elevations, and the season makes quite a difference. When I gave the same exposure on the snow in winter that I gave under like conditions in summer, I found the photographs were under-exposed. Glacial ice requires more exposure than snow, and the early morning views require a much longer exposure than mid-day. To give concrete examples, using a 3-A camera with a Goerz lens, I found a 32 stop and one-fiftieth of a second gave the best results on snow-fields in summer, or taking distant views or cloud effects. The early morning view of snow peaks above camp requires often a fifth of a second or more, although the light seemed perfect. My winter views on snow are generally 16 stop and one-fiftieth of a second in a fair light, and a 4 stop and twenty-fifth on a dull day. Timber views in winter snow are about the hardest views to take, and 32 stop and half a second or even a second give the best results.

Excellent cloud effects can be secured by turning the camera towards the sun and shielding the lens, using a 32 stop and one-fiftieth of a second, or 32 and twenty-fifth. Glacial ice or crevasses require an 8 and one-fiftieth second, or 4 and twenty-fifth, if the sun is not bright. For rock work, 16 and a twenty-fifth give satisfactory detail unless the sun is bright, when a quicker exposure can be made. In the forest below the snow-line, it is nearly impossible to over-expose, and stopping down to 32, I give ten to twenty seconds. This brings out detail in light and shadow. An overcast or dull sky is preferable for photographs in the forest.

Excellent night views of close-by snow peaks or other objects in the moonlight can be secured with a wide open lens and one to three hours exposure. I took a daylight and later a moonlight view of Mt. Hood from timber-line. In the daylight view the snow outline of the mountain blended with the sky, while in the moonlight view the mountain was much whiter than the sky and strongly outlined. A moonlight view of a snow-covered cabin with lighted windows was secured with an hour's exposure.

A ray filter of moderate density may give distant snow peaks and detail not otherwise procurable, but the filter has to be used with caution and often stopping the lens down will give the same results and will prove much more reliable than the filter.

In photographing waterfalls, one can secure life in the falling water and detail in the side rock walls by several successive exposures of a fiftieth or twenty-fifth of a second with a fairly wide stop. For years I did not use a tripod and consequently failed to secure satisfactory views of waterfalls, trees, flowers, trails, etc. I always take one into camp now, but to economize weight rarely take it on the ascent of the mountains. The views taken on the ascent are generally instantaneous of snow, glaciers, moving or distant objects, and generally there is no place or time to set up a tripod. If one uses a ray filter for the distant views on a climb or on the summit, a convenient rock is generally available on which to rest the camera.

One of the greatest difficulties is to secure a photograph of distant snow peaks on the horizon, and they can only be secured by using a filter or by stopping down and giving a quick exposure. Even then the failures are in the majority. Often one photograph fails to take in sufficient scenery and excellent panoramas can be secured by setting the camera on some flat surface and taking two or even three views, turning the camera each time to take in an entirely new adjacent section of the horizon. When the camera is in position to take the right half of the view, for example, make a mark or place any small object on the flat surface just under the center of the lens. Then swing the back of the camera to take in the next section of the scenery on the left with the lens as a pivot always over the mark and keeping the same amount of sky, which is easily done by noticing how high up on the side of the finder the edge of the scenery comes. Two or three photographs can thus be secured which will take in more than half of the horizon and should match perfectly and give a most valuable panorama. Your photographer can enlarge these views, using the overlap in printing so as to blend and blot out the division line.

These remarks are intended for the beginner, and to him I make no apology for their crudeness.

△△△

Far overhead against the arching blue
Gray ledges overhang from dizzy heights,
Scarred by a thousand winters and untamed.
—*Bliss Carman*

Address of Retiring President

WILLIAM P. HARDESTY

(Delivered at Annual Meeting Held in the Club-Rooms, October 1, 1917)

FELLOW MEMBERS:

It seems to be an accepted and befitting custom that your retiring president should give a brief review of the events of the Mazama year, together with a discussion of our plans and policies. In accordance therewith I shall endeavor to give a plain, unvarnished statement of our doings for the year, relating our difficulties and failures as well as our achievements.

It will occur to you at once that this has been an abnormal year. The entry of the United States into the world war occurred in the very middle of our official year. The thoughts of our people have naturally been diverted to graver matters than Mazama recreational affairs, so that interest in our various activities could not be so intense as before. Some of our young men have entered the service of their country and so are no longer with us. The rising cost of living and the call for everyone to be kept at work, each to do his or her share, have operated to reduce attendance on our outings (at least on the annual outing), and many have also lapsed in membership.

Notwithstanding these adverse circumstances, I think we have had a fairly successful year. Our educational courses were continued during last winter, with lectures at our rooms; and we have had several good lectures at the Central Library, one quite noteworthy. Our annual magazine maintained the high standard of excellence set in past years. An echo of war times was found in a Red Cross course in first aid, taken by our members in these rooms.

Our local walks and outings have had an attendance about up to the normal and have been the source of much enjoyable recreation. Our annual Mt. Hood short outing was this year taken later than usual, to afford our visitors in attendance at the N. E. A. convention an opportunity to climb a real mountain. Though successful, the attendance on this outing was not what it would have been in normal times.

Our annual two-weeks' outing took us this year to rugged Mt. Jefferson, an Oregon peak until this year almost unknown to the present generation of Mazamas, except from afar. The outing was most successful and enjoyable, and we may well congratulate ourselves that the difficult and dangerous ascent was made by the great majority of those in camp, without a single mishap. From a financial standpoint this outing was less successful, there being a deficiency of about $350 in the receipts, as compared with the expenses. The advance by leaps and bounds in the cost of provisions and equipment (after we had fixed our low rates) was largely responsible for this. The war also materiallv reduced the attendance.

For the first time steps were taken towards securing a lodge somewhere in the mountains, a committee (very strong in personnel) having been appointed to investigate and report. Certain difficulties in the way of desirability or success in maintaining a lodge soon became obvious, and the committee made a report on progress and was then continued for another year. The lodge question will evidently have to wait for more auspicious conditions.

One consideration, which I think the committee has recognized, is that a mountaineering *lodge* should be what its name implies, a place of *refuge* against the elements for members on exploring and outing expeditions, *not* a country club nor yet a place for the Owl Club's all-night sessions. My own idea is that a beginning on a small scale, the utilization of one modest log house or cabin at first, with addition of other units from time to time as experience and needs justify, would probably be the safest course.

A fully-equipped and commodious lodge establishment will take much money, to be raised by subscription or possibly by issue of bonds or stocks, as what funds could be spared from our general fund would not go very far.

Coming now to our activities and recreations made possible by our possession of club-rooms and other quarters, I may say a word. We have here, in this one room, over 900 square feet of space, which has been used in the main for purely social and recreational purposes. So far as I know, this is the only mountaineering club in America that has anything like so much space for club-room purposes. Our sister club, the Mountaineers, of Seattle, has lately opened up a fine club-room, but with less than half the space afforded by our room. We have also a cloak-room, which, unfortunately, has been pressed into service for storage of all kinds of impedimenta. For the last six months of the year we have also rented a small room on the third floor of this building to receive the surplus of our really fine collection of books, magazines and other material for a library appropriate for an organization like ours. It was one of my ambitions when taking office to see our library fully organized, with everything classified and card-indexed, but this will have to go over to the new year.

In every other similar club the maintenance of a library devoted to mountaineering, exploration, and dissemination of the knowledge of natural things, has been one of the first and most essential functions. Such things have been a sadly neglected part of our Mazama life, and we must confess that, here in this room particularly, the social features have overshadowed all others. To maintain our library in proper order after it is once arranged, and to do the clerical work that is required, it appears that a salaried assistant secretary and librarian, giving certain hours of his or her time each day, will be a necessity. But even this will not suffice to secure full use of the library, as our present quarters (all in one room) are not conducive to studious habits.

I believe that the addition of the circulating feature to our library, as in others, will be necessary to secure all its benefits. When we consider that out of our large resident membership perhaps not over fifty or seventy-five of us habitually visit the club-room at all, and only by circulation of our books can the remainder fully profit from the library, the need of the circulating annex becomes obvious.

I venture to suggest to the incoming officers, and to all members here tonight, that we should devote more effort to securing for our non-resident membership, and for the large resident membership that cannot avail themselves of our club-room facilities, all the benefits possible for them.

Finally, I wish to express my thanks to all members for their support of the Executive Council, now retiring; and to the other officers and heads of committees for their efforts to keep the Mazama ship sailing on an even keel and with a reasonable amount of canvas to the wind. We will now make way for new officers and crew, with best wishes for their success in handling the affairs of the club for the ensuing year.

△△△

Reports of Lodge Committee

September 6, 1917.

EXECUTIVE COUNCIL OF THE MAZAMAS,
Portland, Oregon.
Gentlemen:

In November, 1916, the Executive Council appointed a committee, consisting of R. L. Glisan, chairman; John A. Lee, R. H. Atkinson, C. B. Woodworth and L. A. Nelson, to investigate the cost of a lodge for the club, site, building, etc. As chairman of that committee, I desire to submit the following report:

The committee earnestly considered various sites, but so far have not found one that combines sufficient advantages to warrant its selection. During the summer months the members can and undoubtedly would prefer to camp out in the open, and the Forest Service has erected a number of attractive cabins and shelters along the forestry trails which are rapidly spreading over the state, and especially in the vicinity of the Columbia highway.

We understood that in making our selection we should give a preference to a lodge site where the members could specialize on winter sports. We examined several locations suggested, giving special attention to the locality near Bonneville on the Highway. Whenever there is any snow-fall along the Columbia highway, it is generally heavier at Bonneville than at any other place. Winter before last the snow-fall there was several feet in depth, but last winter there was practically no snow and one could not count on it there except at very brief intervals. None of the places along the Highway offer attractive slopes for snow-shoes or skis, as most of the trails lead up rather precipitous canyons and do not branch out until some distance back.

A site was suggested on the south slope of Mt. Hood not far from Government Camp. The expense of reaching such a site would be less than on the north side, but would take longer time. Any place on the slope of Mt. Hood would be quite a matter of expense and time, which would prove a severe handicap.

We are indebted to your president, Mr. Hardesty, for considerable data which he has secured by correspondence with the other mountain clubs.

I have personally seen the site of the Parsons lodge which the Sierra Club erected in the Tuolumne Meadows in the Yosemite National Park. This is a location ideal for summer trips, but the expense and time of making a winter trip would be a considerable item. I am familiar with the general locality of the Muir lodge erected by the

Southern California section of the Sierra Club near Sierra Madre, east of Los Angeles.

I have also seen the site of the club lodge of the Alpine Club of Canada at Banff and of their shelter camp at Lake O'Hara in the Canadian Rockies. In connection with the lodge they have a number of tents and can accommodate quite a number of members of the club who stop over at Banff. The surroundings, however, are not particularly attractive from a mountaineering standpoint and it is more as a matter of convenience and a help to the members on their mountain trips. Lake O'Hara is about a day's tramp from the railroad and the shelter cabin there would undoubtedly be a great help and obviate the necessity of packing in blankets and cooking utensils.

I intend this year to visit the Mountaineers' lodge in the Cascade mountains, fifty-eight miles east of Seattle and one and one-half miles from the Chicago, Milwaukee & St. Paul Railroad. The round trip rate on the railroad is very low and the lodge is quite accessible. They can depend on snow for about six months in the year and snow-shoeing is good and they are developing slopes for skiing and tobogganing.

Unfortunately for the Mazamas we have no place near any of the railroads that can be reached within a reasonable time at a reasonable expense.

The Mountaineers also have seventy-four acres of land near Puget Sound which they are keeping as a rhododendron preserve and have named this place "Kitsap Lodge." They have an old log cabin on the property now but otherwise have not developed this site.

The College Club Outing Association of Seattle have quite an attractive log cabin which they call "Roaring Camp." We have also received through Mr. Hardesty a bulletin of the Prairie Club of Chicago, with an account of their Beach House.

If desired by the Council, the committee will continue its investigations for another season, but cannot offer much encouragement in the selection of a site which would meet with universal approval.

Respectfully submitted.

R. L. GLISAN, Chairman.

November 20, 1917.

EXECUTIVE COUNCIL OF THE MAZAMAS,
Portland, Oregon.
Gentlemen:

The Lodge Committee submits herewith supplementary and final report.

After careful study and investigation, it is our belief that the proper site for a lodge is on the Mt. Hood high park line, the site to be selected preferably after a permanent road has been completed around the east side connecting the Barlow and Hood River roads, and certainly not before the present roads are hard-surfaced so the site selected can be reached with reasonable expenditure of time and money.

We do not believe any forest lodge advisable at present. By forest lodge we refer to a building erected at some point within easy access from Portland for the benefit of members who would spend week-ends there. Such a building would have to be extra commodious, would cost a large sum, and we do not believe would appeal to nature-lovers or carry out the mountaineering idea which the club must ever keep in mind.

We believe the club should lend its aid in encouraging the Forest Service to establish shelter huts at convenient points in the Columbia National Park.

Respectfully submitted.

R. L. GLISAN, Chairman.

Report of Local Walks Committee

The local walks for the Mazama year 1916-1917, were very successful, the total attendance on the fifty-four trips being 3242, making an average of 60.

The outing to Silver Star mountain was especially attractive. This trip had been scheduled twice before but had never been taken. The annual Mt. Hood outing was also successful, 83 persons succeeding in reaching the summit.

The Larch mountain winter trip continues to be the most popular of any in the schedule and seems to have an unfailing attraction.

It is becoming more difficult each year to secure absolutely new places to visit.

The following is a complete list of local walks taken during the year:

Date	*Time*	*Places Visited*	*Leader*	*Attendance*
1916				
Oct. 7-8	1½ Days	Eagle creek	J. C. Bush; Roy W. Ayer	100
Oct. 11	½ Day	Moonlight walk, Cornell road—Westover road	F. P. Luetters	130
Oct. 15	1 Day	Wauna peak	R. H. Atkinson; A. B. Williams	80
Oct. 22	1 Day	Carlton—Mt. Pisgah	Wm. W. Evans	32
Oct. 29	1 Day	Holbrook—Hillsboro	Agnes Lawson	6
Nov. 5	1 Day	Baker's Bridge	W. W. Ross	56
Nov. 12	½ Day	Parkrose—Columbia river	J. I. Teesdale	72
Nov. 19	½ Day	Council Crest—Oswego	Guy Thatcher	74
Nov. 26	½ Day	Beaverton—Westover Terrace	Dr. Wm. F. Amos	58
Dec. 3	½ Day	Mt. Scott	Margaret A. Griffin	47
Dec. 10	½ Day	Canyon road—Barnes road	Jean Richardson	79
Dec. 17	½ Day	Oregon City—Oswego	Helen Herman; Martha M. Gasch	52
Dec. 24	½ Day	Mt. Tabor—Errol Heights	Harriett E. Monroe	35
Dec. 30—Jan. 1	3 Days	Bull Run—Marmot	A. B. Williams	39
1917				
Jan. 7	½ Day	Riverside—Elk Rock	Forrest L. Foster	74
Jan. 13-14	1½ Days	Larch mountain	Roy W. Ayer; J. C. Bush	187
Jan. 21	½ Day	Fairmount boulevard—Barnes road	Elaine Ewell	93
Jan. 28	½ Day	"Mystery walk"	L. A. Nelson	54
Feb. 4	½ Day	Rose City Park—Elwood	Mary L. Knapp	86
Feb. 11	½ Day	Macleay Park—Blasted Butte	W. W. Evans	55
Feb. 18	1 Day	Columbia highway—Multnomah falls to Rooster Rock	E. C. Sammons; R. H. Atkinson	100
Feb. 25	½ Day	Portland Heights—Marquam hill	W. W. Ross	40
Mar. 4	1 Day	Vancouver river road	Sue McCready; J. G. Edwards	51
Mar. 7	½ Day	Moonlight walk, Washington Park—Arlington Heights	Nettie G. Richardson	6
Mar. 11	½ Day	Parkplace—Clackamas river	Rhoda Ross	86

Date	Time	Places Visited	Leader	Attendance
1917				
Mar. 18	½ Day	Whitwood Court—St. Johns	Coloma Wagnon, Cinita Nunan	78
Mar. 25	½ Day	Oswego—Oregon City	Lola Creighton	102
Apr. 1	1 Day	Moffat Creek falls	H. H. Riddell	55
Apr. 8	½ Day	Kings Heights—Mt. Calvary	W. P. Hardesty	28
Apr. 15	1 Day	Eagle creek	John A. Lee	75
Apr. 22	1 Day	Troutdale—Auto Club	P. G. Payton	47
Apr. 29	1 Day	Champoeg	Jacques Letz	63
May 5–6	1½ Days	Greenleaf peak—Rock Creek canyon	J. C. Bush	18
May 12–13	1½ Days	Cherry Grove—Coast range	J. A. Ormandy	42
May 19–20	1½ Days	Molalla—Wilhoit Springs	C. B. Woodworth	22
May 27	1 Day	Rooster Rock—Bull Run	Harry Wolbers	42
May 30	½ Day	Teuberry hill	Byron J. Beattie	82
June 3–4	1½ Days	Cook's hill—Bald mountain	Frank I. Jones	40
June 10	1 Day	Katani point	Guy Thatcher	16
June 16–17	1½ Days	Silver Star mountain	J. G. Edwards, Alfred F. Parker	98
June 23–24	1½ Days	Mt. Defiance	Chas. J. Merten, George F. Allen	27
July 1	1 Day	Estacada—Clear creek	Anne Dillinger	54
July 2	½ Day	Strawberry festival	J. M. Mason	90
July 3–4	1½ Days	Dowling farm	E. H. Dowling	25
July 8	1 Day	Wahkeena and Multnomah falls	A. B. Williams	19
July 14–15	2 Days	Annual Mt. Hood outing	Harriett E. Monroe, Robt. E. Hitch, Roy W. Ayer	92
July 22	½ Day	Oswego lake	Elsie Silver, Nellie Dalcour	25
July 29	1 Day	Steamboat picnic	A. S. Peterson	145
Sept. 1–3	2½ Days	Herman creek—Wahtum lake—Eagle creek	W. P. Forman	50
Sept. 9	1 Day	Wahclella falls	Harry L. Wolbers	20
Sept. 15–16	1½ Days	Table mountain	Raymond Conway	82
Sept. 23	1 Day	Gale's peak—Roderick falls	R. J. Davidson	21
Sept. 27	½ Day	Moonlight walk, Arlington Heights—Canyon road	Edith Nordeen	55
Sept. 29–30	1½ Days	Oneonta gorge	J. C. Bush	59

Local Walks Committee Financial Report

RECEIPTS

Amount collected on local walks		$ 200.46
Profit from Mt. Hood outing		127.30
Profit from steamboat picnic		34.20
Advertising		48.00
Total		$409.96

EXPENDITURES

Schedules	$100.50	
Postage, etc. (estimated)	42.00	
Supplies, etc.	94.80	
		237.30
Balance		$172.66

ROBERT E. HITCH, Chairman

Summit o Sentine peak, Wallowa mountains, Oregon

Photograph by United States Forest Service

Upper—Aneroid lake, Wallowa mountains, Oregon
Middle—One of the many beautiful lakes of the Wallowa region, with Eagle Cap in background.
Lower—East fork of Eagle Creek canyon, with Eagle Cap in background at left.
Photographs by United States Forest Service

Mazama Outing for 1918

FOR their twenty-fifth annual outing, in 1918, the Mazamas have selected the mountains of Wallowa county, Oregon. This is the extreme northeasterly county of the state, bordering upon Washington and Idaho. Its mountains range in elevation from 6000 to 9800 feet, and include the prominent peaks called Eagle Cap, Sentinel peak, Bennett peak and Marble mountain.

Except for the work done by the United States Forest Service, this beautiful and interesting region is practically unexplored, and affords opportunity for much original investigation. In some respects, it is entirely different from any locality or district hitherto visited by the Mazamas. In addition to its mountains, there are numerous lakes, the largest of these being the well-known and much admired Wallowa lake. One of the others (Aneroid lake) is situated at an elevation of about 7000 feet. In what is known as the lake basin are more than a score of crystal-clear lakes which were stocked with trout some years ago, so that excellent fishing may be fairly counted upon as one of the possibilities of the trip.

The date of the outing has not yet been definitely determined, but it will probably be July 13 to 28. A prospectus giving all necessary details will be issued well in advance of the outing—probably in February. Any information desired may be obtained from MR. ROBERT E. HITCH, *Chairman Mazama Outing Committee,* 503 Fenton Building, Portland, Oregon.

Report of the Certified Public Accountant who Examined the Financial Affairs of the Mazamas

INCOME AND PROFIT & LOSS ACCOUNT

For the period October 1, 1916 to September 29, 1917

Income:			
Members' Dues			$1,134.00
Miscellaneous:			
Balance of 1916 Rose Carnival Prize		$ 12.50	
Badges		6.25	
Stationery		.25	19.00
			$1,153.00
Less:			
Loss in Committee's Transactions:			
Losses:			
Magazine Publication	$173.06		
Mt. Jefferson Outing	371.97	$545.03	
Profits:			
Mt. Hood Outing—1916	50.25		
Mt. Hood Outing—1917	131.55		
Local Walks	206.21	388.01	157.02
Gross Income			$995.98
Expenses:			
Club Room Rent		$370.00	
Telephone Rent and Tolls		63.03	
Printing and Stationery General		342.04	
Furniture and Fixture Expense		87.50	
Parade Expense		19.80	
Entertainment Expense		56.50	
Lecture Expense		68.50	
Library Expense		6.65	
Keys		2.70	
Insurance		8.25	
Sundries		83.64	1,108.61
Net Loss for Year			$112.63

Balance Sheet

As at September 29, 1917

ASSETS

Cash at Banks:		
U. S. National Bank—Checking Fund	$1,042.87	
U. S. National Bank—Savings Deposit	500.00	1,542.87
Club Room Furniture and Camp Equipment . .		1,000.00
Insurance Premium Unexpired		7.15
		$2,550.02

LIABILITIES

Accounts Payable	$ 23.50
Surplus	2,526.52
	$2,550,02

Portland Oregon November 12, 1917.

To the Members of
THE MAZAMA COUNCIL
Portland, Oregon.

Dear Sirs:

In accordance with your instructions I have audited the accounts of the Mazamas for the fiscal year ended September 29, 1917, and in my opinion the accompanying Balance Sheet and Income and Profit & Loss Statements reflect the transactions for the period under review and the financial condition of the Mazamas as at September 29, 1917.

The valuation placed upon the club-room furniture and camp equipment is considered to be conservative. This asset is protected by insurance for a like amount.

Cash funds on deposit were reconciled with the treasurer's statements and verified by a certificate received from the bank.

I examined receipted invoices, cancelled checks and bank statements produced by the treasurer and the various committees and found them to be in order.

Yours truly,

ROBERT F. RISELING,
Certified Public Accountant.

What "Mazama" Stands For

The word "Mazama" is derived from the name of the mountain goat which makes its home high up among the pinnacles and glaciers of the Cordilleran ranges of western America.

The Mazamas were organized on the summit of Mount Hood, Oregon, July 19, 1894, 193 persons (155 men and 38 women) reaching the top that day.

The purposes of the club are to explore mountains, to disseminate authoritative and scientific information concerning them and to encourage the preservation of the forests and other features of mountain scenery in their natural beauty.

Any person of good character who has climbed to the summit of a snow peak on which there is at least one living glacier, and the top of which can not be reached by any other means than on foot, is eligible to membership. The annual dues are $3.00; initiation fee $3.00.

Mazama Presidents and Official Ascents

	Presidents	*Official Ascents*
1894	WILL G. STEEL	Mt. Hood, Oregon
1895	WILL G. STEEL	
	L. L. HAWKINS*	Mt. Adams, Washington
1896	C. H. SHOLES	Mt. Mazama (named for Mazamas, 1896), Mt. McLoughlin (Pitt), Crater Lake, Ore.
1897	H. L. PITTOCK	Mt. Rainier, Washington
1898	HON. M. C. GEORGE	Mt. St. Helens, Washington
1899	WILL G. STEEL	Mt. Sahale (named by Mazamas, 1899), Lake Chelan, Washington
1900	T. BROOK WHITE*	Mt. Jefferson, Oregon
1901	MARK O'NEILL	Mt. Hood, Oregon
1902	MARK O'NEILL	Mt. Adams, Washington
1903	R. L. GLISAN	Three Sisters, Oregon
1904	C. H. SHOLES	Mt. Shasta, California
1905	JUDGE H. H. NORTHUP	Mt. Rainier, Washington
1906	C. H. SHOLES	Mt. Baker (northeast side), Washington
1907	C. H. SHOLES	Mt. Jefferson, Oregon
1908	C. H. SHOLES	Mt. St. Helens, Washington
1909	M. W. GORMAN	Mt. Baker (southwest side), and Shuksan, Wash.
1910	JOHN A. LEE	Three Sisters, Oregon
1911	H. H. RIDDELL	Glacier Peak, Lake Chelan, Washington
1912	EDMUND P. SHELDON	Mt. Hood, Oregon
1913	EDMUND P. SHELDON	
	H. H. PROUTY*	Mt. Adams, Washington
1914	H. H. PROUTY*	Mt. Rainier, Washington
1915	J. E. BRONAUGH	Mt. Shasta, California
1916	FRANK B. RILEY	Three Sisters, Oregon
1917	W. P. HARDESTY	Mt. Jefferson, Oregon

*Deceased

Mazama Organization for Year 1917-18

OFFICERS

Roy W. Ayer (689 Everett St.) . . . . *President*
W. P. Hardesty (60 E. 31st St. N.) . . . . *Vice-President*
Alfred F. Parker (330 Northwestern Bank Bldg.) . . . . *Corresponding Secretary*
Miss Nettie G. Richardson (888 E. Washington St.) . . . . *Recording Secretary*
Miss Martha E. Nilsson (320 E. 11th St. N.) . . . . *Financial Secretary*
Leroy E. Anderson (213 Northwestern Bank Bldg.) . . . . *Treasurer*
Miss Lola I. Creighton (920 E. Everett St.) . . . . *Historian*
Robert E. Hitch (503 Fenton Bldg.) . . . . *Chairman Outing Committee*
A. Boyd Williams (54 King St. N.) . . . . *Chairman Local Walks Committee*

COMMITTEES

Local Walks Committee—A. Boyd Williams, Chairman; Miss Agnes G. Lawson, V. L. Ketchum, Harry L. Wolbers, Miss Vera E. Taylor.

Outing Committee—Robert E. Hitch, Chairman; J. C. Bush, Charles J. Merten.

Library Committee—Miss Lola I. Creighton, Chairman; Miss Beulah F. Miller, Leroy E. Anderson, Miss Florence J. McNeil, W. P. Forman.

House Committee—Miss Jean Richardson, Chairman; Mrs. R. J. Davidson, E. F. Peterson, Miss Agnes Plummer, J. A. Ormandy.

Entertainment Committee—Robert D. Searcy, Chairman; Miss Nellie M. Dalcour, Miss Minnic Heath, Miss Edith Nordeen, Arthur Cook.

Educational Committee—G. W. Wilder, Chairman; W. L. Finley, Miss Ella P. Roberts, Miss Harriett E. Monroe, J. E. Bronaugh.

Publication Committee—Alfred F. Parker, Chairman; Miss Pauline Geballe, Miss Beatrice Young.

Membership Promotion Committee—T. Raymond Conway, Chairman; Miss Margaret A. Griffin, C. E. Blakney.

Membership Committee—Robert E. Hitch, Chairman; Miss Lola I. Creighton, Miss Nettie G. Richardson.

Auditing Committee—R. F. Riseling, Chairman; Miss Anne C. Dillinger, L. Van Bebber.

Roll of Honor

Mazamas who are at the present time engaged in the military or naval service of the United States or our allies

FRANK H. BAGLEY
Sixty-Sixth Aero Squadron

C. E. BLAKNEY
Aviation Corps

W. P. BODWAY
Branch of service not known

K. P. CECIL
First Lieutenant, Oregon Coast Artillery

W. W. EVANS
Corporal, Company D, 162nd Regiment, Forty-First Army Division

FORREST L. FOSTER
Company A, 116th Engineers Forty-first Army Division

A. H. S. HAFFENDEN
Sergeant, Battery A, 147th Field Artillery, Forty-First Army Division

E. G. HENRY
Sixty-Sixth Aero Squadron

RALPH S. IVEY
Branch of service not known

E. C. JENNINGS
Civilian Field Corps

HENRY G. JOHNSON
Reserve Officers' Training Camp

THOMAS R. JONES
Lieutenant, Ordnance Department

F. G. KACH
Sergeant, Eighth Company, Oregon Coast Artillery

D. M. G. KERR
Canadian Army Medical Corps

HENRY A. LADD
University of Oregon Base Hospital Unit

F. H. McNEIL
Twenty-Third Engineers

ALAN BROOKS MORKILL
Lieutenant, Seventh Battalion, Canadian Expeditionary Force

B. W. NEWELL
Reserve Officers' Training Camp

W. D. PEASLEE
Captain, 316th Engineers, Ninety-First Army Division

ARTHUR L. ROBERTS
Company D, Third Telegraph Battalion

E. C. SAMMONS
Captain, United States Reserves, Thirty-Fifth Army Division

J. MONROE THORINGTON
American Ambulance Corps

F. B. UPSHAW
Ensign, National Naval Volunteers

DEAN VAN ZANDT
Aviation Corps

Bureau of Associated Mountaineering Clubs of North America

In May, 1916, nine clubs and societies with common aims associated themselves in a Bureau with headquarters in New York. The membership now numbers nineteen, comprising about 16,000 individual members, as follows:

1. American Alpine Club, Philadelphia and New York.
2. American Civic Association, Washington.
3. Appalachian Mountain Club, Boston and New York.
4. British Columbia Mountaineering Club, Vancouver.
5. Colorado Mountain Club, Denver.
6. Explorers' Club, New York.
7. Fresh Air Club, New York.
8. Geographic Society of Chicago.
9. Geographical Society of Philadelphia.
10. Green Mountain Club, Rutland, Vermont.
11. Hawaiian Trail and Mountain Club, Honolulu.
12. Klahhane Club, Port Angeles, Wash.
13. Mazamas, Portland, Oregon.
14. Mountaineers, Seattle and Tacoma.
15. National Association of Audubon Societies, New York.
16. Prairie Club, Chicago.
17. Rocky Mountain Climbers' Club, Boulder, Colorado.
18. Sierra Club, San Francisco and Los Angeles.
19. United States National Parks Service, Washington.

Among the common aims, aside from the exploration and mapping of mountain regions and the ascent of leading peaks, are the creation, protection, and proper development of National Parks and Forest Reservations, the protection of bird and animal life, and of trees and flowers. Many of the clubs and societies issue illustrated publications on mountaineering, exploration and conservation of natural resources, and are educating their members by means of lectures to a deeper appreciation of nature.

The Bureau publishes an annual bulletin giving the officers, membership, dues, publications, lantern slide collections, outings, and other matters of interest of each club. Data on mountains and mountaineering activities is supplied in response to inquiries.

Acquaintance with the literature of a subject is essential to efficient work in the field, and the Bureau sends many important new books on mountaineering and outdoor life to its members free of charge. A large collection of mountaineering literature has been gathered in the central building of the New York Public Library and the American Alpine Club has deposited its books therein, providing a permanent fund for additions. A bibliography of this collection has been published by the library. An extensive collection of photographs of mountain scenery is being formed and is available to anyone wishing to supplement the literature of a region with its scenery.

LE ROY JEFFERS, Secretary,
Librarian American Alpine Club,
476 Fifth Avenue, New York.

December, 1917.

Geographic Progress in the Pacific Northwest

By Lewis A. McArthur

The year 1917 has witnessed very little geographic progress in the Pacific Northwest largely because of the fact that the greater part of the forces of the United States Geoglogical Survey and United States Coast and Geodetic Survey have been taken over by the military organizations. These organizations have done little or no work in Oregon and Washington during the past twelve months and there is a possibility they will do no geographic work here until the war is over. This, of course, has had a very unfortunate effect on our geographic progress.

During the past eighteen months the Geoglogical Survey has engraved and issued the Albany, Diamond Lake, Estacada, Arlington, Condon and Tualatin quadrangles in Oregon, as well as Willamette Valley sheets numbers seven and eight. It is understood that the Salem sheet is now in the hands of the engraver and the Hillsboro sheet will probably be issued some time within the next few months. Willamette Valley sheet number nine has been completed but never issued in its final form.

The field work on the Kerby sheet in southwestern Oregon has been completed and this sheet is available in its advanced stage. The field work on the Oregon portion of the Troutdale sheet has also been completed and about one-half of the Twickenham sheet in Wheeler county has been done in the field.

No topographical work has been done this year by the Geoglogical Survey in Oregon although there is said to be a probability some of the sheets along the coast may be completed for the War Department.

During the latter part of 1916 extensive triangulation nets were completed in Jefferson, Crook, Deschutes and Lane counties. These furnish triangulation control for a large number of quadrangles in the Cascade National Forest and in the Deschutes National Forest.

During the past year the United States Geological Survey has issued in their final engraved form topographical sheets of Chehalis, Priest Rapids and Coyote Rapids in Washington, and has also issued advanced sheets of the Mt. St. Helens, Prosser, Pasco, Wallula and Connell quadrangles. The Walla Walla quadrangle was almost entirely completed in 1916, but no work has been done since. It is understood that the engraved maps for the Prosser, Pasco and Wallula sheets will soon be issued.

One very interesting feature in connection with this work is that the United States Geological Survey has established a new and probably highly accurate elevation for Mt. St. Helens, 9671 feet. The advance sheet for the Mt. St. Helens quadrangle will be very useful to the members of the Mazamas.

During the year the United States Forest Service, together with the Oregon Geographic Board has been sending in considerable new information for new editions of Geological Survey maps. As a result, new printings of the Mt. Hood, Blalock Island, Snoqualmie and Ashland sheets have been published with many new names and other data.

The United States Coast and Geodetic Survey has completed a net of high class triangulation from Utah to the Cascade range and several hundred points and elevations were established. This work has not yet been published but it probably will be within the next few months.

The Forest Service has done a good deal of intensive reconnaissance work during the past year on isolated townships, but these maps are, of course, widely

Mt. Jefferson from the northwest, near head of Gorge Creek, with Jefferson Park in the foreground

Photograph, 1917, by Ira A. Williams

Jefferson Park, looking north

Photograph by R. J. Davidson

scattered. During the year the Forest Service has made new maps of the Crater, Umpqua, Santiam, Malheur, Wenaha, Umatilla, Columbia, Snoqualmie, Wenatchee and Okanogan national forests. These maps are on a scale of one-half inch to the mile and may be examined any time at the Forest Service headquarters in the Beck Building, Portland, Oregon.

Work is now being done on on new maps of the Olympic, Rainier, Colville, Wallowa, Minam, Fremont, Cascade and Siskiyou National Forests. The Forest Service expects within the next year to begin the preparation of three-color large scale atlas sheets of some of the forests. The Oregon Bureau of Mines has issued an interesting photographic reproduction of its relief map of Oregon.

△△△

Book Reviews

Edited by PAULINE GEBALLE

"THE BOYS' BOOK OF HUNTING AND FISHING"

In Mr. Warren H. Miller, (editor of "*Field and Stream*") up-and-coming boys have a real friend. He has not forgotten the days of his youth, and his book discloses a large and kindly understanding of boys and their needs. He has tried to inspire them with a more definite aim in their out-of-door life by giving them a book especially prepared for them, which tells them exactly how to equip themselves, and how to become real sportsmen—not mere target shooters and fishers of sunnies and perch. He has kept the poor boys especially in mind, and has described the best qualified outfits to be obtained at a reasonable expense.

The chapters on angling lay stress on game fishing. Fly casting for trout, and fly and bait casting for bass are specialized; and complete details are given regarding equipment, its cost, and how to handle it. In part two the author deals almost exclusively with wing shooting. He advocates the shot gun in preference to the rifle, as he aims at big game shooting, and considers the quick aiming of the shot gun in wing shooting more akin to rifle aiming at big bounding game, than military practice with a rifle. Here, also, equipment and its cost are thoroughly discussed. The chapters devoted to camping are rich in valuable information covering the many branches of this broad subject. A week's camp for three boys is described in minute detail.

The book will readily inspire confidence in boy readers. It is enhanced by many illustrations, and the presentation of the technical information is relieved by the narrative style. We share in wonderful upland trips in the brown October days, in shore-bird shooting, and late November duck hunts. These we enjoy as enthusiastically as does the "Kid," the author's eleven year old son, to whom we are introduced, and whose success with his line and his little twenty-eight proves his father's point that game fishing and wing shooting are not beyond the average boy of twelve, who is willing to work to become a real sportsman.

MARGARET A. GRIFFIN

MILLER, WARREN H *The Boys' Book of Hunting and Fishing.* 1916. George H Doran Company, New York $1.25.
(Supplied by the Bureau of Associated Mountaineering Clubs of North America)

"TWO SUMMERS IN THE ICE WILDS OF EASTERN KARAKORAM" In the upper regions of Kashmir, the northern province of India beyond the Himalayas, tower the rival peaks of the Karakoram range. In 1911 and 1912, Dr. and Mrs. Workman visited this little exploited region and made extensive explorations of its glaciers and mountains.

Dr. Workman writes of the first summer's exploration of some of the important Karakoram glaciers, and his wife covers the conquest the following year of the great Rose glacier in the extreme eastern portion of the range, and the Doctor concludes by an interesting study of the physiographical features of the various glaciers visited.

To make the trips required previous knowledge of the country, months of preparation, a careful distribution of reserve supplies at selected bases, a large retinue of coolies and a knowledge of their peculiarities and the proper way of handling them.

The Rose glacier is named after the native flowers which outstrip the other plants in creeping up to adorn the base of this frozen leviathan, a glacier having a continuous stretch of forty-six miles with affluent glaciers bearing in from either side, forcing the ice to break under fearful pressure and forming crevasses, schrunds and seracs, barriers to any but the most indomitable explorers. This wonderful frozen river lies between ranges where the peaks are all over 20,000 and most are 22,000 feet high. The Workmans invaded this region with an army of coolies who went often barefoot over the ice with ponderous packs, driving bands of sheep and goats, as a moving meat supply. The fuel, as well as the food, had to be packed in and frequently the supply trains on their way in ate most of the supplies and consumed or cached the precious fagots, while the slightest suspicion of fuel or food shortage gave ample excuse for the ever reluctant natives to rebel and even mutiny.

Countless little incidents keep up the interest. Three crows followed them from below into the bleak frozen regions of the snow and ice and never left them until the party returned to the grassy plains.

The explorers, spending weeks above the 16,000 foot contour, noticed the elevation by an ever lessening appetite, a tendency to insomnia, with increased exhilaration however, and without apparently lessening their vitality.

The book is profusely and superbly illustrated, the photographs showing a chaotic wealth of unsurpassed glacial and mountain scenery. Hours could be spent profitably on the photographs alone. The combination of text, photographs and maps gives the reader a comprehensive idea of what must be the world's most fascinating upper region. The work is an encyclopedia of glacial information presented in such an interesting style that a hurried glance impels a complete perusal. The average reader with a few isolated peaks to his credit is overwhelmed with the immensity of it all.

R. L. GLISAN

WORKMAN, FANNY BULLOCK and WORKMAN, WILLIAM HUNTER. *Two Summers in the Ice Wilds of Eastern Karakoram.* 1917. E. P. Dutton & Co., New York. $8.00.
(Supplied by the Bureau of Associated Mountaineering Clubs of North America)

"THE BOYS' BOOK OF CANOEING AND SAILING" This book is particularly interesting to boys and may prove a valuable means for gaining their companionship. The information is well rendered and could easily be made practical. His invitations through word pictures and suggestions for outdoor recreation appeal to the enthusiast for out-door sport and awaken a natural desire from those who are indifferent. His plans call for inexpensive material. The book is well illustrated.

ALICE V. JOYCE

MILLER, WARREN H. *The Boys' Book of Canoeing and Sailing.* Doran. Illustrated. $1.25.
(Supplied by the Bureau of Associated Mountaineering Clubs of North America)

"THE BOOK OF CAMPING" This work contains much valuable information, for the novice as well as the seasoned camper, and much of which is only learned by the camper after years of experience. To one who essays to plunge into the wilderness and isolate himself from civilization, living entirely on his own resources, it will prove an ever ready and reliable companion. It covers the field comprehensively.

L. E. ANDERSON

VERRILL, A. HYATT. *The Book of Camping.* 1917. Illustrated. Alfred A. Knopf. $1.00. (Supplied by the Bureau of Associated Mountaineering Clubs of North America.)

"IN CANADA'S WONDERFUL NORTHLAND" "In Canada's Wonderful Northland" is an entertaining account of an expedition whose purpose was the investigation of the mineral resources of Canada in the vicinity of Hudson bay.

The authors, W. Tees Curran and H. A. Calkins, organized the expedition, which, starting from Missinaibi station on the National Transcontinental Railway, made its way down the Missinaibi river. Here formidable rapids were negotiated and portages passed where bulky equipment had to be moved under great difficulties.

At Moose Factory, the party divided, the authors with the assistance of an engineman, making the trip up the bay in a motor boat while the others went in a sailboat which had been chartered for the trip. Exceptional hardships in the shape of weather conditions were encountered, which made the trip require eight months, instead of the five planned on, and necessitated the use of snowshoes on the final lap of the journey back to civilization.

The book is well supplied with maps, and an appendix gives, in well tabulated form, desirable information in regard to harbors and camp sites.

ELLA P. ROBERTS

CURRAN, W. TEES and CALKINS, H. A. *In Canada's Wonderful Northland.* Illustrated. 1917. Putnam. $2.50.
(Supplied by the Bureau of Associated Mountaineering Clubs of North America)

TOURING AFOOT" 'Twould more than pay a person going on a hiking trip, to tuck away "TOURING AFOOT" in his pack—even though he had read it through thoroughly beforehand, and had to leave out some other desired article. For this little book contains facts, devices and hints that are invaluable to the hiker, yet cannot be recalled exactly when most needed.

"TOURING AFOOT" is the result of practical experience in regard to clothing, shoes, outfit, shelter and camp making, rations, first aid, and much needed advice on the care of the feet.

ALICE BANFIELD

FORDYCE, DR. C. P. *Touring Afoot* Outing Publishing Co. 1917. $0 80.
(Supplied by the Bureau of Associated Mountaineering Clubs of North America.)

"MOUNT RAINIER" The God-Mountain, Guardian of the Northwest, Watcher over Puget Sound, topped only by Mt. Whitney in the United States, theme of poetic Indian myth and legend, is a worthy subject of a well compiled volume edited by Professor Meany, President of The Mountaineers.

It is a fascinating account we read here of the history of the great mountain. How Vancouver sailing up the coast in 1792 saw a round snowy mountain, which he named after his friend Admiral Rainier; the first approach to the mountain in

1833; Lieutenant Johnson's trip through Naches pass in 1841; Tacoma and the Indian legend of Hamitchou; the attempted ascents, finally successful in 1870, when General Stevens and P. B. Van Trump after eleven hours of unremitted toil reached the summit and passed the night there, saved from perishing by the steam jets in the crater.

The glaciers, the rocks, the flora, McClure's achievement and tragic death, the official height of 14,408 feet, are all given in reprints from the original sources, so that the reader lives again with the explorers stirring days and nights on the snowy slopes.

It is well illustrated with portraits of the various men whose writings are incorporated in the volume, and there is a chapter which gives for the first time in convenient form place names and elevations in the park.

ARTHUR K. TRENHOLME

MEANY, EDMOND S. *Mount Rainier, A Record of Exploration.* 1916. The Macmillan Company. $2.50. Illustrated.
(Supplied by the Bureau of Associated Mountaineering Clubs of North America.)

"CANADA, THE SPELL-BINDER" The title of this wonderful descriptive work of Canadian resources and its scenery, is well chosen. The author attempts to delineate briefly the history of early Canada, to present to the reader a view of Canada beginning from the earliest settlement by the fur trader down to the present day. The book then sets forth the wonderful progress and development of Canada, particularly in the last decade, including its immense potentialities. The work is primarily, however, a compendium of the achievements of resourceful men who have led in the opening of Canadian resources, and the making possible the cultivation of thousands upon thousands of acres hitherto untouched. Many of these lands are now abounding in the production of wheat and other cereals.

The author's theme, as it were, is a rather too exhaustive attempt on her part to present an historical, sociological and mineralogical discussion of all that Canada is and will be. The scenic beauty of the Yellowhead pass, which is traversed by the Grand Trunk Pacific Railroad, the many beautiful national parks, such as Jasper Park; and the beautiful snow-capped peaks of the Canadian Rockies, are all admirably described. The book is embellished with prints of the larger cities and points of scenic interest.

To the tourist and pleasure seeker, this work will prove a valuable addition to his books of travel.

HAROLD V. NEWLIN

WHITING, LILIAN. *Canada, the Spell-Binder.* 1917. Dutton. $2.50.
(Supplied by the Bureau of Associated Mountaineering Clubs of North America)

"YOUR NATIONAL PARKS" As its title might indicate, this volume contains complete descriptions of our national parks. It may surprise some readers to learn that there are sixteen of these national parks and three national monuments. Much interesting historical matter is presented. One chapter is given to the Hawaiian National Park and one to the Canadian National Parks. In another chapter appears a biography of John Muir.

Mr. Mills' genuine love of nature finds most beautiful expression in the chapter on "*The Spirit of the Forest and the Trail.*" "Scenery," he says, "is our most valuable and our noblest resource." "A campfire in the forest marks the most enchanting place on life's highway wherein to have a lodging for the night."

He portrays for our enjoyment wild life in all its interesting forms, and brings to us vivid pictures of our country's beauty spots. The book is well illustrated with original photographs by the author and there are several maps reproduced by permission of the National Park Service of the Department of the Interior. The "*Guide to the National Parks*" by Lawrence F. Schmeckebier contains much detailed information of value to tourists.

On the whole "YOUR NATIONAL PARKS" is a book of great value and one which might well be placed in every American home. JEAN RICHARDSON

MILLS, ENOS A. *Your National Parks* Illustrated. 1917. Houghton, Mifflin & Co. $2.50. (Supplied by the Bureau of Associated Mountaineering Clubs of North America)

"A THOUSAND-MILE WALK TO THE GULF" John Muir's account of a botanical excursion, which he made from Indiana to Florida a half century ago, is reproduced from the original journal. This he wrote at intervals during the journey southward across Kentucky and Tennessee, then southward to Savannah and later across Florida from the Atlantic coast to the Gulf of Mexico.

Autobiographically it bridges the period between "*The Story of My Boyhood and Youth*" and "*My First Summer in the Sierra.*"

By quotations from his journal the reader is made acquainted with details concerning the different species of flowers and trees encountered, and by accounts of the hospitality extended to him, is given a glimpse of the character of the inhabitants of that region. The intimate details make a personal appeal and the reader feels keenly, with the young naturalist, the dangers of the undertaking, which he met so uncomplainingly.

The chapters may be enjoyed separately, each being in itself an interesting account of some phase of his outing. Walking with John Muir from Indiana to the Gulf, then sailing with him to Cuba, and finally enjoying a first glimpse of California's wonders, through his eyes, is a pleasant exercise, worthy of anyone's time.

MINNIE R. HEATH.

MUIR, JOHN. *A Thousand-Mile Walk to the Gulf.* 1916. Illustrated. Houghton, Mifflin Co. $2 50.
(Supplied by the Bureau of Associated Mountaineering Clubs of North America)

"TROUT LORE"

"A feller gets a chance to dream,
Out fishin',
He learns the beauties of a stream,
Out fishin';
And he can wash his soul in air
That isn't foul with selfish care,
An' relish plain and simple fare
Out fishin'."

The stanza quoted above is not from the book we have been called upon to review, but it so nearly embodies the leading spirit running through the work that we have been constrained to use it as our introduction. The book is by O. W. Smith, angling editor of "*Outdoor Life*" and is entitled "TROUT LORE." The work has come out very recently and is published by the Frederick A. Stokes Company. It is one of many books devoted to "ye gentle art" since the time that Izaak Walton wrote his immortal treatise. One would have thought that nothing remained to be said upon the subject, especially since Henry Van Dyke gave us his "*Fisherman's Luck.*" But the true angler, though perhaps not the "mere fisherman for fish" and especially not the "fish hog," will find throughout the text much that strikes a sympathetic chord, as well as much instruction in the technique of the art. The whole subject is covered—where, when and how to fish, the tackle and lures to be used, the togs to be worn, and how to cook the trout—when caught.

With one statement of the author we will have to take issue. At least our experience has been different from his. He says that the eastern brook trout (*Salvelinus fontinalis*) "will never go into the air of his own free will." While fishing Wahtum lake in October of last year (this lake has been stocked with the *fontinalis*) we had the thrilling experience, not once but several times, of hooking these beauties just at the start of the back cast, when the trout would leap clear of the water to take the fly. With still another statement of the author we cannot quite agree. He pronounces the eastern brook trout more delectable as a tickler of the palate than the rainbow. But this, of course, is a matter of taste, and tastes differ. Besides, as a true westerner, maybe we are prejudiced in favor of the native denizens of our local streams, especially since the *fontinalis*, being a char and not a true trout, is a near cousin to the Dolly Varden, which every western fisherman knows is not comparable to the rainbow as a toothsome viand. The author admits that for gameness the rainbow is without a peer.

In his concluding chapter the author inquires: "Why do we fish; wherein is the attractivity of angling?" Henry Van Dyke's reply, which he quotes—"It is the enchantment of uncertainty"—does not satisfy him fully. Then he proceeds to give his own answer, in which we most heartily concur: "Every true angler is an embryonic poet, feeling things which he cannot express, seeing things which he cannot describe. He who fishes for fish is not an angler but a mere fisherman. He who angles that he may become proficient with latest wrinkles of tackle is not an angler but an experimentist. He who seeks to collect samples of everything in tackle is not an angler but a faddist. The true angler partakes somewhat of the natures of the foregoing, but, first of all, he is a lover of God's Out o' Doors."

JOHN A. LEE

SMITH, O. W. *Trout Lore*. Illustrated. Stokes. 1917. $2 00
(Supplied by the Bureau of Associated Mountaineering Clubs of North America.)

"GLACIER NATIONAL PARK—ITS TRAILS AND TREASURES"

Portions of "GLACIER NATIONAL PARK—ITS TRAILS AND TREASURES" smack very much of commercialism. The authors have caught but a glimmer of the joys and inspirations to be found in a visit to one of our greatest American playgrounds. While this guide may be of great service to the so-called "sight-seeing" tourist, it is of very small value to the mountaineer and outdoor enthusiast.

ROBERT E. HITCH

HOLTZ, MATHILDE EDITH and BEMIS, KATHERINE ISABEL. *Glacier National Park—Its Trails and Treasures*. Illustrated. George H. Doran Company. 1917. $2 00.
(Supplied by the Bureau of Associated Mountaineering Clubs of North America.)

Membership

Address is Portland unless otherwise designated.

ABISHER, MARIE, 335 14th St.
ACTON, HARRY W., 519 West 121st St., New York.
ACTON, MRS. HARRY W., 519 West 121st St., New York.
ADAMS, DR. W. CLAUDE, 1010 E. 28th.
AITCHISON, CLYDE B., 504 Real Estate Trust Bldg., Washington, D. C.
AKIN, DR. OTIS F., 919 Corbett Bldg.
ALLEN, ARTHUR A., Portland Rowing Club.
ALLEN, ENID C., 917 Andrus Bldg., Minneapolis, Minn.
ALMY, LOUISA, Box 372, Forsyth, Mont.
AMOS, DR. WM. F., 1016 Selling Bldg.
ANDERSEN, DR. FREDERICK, Merwyn Bldg., Astoria, Ore.
ANDERSON, LEROY E., 213 Northwestern Bank Bldg.
ANDERSON, LOUIS F., 364 Boyer Ave., Walla Walla, Wash.
APPLEGATE, ELMER I., Klamath Falls Oregon.
ASCHOFF, ADOLF, Marmot, Oregon.
ATKINSON, R. H., O.-W. R. & N. Co.
ATLAS, CHAS. E., Hurley-Mason Co. 619 Perkins Bldg., Tacoma, Wash.
AVERILL, MARTHA M., 1144 Hawthorne AVE.
AYER, ROY W., 689 Everett St.
ANDERSON, WM. H., 4464 Fremont Ave. Seattle, Wash.
BABB, HAROLD S., 578 Miller St.
BACKUS, LOUISE, 122 E. 16th St.
BACKUS, MINNA, 122 E. 16th St.
BAILEY, VERNON, 1834 Kalorama Ave., Washington, D. C.
BALLOU, O. B., 80 Broadway.
BALMANNO, JACK H., 5508 50th Ave. S. E.
BANFIELD, ALICE, 570 E. Ash St.
BARCK, DR. C., 205-7 Humbolt Bldg., St. Louis, Mo.
BARNES, M. H., 658 Schuyler St.
BARRINGER, MAUDE, 5th and M Sts., Fredonia, Kansas.
BATES, MYRTLE, 448 E. 7th St.
BEATTIE, BYRON J., 830 Rodney Ave.
BELL, HALLIE, 549 Belmont St.
BENEDICT, LEE, 185 E. 87th St.
BENEFIEL, FRANCIS W., 750 E. Ankeny
BENTALL, MAURICE, General Delivery. Forsyth, Mont.
BENZ, CHAS. A., Eugene, Oregon.
BIGGS, ROSCOE G., 547 E. Pine St.
BLAKNEY, C. E., Milwaukie, Ore., R. R. No. 2, Box 151.
BLUE, WALTER, 1306 E. 32nd St.
BORNT, LULU ADELE, Box 813, Grangeville, Idaho.
BOWEN, M. W., 405 W. Park St.
BOWERS, NATHAN A., 501 Rialto Bldg., San Francisco, Cal.
BOWIE, ANNA, 297 E. 35th St.
BOYCE, EDWARD, 207 St. Clair St.
BREWSTER, WM. L., 808 Lovejoy St.
BROCKMAN, GUS, 128 E. 16th St.
BRONAUGH, JERRY ENGLAND, Title & Trust Bldg.
BRONAUGH, GEORGE, 350 N. 32nd St.
BROWN, ALBERT S., Campbell-Hill Hotel
BROWN, G. T., 500 E. Morrison St.
BULLIVANT, ANNA, 269 13th St.
BUSH, J. C., 386½ E. Morrison St.
BENEDICT, MAE, 185 E. 87th St.
CALHOUN, MRS. HARRIET S., 367 E. 34th St.
CAMPBELL, DAVID, 404 Boyer Ave., Walla Walla, Wash.
CAMPBELL, P. L. 1170 13th Ave., E., Eugene, Oregon.
CASE, GEORGENE M., Foot Miles St.
CHAMBERS, MARY H., 729 11th Ave. E., Eugene, Oregon.
CHENOWETH, MISS MAY, 104 E. 24th St.
CHRISTIANSON, WM. D., 134 Colburn St., Brantford, Ontario.
CHURCHILL, ARTHUR M., 1229 Northwestern Bank Bldg.
CLARK, J. HOMER, 706 Glisan St.
CLARKE, D. D., Water Bureau, City Hall.
COLBORN, MRS. AVIS EDWARDS, 383 Summer St., Buffalo, N. Y.
COLLAMORE, MARY ERNA, E. 6th and Oregon Sts.

COLLINS, W. G., 510 32nd Ave., S., Seattle, Wash.

*COLVILLE, PROF. F. V., Department of Agriculture, Washington, D. C.

CONNELL, DR. E. DEWITT, 628 Salmon.

CONWAY, T. RAYMOND 4705 60th St., S. E.

COOK, ARTHUR, 243 West Park St.

COURSEN, EDGAR E., 658 Lovejoy St.

COURSEN, GERALDINE R., 658 Lovejoy.

COWIE, LILLIAN G., Wellesley Court.

COWPERTHWAITE, JULIA, P.O. Station E.

CRANER, HENRY C., Room 214 M. A. A. C.

CREIGHTON, LOLA I., 920 E. Everett St.

CROUT, NELLE C., 1326 Tillamook.

CURRIER, GEO. H., Leona, Oregon.

*CURTIS, EDWARD S., 614 Second Ave., Seattle, Wash.

DALCOUR, NELLIE MAE, Karl Hotel.

DAVIDSON, PROF. GEO., 530 California St., San Francisco, Calif.

DAVIDSON, R. J., 458 E. 49th St.

DAVIDSON, MRS. R. J., 458 E. 49th St.

DAY, BESSIE, 690 Olive St., Eugene, Oregon.

*DILLER, PROF. JOS. S., U. S. Geological Survey, Washington, D. C.

DILLINGER, ANNE C., 121 E. 11th St.

DILLINGER, MRS. C. E., 121 E. 11th St.

DUFFY, MARGARET C., 467 E. 12th St.

EDWARDS, J. G., N. P. Ry., Vancouver Wash.

ELLIS, PEARL, 454 E. 22nd St. N.

ELLIS, EDITH, Empire, Canal Zone.

ENGLISH, NELSON, 267 Hazel Fern Place

ESTES, MARGARET P., 1063 E. Washington St.

EVANS, WM. W., 246 20th St.

EVERSON, F. L., 210 Chamber of Commerce Bldg.

EWELL, ELAINE, 608 E. Taylor St.

FAGSTAD, THOR, Cathlamet, Wash.

FARRELL, THOS. G , 328 E. 25th St.

FELLOWS, LESTER O., 4319 74th St., S. E.

FINLEY, WM. L., 651 E. Madison St.

FISH, ELMA, 259 E. 46th St.

FITCH, LOUISE R., Tau Delta House, Eugene, Oregon.

FLEMING, MISS M. A., 214 P. O. Bldg.

FLESHER, J. N., Carson, Wash.

FORD, G. L., 309 Stark St.

FORMAN, W. P., Y. M. C. A.

FOSTER, FORREST L., 354 49th St., S. E.

FRANK, ALBERT C., Box 136, San Diego, Calif.

FRANING, ELEANOR, 549 N. Broad St., Galesburg, Ill.

FRIES, SAM L., 691 Flanders St.

FUHRER, HANS, Box 18, Cathcart, Wash.

FULLER, MARGARET E., 409 16th St.

GALUSHA, ORA W., 30 Hillcrest Parkway, Winchester, Mass.

GARRETT, GEO., 646 Cypress St.

GASCH, MARTHA M., 9 E. 15th St. N.

GEBALLE, PAULINE, 782 E. Yamhill St.

GEORGE, MELVIN C., 616 Market Drive.

GETZ, FLORENCE I., 1016 Clackamas St

GILE, ELEANOR, 622 Kearney St.

GILMOUR, W. A., Title & Trust Bldg.

GLISAN, RODNEY L., 612 Spalding Bldg.

GOLDAPP, MARTHA OLGA, 455 E. 12th St.

GORMAN, M. W., Forestry Bldg.

GRAVES, HENRY S., U. S. Forest Service, Washington, D. C.

*GREELEY, GEN. A. W. Washington, D.C

GREY, HARRY C., 1483 Mallory Ave.

GRIFFIN, MARGARET A., 303 Title & Trust Bldg.

GRIFFITH, B. W., 1736 Kane St., Los Angeles, Calif.

HARDESTY, WM. P., 4160 E. 31st St., N.

HARNOIS, PEARLE E., 1278 Williams Ave.

HARRIS, CHARLOTTE M., 1195 E. 29th.

HARZA, L. F., Great Lakes Power Co., Ltd., Sault Ste. Marie, Ont.

HATCH, LAURA, 36 Bedford Terrace, Northampton, Mass.

HATHAWAY, WARREN G., 800 Burwell Ave., Bremerton, Wash.

HAWKINS, E. R., 655 Everett St.

HAZARD, JOSEPH T. 4050 1st Ave. N. E., Seattle, Wash.

HEATH, MINNIE R., Wheeldon Annex 395 Salmon St.

HEDENE, PAUL F., 720 E. 22nd St., N.

HENDERSON, G. P., 1155 E. Yamhill St.

HENRY, E. G., 488 N. 24th St.

HENTHORNE, MARY C., 270 E. 52nd St.

HERMANN, HELEN M., 965 Kirby St.

HEYER, A. L. JR., 253 6th St.

HIGH, AUGUSTUS, 300 W. 13th St., Vancouver, Wash.

HILD, F. W., Denver Tramway Co., Denver, Colo.

HILTON, FRANK H., 504 Fenton Bldg.

HIMES, GEO. H., Auditorium.
HINE, A. R., 955 E. Taylor St.
HITCH, ROBERT E., 503 Fenton Bldg.
HODGSON, CASPER W., Rockland Ave., Park Hill, Yonkers, N. Y.
HOGAN, CLARENCE, 591 Borthwick St.
HOLDEN, JAS. E., 1652 Alameda Drive.
HOLMAN, F. C., 558 Lincoln Ave., Palo Alto, Calif.
HOLT, DR. C. R., 217 Failing Bldg.
HORN, C. L., Wheeldon Annex.
HOWARD, ERNEST E., 1012 Baltimore Ave., Kansas City, Mo.
IVANAKEFF, PASHO, 246 Clackamas St.
IVEY, RALPH S., R. F. D., Milwaukie, Oregon.
JAEGER, J. P., 131 6th St.
JONES, F. I., 307 Davis St.
JOYCE, ALICE V., 595 Lovejoy St.
KERN, EMMA B., 335 14th St.
KERR, DR. D. T., 556 Morgan Bldg.
KETCHUM, VERNE L., 742 Y. M. C. A.
KNAPP, MARY L., 656 Flanders St.
KOERNER, BERTHA, 481 E. 45th St., N.
KOOL, JAN, 1308 Yeon Bldg.
KREBS, H. M., 582 Main St.
KUENEKA, ALMA R., 869 Clinton St.
KUNKEL, HARRIET, 405 Larch St.
KUNKEL, KATHERINE 113½ Russell St
LADD, HENRY A., Ladd & Tilton Bank.
LADD, WM. M., Ladd & Tilton Bank.
LANGLEY, MANCHE IRENE, Forest Grove, Oregon.
LAWFFER, G. A., 309 Stark St.
LAWSON, AGNES G., 767 Montgomery Drive.
LEADBETTER, F. W., 795 Park Ave.
LEE, JOHN A., 505-6 Concord Bldg.
LETZ, JACQUES, State Bank of Portland.
LIND, ARTHUR, U. S. National Bank.
LOUCKS, ETHEL MAE, 466 E. 8th St.
LUETTERS, F. P., 689 Everett St.
LUND, WALTER, 191 Grand Ave., N.
LUTHER, DR. C. V., E. 34th and Belmont Sts.
MCARTHUR, LEWIS A., 407 Clay St.
MCBRIDE, AGNES, 1764 E. Yamhill St.
MCCLELLAND, ELIZABETH, 2187 East Washington St.
MCCOLLOM, DR. J. W., 553-557 Morgan Bldg.
MCCREADY, SUE O., 512 U. S. Bank Bldg. Vancouver, Wash.
MCCULLOCH, CHAS. E., 1410 Yeon Bldg.
MCDONALD, LAURA, 354 E. 49th St.
MCISAAC, R. J., Parkdale, Oregon.
MCLENNAN, MARGARET, Box 324, Honolulu, T. H.
MCNEIL, FLORENCE, 607 Orange St.
MCNEIL, FRED H., The Journal.
MACKENZIE, WM. R., 1002 Wilcox Bldg.
MARBLE, W. B., 3147 Indiana Ave., Chicago, Ill.
MARCOTTE, HENRY, D. D., 218 E. 56th St., Kansas City, Mo.
MARKHAM, B. C., 343½ Washington St.
MARSH, J. WHEELOCK, Underwood, Wash.
MARSHALL, BERTHA, Station A, Vancouver, Wash.
MASON, J. M., Box 52, R. R. No. 2, Milwaukie, Oregon.
MAY, SAMUEL C., May Apts.
MEARS, HENRY T., 494 Northrup St.
MEARS, S. M., 721 Flanders St.
MEREDITH, JOHN D., 329 Washington St.
MEREDITH, GEORGE, 8½ N. 11th St.
MERRIAM, DR. C. HART, 1919 19th St., N. W., Washington, D. C.
MERTEN, CHAS. J., 307 Davis St.
METCALF, ALICE K., 531 E. Couch St.
METCALF, EDNA, 531 E. Couch St.
MILES, S., 32 E. State St. Albion, N. Y.
MILLAIS, JAMES A., 415 10th St.
MILLAIS, MRS. ADA M., 415 10th St.
MILLER, BEULAH F., 629 E. Ash St.
MILLER, MAUDE ETHLYN, 767 15th St., E., Eugene, Oregon.
MILLER, JESSE, 726 E. 20th St.
MILLS, ENOS A., Longs Peak, Estes Park, Colorado.
MONROE, HARRIETT E., 1431 E. Salmon St.
MONTAG, JOHN W., 883 Commercial St.
MONTAGUE, JACK R., 1310 Yeon Bldg.
MONTAGUE, RICHARD W., 1310 Yeon Bldg.
MONTGOMERY, ANDREW J., 374 E. 56th.
MOORE, EDITH, Knickerbocker Apts.
MORGAN, CHRISTINE N., Box 144, Palms, Calif.
MORKILL, ALAN BROOKS, Canadian Bank of Commerce, Vancouver, B. C.

MUNGER, A. R., 112 W. 28th St., Vancouver, Wash.
MYERS EARL, Box 355, Sunnyside, Wash.
NELSON, L. A., 410 Beck Bldg.
NEWELL, B. W., Ladd & Tilton Bank.
NEWTON, JOSEPHINE, 1350 Pine St. Philadelphia, Pa.
NICHOLS, DR. HERBERT S., 802 Corbett Bldg.
NICKELL, ANNA, 304 College St.
NILSSON, MARTHA E., 320 E. 11th St., N.
NISSEN, IRENE, 969 E. 23rd St., N.
NORDEEN, EDITH, 361 Graham Ave.
NORMAN, OSCAR M., 499 E. 9th St., N.
NUNAN, GINITA, 489 W. Park St.
**O'NEILL, MARK, Worcester Bldg.
O'BRYAN, HARVEY, 602 McKay Bldg.
OGLESBY, ETTA M., Baron Apts., 14th and Columbia Sts.
ORMANDY, HARRY M., 501 Weidler St.
ORMANDY, JAMES A., 501 Weidler St.
PARKER, ALFRED F., 374 E. 51st St.
PARSONS, MRS. M. R., Mosswood Road University Hill, Berkeley, Calif.
PASTORIZA, HUGH, The Technology Club of New York, 17 Gramercy Park, New York City.
PATTULLO, A. S., 500 Concord Bldg.
PAUER, JOHN, 485 E. 20th St., N.
PAYTON, PERLEE G., 3916 64th St. S. E.
PEASLEE, W. D., 125 E. 11th St.
PENLAND, J. R., Box 345, Albany, Ore.
PENWELL, ESTHER, 95 E. 74th St.
PETERSON, ARTHUR S., 780 Williams Ave
PETERSON, E. F., 780 Williams Ave.
PETERSON, LAURA H., 309 College St.
PHILLIPS, MABEL F., 335 14th St.
PILKINGTON, THOS. J., Sebastopol, Calif.
**PITTOCK, H. L., Imperial Heights.
PLATT, ARTHUR D., 16 Hilliard St., Cambridge, Mass.
PLUMMER, AGNES, 3rd and Madison Sts.
POWELL, MARY E., 1330 E. Taylor St.
PREVOST, FLORENCE, 689 Everett St.
REA, R. W., Ochoco Irrigation District, Prineville, Oregon
REED, MRS. ROSE COURSEN, 208 Eilers Bldg.
*REID, PROF. HARRY FIELDING, Johns Hopkins University, Baltimore, Md.
REIST, LINN L., 600 Chamber of Commerce Bldg.
RHODES, EDITH G., 5005 43rd Ave., S. E.
RICE, EDWIN L., 1191 E. Yamhill St.
RICHARDSON, JEAN, 888 E. Washington Street.
RICHARDSON, NETTIE G., 888 E. Washington St.
RIDDELL, GEO. X., 705 Alaska Bldg., Seattle, Wash.
RIDDELL, H. H., 415 E. 19th St.
RILEY, FRANK BRANCH, Chamber of Commerce Bldg.
RILEY, MRS. FRANK BRANCH, 61 Lucretia St.
RISELING, ROBT. F., 1426 Northwestern Bank Bldg.
ROBERTS, ELLA PRISCILLA, 109 E. 48th Street.
ROBINSON, DR. EARL C., 658 Morgan Bldg.
ROGERS, HOMER A., Parkdale, Oregon.
*ROOSEVELT, THEODORE, Oyster Bay. N. Y.
ROOT, BELLE OTT, 302 Wilcox Bldg.
ROSENKRANS, F. A., 281 E. Morrison St.
ROSS, RHODA, 1516 E. Oak St.
ROSS, WILLIS W., 494 Yamhill St.
ROUTLEDGE, FRED A., 159 E. 67 St., N.
ROYAL, OSMON, 207½ 4th St.
RUCKER, WILLARD, 681 E. Ash St.
SAMMONS, E. C., 69 E. 18th St.
SAMMONS, RETA, 69 E. 18th St.
SAMPSON, ALDON, Hotel Bon Air., Augusta, Ga.
SCHMIT, LUCIE, S. 667 Everett St.
SCHNEIDER, MARION, 260 Hamilton Ave.
SCHUYLER, JAMES T., 205 Broadway.
SCOTT, ELISE, 1565 Knowles St.
SEARCY, ROBERT D., 616 Chamber of Commerce Bldg.
SHARP, J. C., 1360 E. 17th St.'
SHARP, MRS. J. C., 1360 E. 17th St.
SHELTON, ALFRED C., 1390 Emerald St. Eugene, Oregon.
SHERMAN, MINET E., 229 E. 23rd St.
SHIPLEY, J. W., Underwood Wash.
**SHOLES, C. H., Box 243.
SHOLES, MRS. C. H., 1530 Hawthorne Avenue.
SIEBERTS, CONRAD J., 728½ Milwaukee Street.
SIEBERTS, MRS. C. J., 728½ Milwaukee Street.
SILL, J. G., 539 Vancouver Ave.

SILVER, ELSIE M., 100 6th St.
SMEDLEY, GEORGIAN E., 262 E. 16th St.
SMITH, KAN, Ketchikan, Alaska.
SMITH, LEOTTA, Palatine Hill, Oswego, Oregon.
SMITH, PROF. WARREN D., 941 E. 19th St. Eugene, Oregon.
SMITH, W. E., 589 E. 12th St., N.
SNEAD, J. L. S., 572 E. Broadway.
SPAETH, DR. J. DUNCAN, Princeton, N. J.
SPURCK, NELL I., Campbell Hotel.
STARKWEATHER, H. G., R. R. No. 1 Milwaukie, Oregon.
STARR, NELLIE S., Whitney Apts.
STEARNS, LULU, 553 Belmont St.
STONE, DR. W. E., Purdue University, Lafayette, Indiana.
STONE, MRS. W. E., 146 N. Grant St., Lafayette, Indiana.
STUDER, GEO. A., 1114 Williams Ave.
SMITH, MARY GENE, Campbell-Hill Hotel
STURGES, DANIELA R., 648 Gerald Ave.
TAYLOR, VERA E., 604 Spalding Bldg.
TENNESON, ALICE M., 5 E. 71st St.
THATCHER, GUY W., 302 Sacramento St.
THORINGTON, J. MONROE, 2031 Chestnut St., Philadelphia, Pa.
THORNE, H. J., 452 E. 10th St., N.
THURSTON, BLANCHE M., 1384 E. Lincoln St.
TINDOLPH, A. G., 621 Empire Bldg. Boise, Idaho.
TREICHEL, GERTRUDE, 535 Mall St.
TREICHEL, Chester H., 501 Y. M. C. A. Bldg., San Diego, Calif.
TRENHOLME, ARTHUR K., 333 E. 44th St.
UPSHAW, F. B., 401 Concord Bldg.
VAN BEBBER, L., 601 Fenton Bldg.
VAN ZANDT, DEAN, 849 Front St.
VEAZIE, A. L., 695 Hoyt St.
VERNON, HOWARD W., 22 Reade St., New York City.
VESSEY, ETHYLE, 1207 W. 18th St., Vancouver, Wash.
VIAL, LOUISE ONA, 580 E. Main St.
WALDORF, LOUIS W., Western, Neb.
WALTER, WILLIAM S., 55 N. 21st St.
WARNER, CHAS. E., 248 Nartilla St.
WEER, J. H., P. O. Box 1563, Tacoma, Wash.
WEICHELT, O. H., 6015 Hillegon Ave., Oakland, Calif.
WEISTER, G. M., 653 E. 15th St., N.
WENNER, B. F., 675 Glisan St.
WHITE, WM. JR., 1302 Commonwealth Trust Bldg., Philadelphia Pa.
WILBURN, VESTA, Richmond Paper Co. Seattle, Wash.
WILDER, GEO. W., 226 14th St.
WILLARD, CLARA, 112 W. 10th St., Vancouver, Wash.
WILLIAMS, A. BOYD, 54 King St., N.
WILLIAMS, MRS. A. BOYD, 54 King St., N.
WILLIAMS, GEO. M., 713 F St., Centralia, Wash.
**WILLIAMS, JOHN H., 2671 Filbert St., San Francisco, Calif.
WILSON, A., 1324 Cascade St., Hood River, Oregon.
WILSON, CHAS. W., Bellevue, Idaho.
WILSON, RONALD M., United States Geological Survey, King City, Calif.
WINTER, C. L., 240 E. 32nd St.
WOLBERS, HARRY L., 577 Kirby St.
WOODWORTH, C. B., Ladd & Tilton Bank.
WYNN, Dr. FRANK B., 421 Hume-Mansur Bldg., Indianapolis, Indiana.
YORAN, W. C., 912 Lawrence St., Eugene, Oregon.
YOUMANS, W. J., 687 E. 9th St.
YOUMANS, MRS. W. J., 687 E. 9th St.
YOUNG, BEATRICE, 652 E. 73rd St., N.
ZIEGLER, MAE, 89 Mason St.

Active members	356
*Honorary members	6
**Life members	4
TOTAL	366

Saddle-Up

for a horseback tour of America's Vacation Paradise. Glacier National Park has wide, safe trails through some of the most romantic scenes on the continent.

Visit the picturesque passes of "the roof of America"—Piegan Pass—Two Medicine and Many Glacier. Stop at modern hotels or Swiss Chalets. 'Visit' with the picturesque Blackfeet Indians.

Glacier is on the main line of the Great Northern—and "right on your way"—no matter what your Eastern destination.

Plan to spend next summer's vacation in GLACIER NATIONAL PARK

"See America First"
GREAT NORTHERN RAILWAY
Glacier National Park

C. E. STONE	M. J. COSTELLO	C. W. MELDRUM
Pass. Traffic Mgr.	Asst. Traffic Mgr.	Asst. Gen. Pass. Agt.
St. Paul, Minn.	Seattle, Wash.	Seattle, Wash.

HENRY DICKSON
City Pass. & Tkt. Agt.

To the Lover of the Great Outdoors

The Wilderness and the solitary places speak to you with tongues of silver
The pearly white of a glistening summit carries a call and a challenge
The massive crag and the cold green glacier depths bear the message of the ages
The moaning tide and the flickering camp-fire whisper sweet lullabies
To you there is music in the waterfall and the flowers of the uplands smile and bid you welcome

You Lover of the Great Outdoors

will find among the mountains of Western Oregon ideal hunting grounds

Our booklets "Camping, Hunting and Fishing in Western Oregon" "Oregon Outdoors" and our new map of Oregon will point the way

Inquire of any agency or address

John M. Scott
GENERAL PASSENGER AGENT
PORTLAND OREGON

SOUTHERN PACIFIC LINES

PRINTED BY JAMES, KERNS & ABBOTT CO., PORTLAND

DECEMBER, 1918 NUMBER 3

MAZAMA

A Record of Mountaineering in the Pacific Northwest

NESIKA KLATAWA SAHALE

Published by THE MAZAMAS
213 NORTHWESTERN BANK BUILDING
PORTLAND, OREGON, U. S. A.

Fifty Cents

Mazama Organization for the Year 1918-1919

OFFICERS

EDGAR E. COURSEN (658 Lovejoy St.) ....President

MISS HARRIET E. MONROE (1431 E. Salmon St.) ....Vice-President

MRS. BEULAH MILLER CARL (629 E. Ash St.) ....Corresponding-Secretary

MISS JEAN RICHARDSON (131 E. Nineteenth St.) ....Recording-Secretary

MISS MARTHA E. NILSSON (320 E. Eleventh St., N.) ....Financial Secretary

MISS LOLA I. CREIGHTON (920 E. Everett St.)................................Historian

MISS MARION SCHNEIDER (260 Hamilton Ave.) ....Treasurer

ROY W. AYER (689 Everett St.)............................Chairman Outing Committee

JOHN A. LEE (Multnomah Club)................Chairman Local Walks Committee

---

COMMITTEES

OUTING COMMITTEE—Roy W. Ayer, Chairman; Charles J. Merten, Miss Martha E. Nilsson.

LOCAL WALKS COMMITTEE—John A. Lee, Chairman; W. P. Hardesty, Vice Chairman; E. H. Dowling, Miss Minna Backus, Guy W. Thatcher, Miss Olga Hallingby.

HOUSE COMMITTEE—Mrs. Laura McDonald, Chairman; Miss Amy Johnston, Miss Althea Lee, W. A. Gilmour, A. A. Bailey, Jr.

ENTERTAINMENT COMMITTEE—Jan Kool, Chairman; Miss Katherine Schneider, Miss Ethel Loucks, E. F. Peterson, Miss Anna Bowie.

PUBLICATION COMMITTEE—G. W. Wilder, Chairman; Miss Laura Peterson, Miss Lola I. Creighton.

EDUCATIONAL COMMITTEE—L. A. Nelson, Chairman; Miss Agnes Plummer, M. W. Gorman, Mrs. C. N. Morgan, R. J. Davidson.

LIBRARY COMMITTEE—Miss Lola I. Creighton, Chairman; Miss Florence J. McNeil, L. E. Anderson, Miss Anna Bullivant, Mrs. Beulah Miller Carl.

MEMBERSHIP PROMOTION COMMITTEE—W. P. Hardesty, Chairman; Miss Mary Gene Smith, Frank I. Jones.

MEMBERSHIP COMMITTEE—Roy W. Ayer, Chairman; Miss Marion Schneider, Miss Lola I. Creighton.

AUDITING COMMITTEE—Robert F. Riseling, Chairman; Miss Anne C. Dillinger, L. Van Bebber.

ANEROID LAKE

Photo by Boychuk

MAZAMA

A RECORD OF MOUNTAINEERING IN THE PACIFIC NORTHWEST

Publication Committee

Lola Creighton | G. W. WILDER, Editor | Laura Peterson

VOLUME V | PORTLAND, OREGON, DECEMBER, 1918 | NUMBER 3

Contents

MAZAMA

VOLUME V DECEMBER, 1918 NUMBER 3

The Wallowa Outing, 1918

LOLA CREIGHTON

On Saturday evening, July 19, a party of twenty-two left the Union Station, Portland, on the O. W. R. & N. train for the Mazama outing in the Wallowa Mountains. As all previous annual outings have been confined to the well-watered western portions of the Pacific states, this trip into the National Forest Reserve of the extreme northeastern part of Oregon introduced a new field for Mazama outings.

Sunday morning breakfast was eaten in La Grande, and the luggage was transferred to a branch line train for Joseph. This train route follows first the Grande Ronde River to the northeast and then turning at the junction of the Wallowa River follows it to the southwest into the beautiful Wallowa Valley.

Those of the party who had been somewhat discouraged of a pleasant trip the evening before, as each turn of the swiftly pounding wheels took the train farther into what appeared a tractless waste of sagebrush and sand, were reassured when near Enterprise, for first real view of the mountains impressed the eye with their sheer height and massiveness, as they rise abruptly from the comparatively level floor of the valley. The many crags and peaks, shelving patches of snow, the bright blue sky, and warm sunshine, caused many heads to be withdrawn from windows with expressions of surprise and delight and a general feeling of satisfaction that here indeed was a region worthy of a prolonged visit.

When the party climbed down from the hot, dusty coach, in Joseph, members of the Joseph Commercial Club were immediately introduced. These hospitable folks took the party to lunch at the restaurant and then for the six mile automobile ride to permanent camp, situated one and one-half miles above Wallowa Lake, on the west fork of the Wallowa River. Very unusual is it to reach a permanent camp without a hike of many miles over a mountain trail. The ride along the lake and through the woods seemed so short that no one could believe that the white tent, in front of which the machines halted, could really be the abode of the commissary department. However, the sight of Mr. Charles J. Merten, the member of the outing committee who had arrived two days before, to establish camp, greeting all the newcomers, put an end to all debates on the question.

Soon the many attractions of this camp, at an elevation of 4500 feet, were appreciated. It was situated at the junction of many streams, up which trails led into the regions to be explored. Yet it was such a short distance from Joseph that mail and groceries were daily brought in by auto delivery. The cook tent was pitched upon the banks of the streams which furnished excellent water for all camp purposes and for cold showers for dusty hikers. Enthusiastic fishermen caught trout within a stone's throw of the camp frying pan. A five minute's walk up the trail led to a view of the beautiful Wallowa Falls. Tall pines furnished wood for fires and welcome shade for warm afternoons. Willow and spirca bushes formed the lesser growth. The table, neatly constructed of level boards, was far too stylish for a Mazama camp. Delightful it was to sit with elbows resting on it and watch the rushing waters of the streams or the sunlight on the crags which towered upward on every hand. And, best of all, there were no mosquitos—the bane of all lake or meadow camps—to mar the enjoyment of camp life.

The afternoon and next day were spent in the near vicinity of camp—establishing comfortable individual camps, hanging the flag and banner of Camp Roy Ayer, hiking, fishing and swimming in Wallowa Lake.

The following day, Tuesday, camp was early astir and by seven o'clock a party of nineteen, under the leadership of Mr. Robert E. Hitch, were off on the trail for Aneroid Lake. It proved to be a veritable gem of blue waters, rimmed in on three sides by sheer mountain walls. Fishing, swimming and rowing were enjoyed before lunch was eaten. Seven of the more strenuous members, during the afternoon, climbed the highest points to be seen from the lake—Petes Peak and Aneroid Point. The others explored the many points of interest along the lake shore.

On Wednesday Mr. E. F. Peterson and Charles J. Merten started on a scouting trip to Lake Basin to fish and map out the most feasible route for the official climb to the summit of Eagle Cap, elevation 9,860 feet.

Seventeen were members of the party that at 10:45 Thursday morning reviewed Ice Lake after five hours of steady climbing.

The trail, after crossing the west fork about four miles above camp, leads through groves and meadows carpeted with beautiful flowers, and then up for a thousand feet or more by a succession of wonderful waterfalls which had been sighted from the valley trail. The lake of ice cold water is far larger than the first view indicates as it stretches in horseshoe shape beyond a projecting point of rocks. It is walled in on the south by a gray weathered mountain which joins Marble Mountain on the westward side. Had time permitted an attempt would have been made to climb Marble Mountain, as it appeared to present few difficulties. To the west and north stretches the long ridge of a barren brown mountain. A party, under the leadership of Mr. W. H. Harris, supervisor of the Wallowa National Forest, climbed this ridge during the afternoon, and by following it and then dropping down to Big Creek they arrived at camp but shortly after the main party. During the remaining days in camp many discussions were held over the rival beauties of Aneroid Lake and Ice Lake. Many expressed that these two lakes, with their surrounding peaks, would easily repay any outing trip.

Friday evening the camp fire session had the unusual pleasure of celebrating a Mazama birthday in camp. Mr. Adolf Aschoff, who was one of the party that organized the club on the summit of Mt. Hood, on July 19, 1894, gave a very interesting account of that eventful day in Mazama history. The four ex-presidents, Messrs. C. H. Sholes, R. L. Glisan, John A. Lee, and Jerry Bronaugh were also speakers of the evening. After speeches and songs, a large birthday cake, lighted by twenty-four candles, was carried into the bright light of the campfire. This cake was so artistically decorated and proved of such delicious flavor that Chef Thompson received many rounds of praise for it.

Farewell was said next morning to the four ex-presidents, who resumed their automobile trip. Their fish stories, accounts of daily trips, and patriotic speeches had been enjoyed at the camp fires.

The early morning sunshine of the following day found twenty-one hikers swinging along the trail in Hurricane Canyon.

They were indeed fortunate for machines furnished by the people of Joseph had taken them to the beginning of the trail where the commissary and bedding was safely stowed on three pack horses. The open glade soon disclosed a view of Eagle Cap. Marble Mountain served as a gauge of distance traveled. A delightful day was spent as each hour brought forth some new wonder. Cascades on the precipitous canyon wall, the rock formations, the beautiful stretches of forests, and the clear, winding streams and bright flowers of the upland meadows made this a paradise for those carrying cameras.

Departing from this easy valley trail a steep climb over an eastward ridge brought the hikers, after a thirteen mile trip, to bivouac camp, near Mirror Lake, with the summit of Eagle Cap so near at hand that after an excellent supper many expressed the wish to climb it by moonlight. With the trees and surrounding mountains clearly reflected in the many beautiful lakes which abound in this basin, it was hard to be forced to retire to sleeping bags for the necessary rest before the next day's long trip.

Next morning, July 21, under the leadership of Charles J. Merten, every member on this, the official climb of the outing, reached the summit of Eagle Cap at 8:15, after a two hour's easy climb, following the northwest ridge. The extent of these mountains was realized by the view from the top. It disclosed ridge after ridge in every direction, with more than eighteen lakes showing as blue splotches in the valleys. Thoughts went to the Mazama boys in service, with the wish that they were there to enjoy and explore this wonderful country—a new peak, or ridge, or valley could be chosen for each day, and at the end of two weeks much would still remain unseen. Names were signed in the Mazama book and great monument of rocks built above the original small pile. Turquoise Lake held the interest of many and had the party but known the route to camp as planned by the packer, many miles could have been cut from the morning's trip by going from the top to the lake. Great fun was derived on the journey back to bivouac camp by sliding down the snowfields which lie on the northern steep slope of Eagle Cap. A circular detour was made. After a climb over a ridge that seemed the height of Eagle Cap to the tired and hungry hikers, lunch was served at 1:45 on the stream, but a short distance from its source in Turquoise Lake. Everyone was sorry that

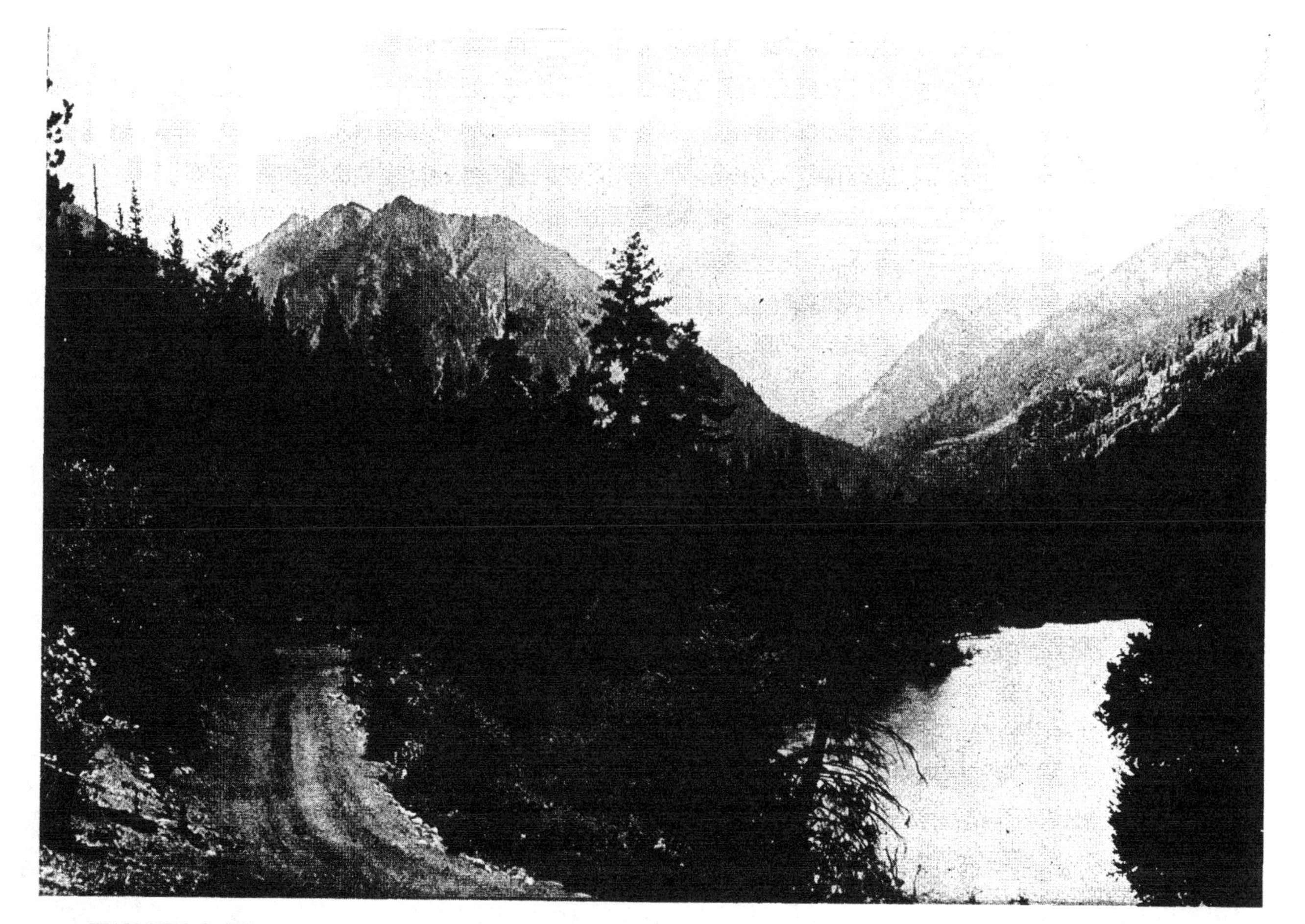

WALLOWA LAKE

Photo by Boychuk

WEST FORK CANYON

Photo by Boychuk

the afternoon's trip had to be taken so hurriedly as it led through one of the most interesting valleys. The lake near the prospector's cabin was a beautiful sheet of water. A hillslope near the West Fork crossing deserves special mention for its great variety of flowers that are indigenous to this region. A columbine, of a yellow shade, attracted much attention. Some of the faster walkers reached camp by 6:30. The entire walk, including the climb of Eagle Cap, was about twenty-two miles.

Tuesday the camp heartily welcomed the arrival of Dr. W. D. Smith, and in the afternoon he led a party to the top of the eastern part of the long, treeless, old moraine ridge, which so completely banks in the waters of Wallowa Lake. The bright afternoon sunshine brightened the golden fields of grain on the many farms that stretch in every direction from this view point.

During the remainder of the stay scarcely a day passed that some new trip was not taken, the more strenuous hikers daily added some new canyon, valley, ridge, or peak to their list. A special trip was made to Ice Lake that Dr. W. D. Smith might study the geology of that region, and on Thursday he and Mr. Merten climbed Eagle Cap from the Turquoise Lake side. Fog obscured distant views on both trips.

This outing included all the usual jolly good times of camp life. No one missed the camp fire sessions where the Ayergronian, the daily newspaper, was read; fish stories told; plays staged; songs sung, poems recited and music played to the great enjoyment of the tired but happy hikers. The afternoon teas were well attended, especially the one at which ice cream, a rare dish in a mountain camp, was served. President Roy W. Ayer and everyone in camp lined up on the evening of July 24, to give a hearty welcome to about forty of the people of Joseph to whom dinner was served in camp style. Later they took part in the camp fire session and their speeches urged that the Mazamas should come soon for another outing in the Wallowa Mountains.

On every mountain camp some rain must fall, and this was no exception to the rule. Friday night's rain seemed sent to test the sleeping bags of all those who dared sleep out with no protecting tent for shelter and next morning saw a number holding blankets before a big fire. Tea, cards and the stuffing of dunnage bags helped to pass the rainy afternoon hours.

Next morning the last of the dunnage was packed, the last hurried visit made to the falls; and farewell said to all the favorite spots about camp. At seven o'clock automobiles again arrived in camp and the entire party was taken for a ride, until noon, through the beautiful Wallowa Valley. All were amazed at the crops which that soil will produce. No words can give any idea of the hospitality of the people of Joseph to those who have never visited their valley. Their generosity to the outing party culminated in a most delicious banquet. Speeches, songs and musical selections made the hour until train time seem very short.

Portland was reached at one o'clock of the following afternoon. So was ended a most successful outing, with every member greatly strengthened for more effective service during the coming year.

THOSE WHO CLIMBED EAGLE CAP ON THE MAZAMA OUTING OF 1918

Aschoff, Adolf
Ayer, Roy W.
Ayer, Leroy, Jr.
Boychuk, Walter
Creighton, Lola
Griffin, Margaret A.
Hallingby, Olga
Hitch, Robert E.
Kerr, Dr. D. F.
Knapp, Mary L.
Kool, Jan
Kress, Charlotte
Lawson, Agnes G.
Merten, Charles J.
Miller, Jessie
Morgan, Mrs. C. N.
Nilsson, Martha E.
Pendleton, Cecil
Penwell, Esther
Peterson, E. F.
Prenbys, R. P.
Smith, Dr. Warren D.
Taylor, Vera
Taylor, Zella
Young, Crissie

The Wallowa Mountains—Geology and Economic Geography

By WARREN DU PRE SMITH

Just as "All Gaul is divided into three parts," so they say Oregon is divided into three parts; Western Oregon, Eastern Oregon and Wallowa County. Whoever "they" are, they are right, for certainly Wallowa County comprising chiefly mountain country, is unlike anything else the writer has seen in Oregon.

Before launching into the pleasant task of describing this wonderful, and we fear to most Oregonians, unknown country, the writer should tell how he came to know it. Though he had read some geological notes relative to it from one or two fellow geologists and had listened to an enthusiastic student paint the glories of his native haunts, he had little conception of it until that progressive body, the Mazamas, held their annual outing in that region, July, 1918, and he was a privileged member of their camp.

We almost wish we could forget for the moment the vocabulary of geology and be permitted to draw upon the words of Wordsworth, or that we had the power of description of a Tyndall, or a Winchell, for truly the region we are about to tell you of merits the noblest words ever coined and the finest phrases in our language. As we stood more than once on the top of some natural minaret in the flush of triumph, we wished we might, in the words of Kipling, "take hold of the wings of the morning and flop around the earth 'till we're dead," but this is permitted to the poet, the geologist must keep to the earth and the things thereof.

To the following persons we would like to make especial acknowledgment for assistance in the acquiring of data for this article: Mr. A. Bodmer, Manager McCully Mercantile Co., Joseph, Oregon, who furnished transportation, samples of mineral and much information; Mr. H. W. Harris, Forest Supervisor of Wallowa National Forest, for the use of the excellent topographic folio of this Reserve; Messrs. Richardson and Reade, Engineers of the Baker Mines, Cornucopia; Waldemar Lindgren, whose articles on the Blue Mountains of Oregon, published years ago, while it does not include our particular territory, yet is invaluable to one making a geological study of this portion of the state; R. M. Swartley, formerly Mining Engineer of the State Bureau of Mines and Geology, who has published some notes on the geology of this region and who has perhaps made the only detailed examination of the mines of this district.

Geographical Position

Situated in the extreme northwestern corner of the state, near the junction of the States of Washington, Oregon and Idaho, bordered on the one side by the famous Snake River, the country we are delineating lies in the heart of the great "Inland Empire," with fair transportation it is in contact with such important cities as Walla Walla, Lewiston, Baker City and Boise, it is more than a county of Oregon, it is part of a more cosmopolitan region. This is perhaps the most striking characteristic about Wallowa County, it belongs politically to one state, yet geographically, commercially, socially and in every other respect it is not bounded by that state.

Second, it is ideally replete with mountains, plains, rivers, lakes and forest. Every variety of climate, of scenery, of soil and resources can be found in it. This fact is of paramount importance. The beneficial effect upon the prosperity, wealth and mentality of its citizens cannot be gainsaid. The region, though old geologically, is but just emerged from the frontier stage and so has not had time to acquire much history or tradition, and now that the childhood of the race has gone never to return, it has no chance to become like Switzerland, though the natural setting is there.

Although the Wallowa County is far from the sea, some of its area is rolling lava desert, which in some of its aspects and moods, is not unlike the sea. The motor car has supplanted the old prairie schooner which aforetime rocked and plunged from one lava hummock to another, and prosperous cities are located where once old camp fires burned.

One thing which is very confusing to the traveler is the way you come into Wallowa, by the back door as it were. You go north-by-east, and then east-by-south, because the Wallowa Range extends from west to east and the one railroad which taps this region must make an end run, so to speak, up the Grande Ronde, and thence along the Wallowa to Joseph at the lower end of Wallowa Lake.

TOPOGRAPHY

The general topographical scheme comprises an extensive undulating plan or rather low plateau of basalt sloping up to an almost precipitous wall of mountains, so abrupt as to be explained in one way only, namely, the fault scarp. These mountains are variously known as the Wallowas, The Eagle Creek Mountains or the Powder River Mountains.

This front mountain wall varies from two to four or five thousand feet high, and is cut through in several places by deep canyons by way of which foaming torrents hurdle their way as if glad to be released from their "high mountain cradles" and join the older and more sedate Wallowa, whose course through the lava fields is denoted by a great velvet band thrown carelessly out across the landscape. And back of this rock rampart of marble and porphyry rise other rocky ranks with here and there taller sun-kissed and cloud capped leaders, and one among all supreme, "Old Eagle Cap."

Right in the largest of the openings through the range where it seems a giant axe wielded by some god has cleaved asunder the mass of stone is one of Oregon's and Wallowa's chiefest jewels, Lake Wallowa. Hemmed in, in part by the canon walls, in part by huge moraines extending far beyond the mountain portal, it lies there, a beautiful sight, making a natural and ideal reservoir of water for the use of man, beast and vegetation.

The principal streams debouching from these mountains are the Wallowa and the Minam, on the north side, and on the south side Pine and Eagle Creeks; on the west side, Catherine Creek, and on the east the Imnaha. None of these is navigable, but they are all a source of power and some, except those which flow out of the mining districts, are stocked with fish.

Joseph, the railroad terminus and an enterprising town of about 1,000 people, at the gateway of the mountains, is 4100 feet in elevation. The Wallowa Valley bottom is anywhere from 2,500 to 3,700 feet and the summits of the passes across the ranges are all close to 8,000 feet. The mountains are serrated, rugged, almost bare of their timber or snow in summer. One dwindling glacier still clings to the slopes between Sentinel Peak and Eagle Cap, a pitiful remnant of its former self.

To one who has visited Glacier National Park, up in the corner of Montana, a striking resemblance between the two regions will be at once apparent. The scenery of the Wallowa is just as fine, though there is not as much of it.

Climate and Vegetation

To the dweller in Western Oregon the land of rain and mist, of vegetation almost tropical in its rankness, of color tones dominantly green and cobalt and purple, the climate, atmosphere and vegetation of this region offer almost startling and yet not displeasing contrast. Perhaps the best way to show this is by tabular arrangement as follows:

Western Oregon	Wallowa
Vegetation — Large trees — firs, dense underbrush.	*Vegetation* — Small trees — pines and tamaracks, little or no underbrush.
Rainfall — Abundant to excess, 40-100 inches, distributed over many months. Slow discharge. *Humid.*	*Rainfall* — Light precipitation, 10-25 inches, but concentrated in short periods, thunder showers, rapid run-off. *Dry.*
Clouds — Excessive cloudiness.	*Clouds*—Excessive sunlight.
Temperature—Moderate, 10 to 100 degrees F.	*Temperature* — Extremes, —17 to 110 degrees F.
Climate—Marine.	*Climate*—Continental.

Fauna and Flora

Though it is not within my province to discuss at any length in this article these two subjects, for completeness and in order to show the interdependence of the various branches of scientific knowledge of any region, some allusion must be made to them here. Geographical position, physiography and geology, as all know, have a large share in the distribution, grouping and welfare of plants and animals.

The lighter rainfall, the steepness of the mountain slopes, the greater amount and force of the winds, the light soil covering all have influenced the character of the vegetation. The lack of undergrowth has, of course, been one of, if not the chief cause, for the scarcity of animals. It is said that mountain sheep are still occasionally to be seen on some of the more inaccessible ridges. Undoubtedly this region was once well stocked with game of the kind now to be found in the wilder portions of the Rocky Mountains and formerly, too, there must have been a plentiful supply of birds, but in ten days or more roaming through the woods and over the rocks of Wallowa, only one or two blue birds were seen, and no beast, save domesticated animals from the lowlands. Undoubtedly a field zoologist working at night or with traps would see things which the layman would pass unnoticed.

Not so with plants, for in this domain even the untrained can note, in passing, their profusion and the great contrast to those in the coastal area. On the north side of the Wallowas one finds among the trees mainly yellow pine, tamaracks and lodge pole pine in much denser stands than on the south side. On the divides there is scarcely any timber at all, or if any, a sort of stunted pine.

The flowers of the lowland reaches everyone knows, but comparatively few know the rich carpets of flowers on the upland meadows. Their number and variety are legion. Of course, we know the names of a few of these, but to know their beauty is still better and so we shall leave the more scientific discussion of them to some one better informed.

POPULATION

The census of 1910 shows that there were 7,863 people in this county; of these 502 only were foreign born. This fact alone may argue much in favor of the enterprise of this region. Regions of variable and extreme temperature, with storms, plenty of sunlight, high barometer and moderate rainfall, coupled with varied topography are energy producers. Wallowa County has all of these. No, I shall not be too positive about the matter of the atmospheric pressure, there being insufficient data on this point.

GENERAL GEOLOGY

Just as the medical student must spend long hours in the dissecting room, so must the student who would know the earth, its component parts and how the various parts function, go into Nature's great laboratory where old Earth has been rent asunder, disembowelled as it were. The story of the earth cannot be deciphered from a mere examination of its undisturbed surface.

The Wallowa Mountains offer a splendid opportunity to make these necessary studies. Here in this gaping canyon we see a clean cut incision through the tissue of the outer integument of our old world; in another place a terrible, hardy healed-over wound where a side of the mountain has fallen away; over yonder an old scar where some ancient glacier scratched and gouged its way across its face, and there in the center of the range some convulsion has torn open its side, bringing the very entrails out into the light of day.

Now only the quieter process of erosion and weathering are going on, but once in the morning of Time, there was heaving and grinding, writhing and slipping, twisting and breaking, as a result of tremendous surging from within. And now we come in the noon-time and classify and theorize. What then do we find? First let us list all the different kinds of formations. Beginning with the oldest, perhaps, there are some slates and quartzites with impressions of some primitive clams, called for want of better names, Halobias, and Daonellas, which indicate that comparatively remote time in the world's history known as the Triassic Period. Mingled with these are some lenses of limestone now changed to marble containing almost indeterminable corals. These certainly tell us of warm tropical seas and strange little industrious animals working away to construct fantastic dwelling places for themselves beneath the surface of the sea. We find them now thrust up and far away from their ancient home, their delicate cups rudely torn and crushed by the ruthless ice stream which ground over them.

If we look around pretty carefully we might find some of the cause of the disturbance. There! you are looking right at it, though you don't realize that it did all the mischief, miles and miles of gleaming white granite, or granodiorite, to be more exact. This formation is part of the great Post-Jurassic bathylith which is found extensively all the way from Alaska down into California, but is not very evident in Western Oregon. You will find it there, too, but away back in some of the canyons where erosion has removed the ever present basalt cover. This rock in the Cornucopia district is, according to Lindgren, a more acid type than the California rock, and has more quartz in it.

On the north and south flanks of this perhaps younger, perhaps older (we are uncertain which) lies an old, (I say old here

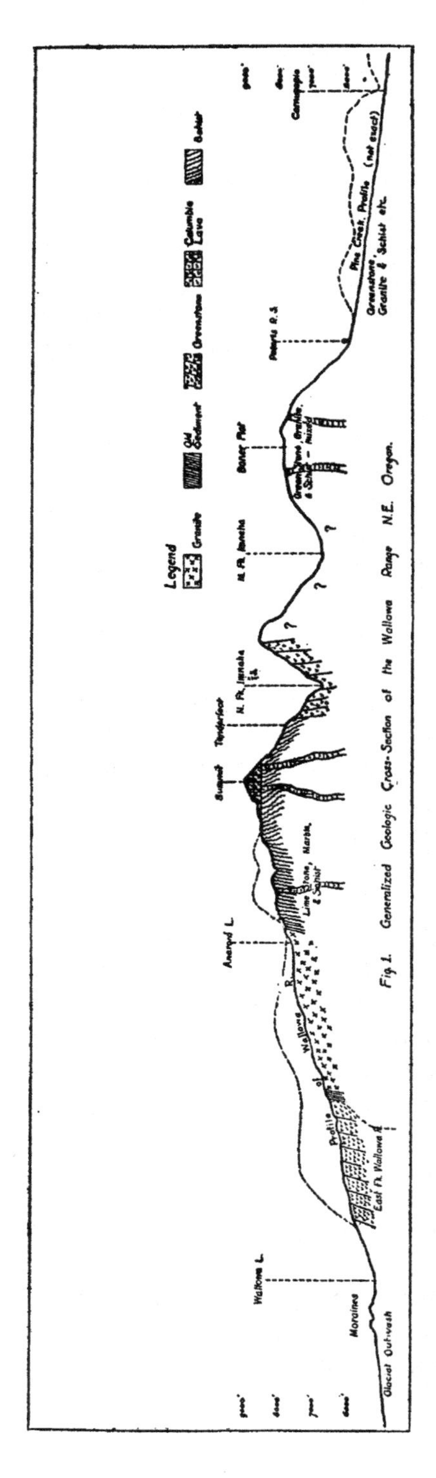

Fig. 1. Generalized Geologic Cross-Section of the Wallowa Range N.E. Oregon.

because there is a still younger one) which from its field appearance and for want of a more exact name, we call a greenstone. This is an extremely hard, dense rock, which emits a ringing sound when struck sharply with the hammer. In places it is characterized by great white felspars an inch or more in length in a green matrix, like plums in a pudding. Overlying the whole mass is a series of brownish layers of more porous lava resting more or less horizontally on the upturned members of this series. This we at once recognized as the well known Columbia lava of the Cascades which also spreads out over the lower plateau regions of Eastern Oregon. Cutting all the older rocks and leading up into these overlying lava beds are great, wide (20-100 feet) sinuous dikes of basalt by way of which the once molten rocks reached the top from the hot viscous reservoir of magma below the "roots of the mountains." Still later than these, though not always above, we find the products of degradation, moraines and talus from glacier, river and landslide.

FIG. I. (Text Figure.)

In Fig. I is presented a generalized scheme of the formations as one finds them in crossing the range.

There are rocks in this region at least as old as the Triassic and probably much older, perhaps Archean, though the writer doubts this very much. Not a scrap of evidence in this part of the Blue Mountains has been seen to support this assumption. It should be distinctly borne in mind that the writer has not seen all this region nor has he spent very much time in it. There are Tertiary lavas and Pleistocene moraines and recent talus deposits. The Tertiary sediments so characterictic of Western Oregon are here either hopelessly buried, unrecognizably metamorphosed, or else missing altogether. Fossils, the time markers of Nature, are few and hard to find and when found are nearly always scarcely recognizable. Of course the granites and lavas are almost never found to contain fossils. When they do it is only an accident.

Principal Geological Events

The principal acts in the geological drama in those far reaches of time long, long before the advent of man were something like these:

Act I.

The deposition of the oldest sediments. This process, of course, presupposes somewhere and somewhen primordial igneous rock from which such sediments might have been derived, but all this we'll relegate to the prologue. Act I, then, we can look upon as a quiet time of preparation.

Act II.

This was the beginning of a period of storm and stress with plenty of stress, but the storm probably not coming until later. In this act the heavy villian, a sleeping giant, underneath, beginning to turn over and push up the covers, asserts himself; this is the granite mass which forced its way up in the post-Jurassic Period. However, he did not emerge in this act, though he made himself felt.

Act III.

In the third act again, a time of comparative quiet, events of ultimate tremendous import to the future men who were to inhabit this region transpired. At this time the circulating ground waters both the meteroic (that which comes from the atmosphere and the magmatic (that which comes from more deep-seated source in the lithosphere) began their slow patient accumulation of the metals. Collecting here and there tiny particles, sometimes dissolving and precipitating them, sometimes transporting them bodily and concentrating them in the crevices in the rocks, where ages after man with his drills and picks and sweat could dig them out to buy and sell with, to gamble, perhaps, a bit, and not least, to adorn his women folk with.

ACT IV.

Another time of storm and stress follows when the pent-up energy below must find release and huge tongues of red molten rock worked their way out to the surface and spread waste rock and destruction over the surface. Contemporaneously there was probably more lifting and heaving and the Wallowas rose to their full height. At the same time, perhaps, occurred a large fracturing and dropping down of the valley portion away from the mountain block and we have what is known as a fault. This block faulting is characteristic of Eastern Oregon, evidence of it being seen near Baker, along the front of the Elkhorn Range, in the Stein Mountains in Southwestern as well as in South Central Oregon.

In the ensuing act the forces which brought about excessive accumulation and concentration of temperature abated and the old giant of Wallowa cooled his wrath to such an extent that his flushed visage became paler and paler and finally his head and shoulders were completely buried in a white canopy, this time not of rock but of snow and ice.

What these ice streams did in their heyday we shall next consider.

One afternoon was spent by the Mazamas exploring the long curved finger-like ridge running along the east side of the Wallowa Lake. Plate No. 1. This ridge is very remarkable and very noticeable as a topographical feature of the region. It is between five and six miles long and seems from the lake shore to be two or three hundred feet high and single crested. It is absolutely bare of trees. On climbing it one finds that near the upper end of the lake, its top is 600 feet above the waters of the Wallowa, and at the upper end is a single ridge of a quarter of a mile in width, in a short distance it bifurcates and further on it becomes five parallel ridges with shallow swales in between. At the extreme lower end this five-fold lateral moraine of sand, clay and boulders, passes imperceptibly into the characteristic irregular terminal moraine. The moraine on the west side is shorter and less interesting. Were it not for the fact of the glaciation we would be absolutely at a loss to account for these topographical freaks.

The effect of the character of the rock in relation to the work of the glaciers is nicely demonstrated in the Wallowas. Where limestone and marble were passed over by the ice we get rounded surfaces, gentle curves, but granites and lavas were not so easily polished off. At Ice Lake the *roches moutonnees* or "sheeps backs" due to glacial polishing are particularly well developed.

The effects of glaciation are everywhere seen in the mountainous portions of the Wallowa County. Wallowa Lake owes its existence to glaciation, it being merely a river gorge widened and deepened, dammed up on the sides and at the end by lateral and terminal moraines respectively. These "finger lakes," of which Wallowa is a fine, and the only example in this region, are numerous in Glacier National Park.

The glacial cirque with amphitheatric form and the little jewel of a lake always to be found in it is one of the topographic features also of this region and one of its choicest bits of scenery. The wonderful curves and colors of these mountain rock bowls can only be appreciated by those who return again and again to them. From Eagle Cap, on a clear day, it is possible to see as many as sixteen of these tiny mountain jewels. Some of these are turquoise, some emerald and others mauve and gray colored, depending upon the nature of the rocky basins which contain them and the particular state of the sky overhead. Besides the cirque ampitheatres with the snow fields above them are all the rest of the glacial phenomena, such as *roches moutonnees*, already mentioned, U-shaped valleys, lateral and terminal moraines, valley trains, etc., which one will find in any well-glaciated region. The long, gentle slopes on which the city of Joseph is located are, for the most part, made up of glacial outwash and are wonderfully fertile. Soils of glacial origin, it is hardly necessary to say, are invariably exceptionally rich.

The geology, physiography and glaciology of this region are in many respects like that of the Rocky Mountain regions, in other respects like the Sierras of California, but very little like the Cascades or the Coast Range of Oregon. All this bears out and emphasizes the statement made elsewhere by the writer that fundamentally there is little in common between Eastern and Western Oregon. They belong to separate and totally distinct provinces. If scientific men had the making of political boundaries some of our states would have quite different shapes.

In spite of the fact that there are some comparatively old rocks in the mountains of this country it is in a state of topographic youth and also in development of its resources young. The youthfulness of its topography and exploitation (the best and correct meaning of this word is here employed) are reflected in the youthfulness of its people. Progressive ideas have only to be suggested to be given an immediate trial. Mental inertia is not one of the faults of the inhabitants of Wallowa County. It may seem to some that this statement suggests an invidious comparison. The shoe may be tried on by anyone who wishes, but the point we wish to bring out now is the effect of environment, physiography, climate, the whole assemblage of physical factors, upon man and his development. This effect must not, of course, be overdrawn, neither must it be belittled.

Some of the interesting details of the geology of this region might very properly be noted here.

At Aneroid Lake, on the property of Mr. Sieber, is a rather unusual development of garnets varying in size from an eighth of an inch in diameter to an inch or more. In spite of the weathering they have been subjected to they show very distinctly their characteristic dodekahedral shape, having twelve crystal faces, each face being diamond shape. Some of these are greenish color and some a cinnamon brown.

On the right (west side) of the trail leading from Aneroid Lake to the first summit, and at an elevation of about 7,500 feet, is a fine exposure of old slate, limestone lenses, schists, etc., dipping to the northwest. These slates, which are extremely hard, emit a ringing sound when struck with a hammer, and contain index fossils of the Triassic Period in geological history.

Over on the west fork of the Wallowa, well up to the source of the stream, on the claim called the "Opal," and owned by Manuel Lopez, was seen a very interesting grouping of minerals. In a pocket in the quartz vein which lay between foot wall and hanging walls of granite, we found scheelite (the ore of tungsten) almost every copper mineral known, including the principal ores of copper, pyrite and bornite, molybdenite (the ore of molybdenum and one of the chief substances used to harden steel) and phlogopite mica. Between this pocket and the footwall lay a band a foot wide of fine, green mica, in minute flakes.

Over the divide, in the Cornucopia district, one of the most distinctive geological features is the Aplite dike, known as the "forest dike." Aplite is a variety of fine grained granite, consisting chiefly of the minerals—quartz and feldspar. The local name of forest dike has an interesting derivation. Along the cracks are fine markings, arborescent in shape, due to fine deposits of manganese oxide dissolved out of the rocks and deposited in the cracks and joints. These are not fossil moss or vegetation of any kind as some have supposed, but the likeness is so marked that the name "Forest Dike" is a very appropriate name. The most interesting fact in connection with this dike is that the principal gold deposits so far located on the Cornucopia side of the range are found right up against this dike either on one side or the other. Some gold values even penetrate the dike itself. The dike is several feet wide and can be plainly followed on the surface for a half mile or more.

Economic Geography

It would be only natural to find in a region so geologically and climatically diversified all manner of resources, some well developed, others just beginning to be opened up, and still others unsuspected or little availed of.

Some of these are listed below in the order of their present development:

1. Agriculture.
2. Hogs and sheep.
3. Minerals (gold, limestone and decorative stone).
4. Forests.
5. Water power.
6. Scenery.

In 1915 this country ranked first in rye production, third in barley and ninth in wheat, although in area it comes tenth of all the counties of Oregon. The writer is no farmer so he will offer no suggestions relating to this branch of industry, save this: It seems that some device, such as hydraulic rams, might be employed to get water up onto the more elevated lava plateaus where dry farming is now being used. Dry-farming, in spite of all that has been written about it, is a poor substitute for the old, old method of using water. It will, of course, have to suffice when water is not available, but in Wallowa County there is no shortage of water. All that is necessary is to devise a practicable method of bringing water to this land. It may surprise some to know that this county is first in the raising of hogs, a very patriotic kind of crop just now, as well as profitable.

As for minerals—on the Cornucopia side much prospecting and development work has been done, with some good, substantion producers as a result, but at the present time there are only two mines worthy of the name operating in a district which ought to have a score. On the Wallowa side of the range, with even better surface indications, perhaps, there is not a single metal mine in operation. Gold is the only metal now being won, but the veins show copper, lead, molybdenium, tungsten, etc., besides gold, silver and tellurium. What are the reasons for this lack of development? They are several, as follows: (the writer here is not fault-finding, but making an honest effort to present conditions in their true light so as to help the country. Scientific promotion and not boosting is what this country, as well as the rest of Oregon needs).

1. Ignorance about the region on the part of the outside world.

2. Early ill-advised ventures and downright swindles such as the notorious Tenderfoot.

3. Absence of scientific prospecting and development in the beginning.

4. Lack of capital and backing for those who have faith in the country.

Now these are not peculiar to any one camp, it has been the history of every new region in the world, certainly of every new mining field.

There is one mineral, or rather rock, which is found in the Wallowa country which is almost unique, certainly very rare, namely, the black marble, which is found, plastered, as it were, up against the face of the escarpment on the side toward and nearest Enterprise, on Silver Creek. The overlying waste material is now being burned to make a first-class grade of lime. The new kiln recently erected for the company by W. A. Gossett, of Baker City, appeared in every way a very efficient one. The lime is only a by-product from this quarry, the principal asset being the exceptionally high grade decorative marble. A table made of this black polished stone, the blackness here and there relieved by a spot of snowy whiteness, was on exhibit at the Panama Pacific Exposition in 1915, and struck many as being the finest single article on exhibit in that great galaxy of rare and wonderful objects. As far as we know Belgium is the only other country which has produced as fine a quality of stone. There appears to be an almost inexhaustible supply of the Wallowa stone, but lack of a market will for some time work against it. Oregon architects need not go elsewhere for decorative stone, for Wallowa furnishes a superior product. The blackness of the rock has puzzled some who have difficulty in realizing that it is marble. It is due to included carbonaceous (organic) matter which disappears as soon as it is subjected to a fair amount of heat.

Materials for making cement are probably also to be had. The above mentioned limestone and glacial clays in the moraines could, in all probability be utilized to this end, though no detailed investigations have been made to our knowledge.

The war, if it continues much longer, will cause the exploitation of the tungsten and molybdenum bearing veins, just as it has developed the chromite industry in other parts of Eastern Oregon.

If all the mineral territory could be brought into responsible and efficient hands, as is the case of the properties controlled by the Baker mines, there would be an end of waste and muddle and the whole mineral industry would go forward with steady strides.

Forests

As others can speak with more authority and from better information about timber resources of this region, I shall in passing, merely give the impressions of one who has seen forests in many parts of the world, and who has been trained to take rapid inventories of natural resources of new countries. There is an abundance of forested country in the Wallowa region, especially on the north side of the range. A great deal of the timber is small, more like lodge-pole pine, tameracks, etc., excellent for certain purposes such as mine stulls, and interspersed with it a

MIRROR LAKE

Photo by Boychuk

CE LAKE—MARBLE MOUNTAIN ON THE LEFT

Photo by Boychuk

fair supply of yellow pine. The forests are very similar to those of Idaho and Montana, but, of course, much inferior to those of Western Oregon and Washington. The lack of undergrowth is a conspicuous and pleasant point about these forests.

On the south slopes of the Wallowa Mountains the quality and stands of timber are both, as far as we encountered them, disappointing. There is still in this region all the timber needed for all reasonable local uses. The Government is taking good care now that this supply is not wasted.

Water Power

It goes, almost without saying, that a region of the kind we have under discussion, should be amply supplied with water power. The elevation, the latitude, the precipitation, all insure deep and lingering snow fields. The melting of these, with the sharp declivity, furnish abundant and swift streams, in which the power is enhanced by numerous water falls. Add to all these the cirque lakes, which insure a permanent and steady flow of water, we have ideal conditions. The writer has not made nor seen any estimate of the horsepower available, but is firmly convinced that there is plenty in these mountains to enable the industrial wheels of Wallowa County to turn, for centuries to come.

Scenery

As yet little or nothing has been said on this subject from the commercial standpoint. Here is a very profitable commercial asset of the country, which some have realized, but which has not yet borne full fruit. The chief reason being the lack of advertising of the right kind. No organization in Oregon can supply this deficiency as can the Mazamas, and that is the reason why in this article the writer has covered some subjects which are not strictly geological. However this may strike the reader, nothing so far has been said, which is not as far as we now know true, nor which does not follow legitimately from a consideration of the geological and physiographical features of this region. Personally the writer regards Wallowa Lake as one of the most attractive spots in Oregon. Build a good, attractive hotel at the head of the Wallowa Lake, another one at Cornucopia, place a few Swiss Chalets here and there in the mountains between, put some motor boats on the Lake and wait for returns. You will not have to wait long.

What with abundant water, rich and diversified soil, both volcanic and glacial, in origin, in addition to other resources, plenty of sunlight, varied scenery, some of it meriting the adjective "grand," favorable climate, extensive mineralization, a topography in no way monotonous, and last, a class of people energetic and skilled, there is a bright future ahead for this section of Oregon.

EPILOGUE

The following few words have little direct relation to the foregoing descriptive matter and are appended here for the perusal of the writer's Mazama friends, and the casual reader is forewarned that he reads them at his own risk.

We have, in the above descriptive matter, been dealing entirely with concrete facts about purely material things. There are some other things, less tangible, ideas, impressions, sentiments, whatever you are pleased to call them, which came to us very forcibly on this last outing. Perhaps others felt and saw the same as the writer but on the chance that some of this may find an answer in the hearts of our Mazama friends, we put these thoughts forth for what they may be worth.

As we climbed on our way to Eagle Cap the crowning point of the Wallowa Range, we began to compare the long toiling upward with the long march of humanity from the time of the anthropoids up to the present time. At the start we floundered around a bit, looking for the trail. There was some confusion as some of these trails were blind ones. Undoubtedly our ancestors made many false starts and may not the monkeys, gorillas, etc., represent these false starts of humanity? In our climb up the mountain we encountered long, gentle slopes, sunlit and flower strewn, where all seemed well with the world. These correspond to the years of peace when some progress upward, though slow, is made. And then comes a temporary reverse grade, where we seem to be losing all we had gained before. How often do we hear persons pessimistically inclined bemoan the fact that the human race is degenerating, and assert that civilization is not on as exalted a plane as in the time of the Greeks. These persons lack perspective, they are too close to the thing they are looking at and do not see enough. Following these short reverse grades we strike sharp upward slopes, rough, to be sure, but away we go higher and higher, and we "make elevation" very rapidly. At the top of some of these steep ascents we stop all out of breath and almost exhausted. May we not liken these times to war times in the life of the race, when in spite of the temporary exhaustion we find that great progress has been made? Some times as the column of climbers goes plodding along discouragement is plainly evident and doubts are expressed as to whether progress is being made, or as to the objective, and some say, "Oh, whats the use! Oh, Lord, how long, how long!" And now and then one drops out by the wayside, but the leader up ahead has had a vision of the heights beyond and he sends back word to the toilers to be of good cheer, "to carry on." And so it is in life. We are making progress all the time, but we must have leaders. Once in a while some demagogue, some dangerous theorist, will persuade the people to do away with leaders and the proletariat is enthroned for a while, then the trail is lost and all go floundering in the snowdrifts or come to some yawning chasm, from which all shrink in horror.

But what, some cynic asks, will be gained by getting to the top, what do we do there, what can we see, "where do we go from there?" Perhaps there is nothing more, but we think otherwise. At any rate it is "not the gold, but the finding of the gold" that makes many a man go on.

Have you ever thought, reader, what a long, long, hard, hard trail the human race has come over? Have you ever thought of this that considering what this race has come from and through, we ought not to deplore the meanness of the human species, but rather wonder at the goodness in man? And having come thus far on the upward trail, may he not climb to heights yet undreamed of, to pinnacles of glory still hidden by clouds and foliage?

Even now we are on one of those steep, terribly hard, upward pulls; the rocks slip under our feet, the mist surrounds us, we seem to see darkly. "Patiencia." Some of us will be bruised and broken, the packs are too heavy, some will not come out of the shadow and a stone or two crossed sticks will mark where this one or that one succumbed, but of one thing, reader, have no doubt, the main column will pull through—this old world may be, as one writer says, a disorder of forces, but there is in man a spirit which is resistlessly curbing the elements—bringing order out of disorder. There are many physical signs that we are nearing the end of an epoch and man and woman released from the old, worn out shackles, will march together as Mazama comrades in a new and better relation to those heights seen only by the leaders, and as we go we sing together the words of the psalmist of old, "I will lift up mine eyes unto the mountains."

Eugene, Oregon, September, 1918.

Out in the Fields

The little cares that fretted me,
I lost them yesterday
Among the fields above the sea,
Among the winds at play,
Among the lowing of the herds,
The rustling of the trees,
Among the singing of the birds,
The humming of the bees;
The foolish fears of what might happen,
I cast them all away
Among the clover-scented grass,
Among the new-mown hay,
Among the hushing of the corn
Where drowsy poppies nod,
Where ill thoughts die and good are born,
Out in the fields with God.

Aboriginal Nomenclature

HENRY SICADE

Of the big peaks or snow covered mountains, in the United States, there is but one which is known and called by its aboriginal name, Mt. Shasta.

There are a few small mountains bearing aboriginal names but nearly all are known by names of explorers or named after famous or prominent persons.

My people, the red race, has never been a party to any disputes as to names of rivers, lakes, bays, mountains or ranges, nor do I desire to take part in any such controversy. With us, more often the naming of tribes, rivers, mountains, etc., are given by other tribes of other sections. For instance, my tribe, the Squallys, now officially known as the Nisquallys; derived their name from the tops of flowers, herbs and grass; waving back and forth when the wind blows. When the first French explorers came to our country, they asked our people their tribal name and were informed Squally, "Oh! Quarre! Nez-quarre." My people have blunt noses, the French, naturally named them, "Nez'-quarre," square noses; the Americans now call us Nisquallys.

My adopted tribe, the Puyallups, got their name from the plains people because they were quite generous and gave more than was needed; "Pough"—pile up, add more, or running over, "Allup," people. The aboriginal people were quite original in their ways.

Another name in dispute; this dispute is not by us or by any inter-tribal disagreement, but by prominent white people of different sections of the state of Washington. It is the name of nearly all of Pierce County, Washington, given us by tribes further north or west; this name is "Tiswauk," Tiswauk means "barely discernable at a great distance;" having the great mountain as a mark to designate the surrounding country. As you are all aware, originally, aboriginal names had meanings, some we know, others have been lost, all having been handed down through countless generations and sometimes only the name survives, but not the meaning. You will also notice that most aboriginal names have significant meanings, suggestive or fitted to signify some thing or having peculiar characteristics. Some of them to my ear are very euphonious, and easy to pronounce, considering that our language is very gutteral.

I am going to tell you in three stories, or to be exact, if you please, three legends, how the name Tacoma originated. My mother tongue is spoken by nearly all of Western Washington and with different tribes there are slight variations and the farther away you get, north, west, or east, there are great variations, in British Columbia or its borders and east of the Cascade Range the dialect or languages are entirely different.

My own people call it "Tacobud." Ta—that or the, Co—water, Bud—where it comes from. The plains people, those living east of the mountain range call it Tahomah. When a Klickitat says Tahomah, he gives that ring in awe; the mountain, the great mountain, which gives thunder and lightning, having great unseen powers. These plains people dominated Western Washington, and like the descendants of the Puritans the Cavaliers, we who are from the plains people are proud of them. But Tacobud or Tahomah, do not tell us how the name Tacoma originated; some of us have tried to delve into our past. In the unrecorded past of the aboriginal people, their authority for names, customs, etc., are legends; myths in your estimation. Legends are our histories, telling of the origin of peoples, lands and natural objects, etc.

There is on the western slope of Mt. Tacoma, a little valley rich in nature's gifts of berries, game and fish. This particular place was the real Tiswauk, inhabited by a small band of natives, called Tiswaukumsh (meaning people of that locality.)

There lived and grew up a beautiful young woman; before the mountain existed. This young woman was betrothed to a brave of the Olympic peninsula,.who already had a wife. The young wife was jealous of the other wife and their rivalry grew bitter, and at last the young woman started something and scratched the face of her rival, and the brave interfered, and he, too, got scratched. The young woman prepared to leave for good, taking her son along, and also a liberal supply of dried fish. When about ready to pull away in their canoe, the young mother exclaimed to the boy "Tacoma," meaning don't forget to take the snow water along. From that exclamation she was called or named "Tacoma." When the young woman and the young son returned to their native land, the Great "Changer," taking the form of a fox (Doupuel-buth) came and handed her as punishment for leaving her husband, the decree that she should be changed into the present, snow covered mountain; and the little son is that sort of a hump on the southwestern part of the peak. In measuring out justice to both man and wife, the husband was also punished and was turned into a part of the Olympic Range and about Jackson's Cove, Hood's Canal, the old Indians point to great fissures as the finger marks of the jealous wife.

Here is another legend, or the second one. For a small area of land between Hood's Canal and the Pacific Ocean, no other place or section had produced so many peaks as Ha-had-hun (the Olympic Range.) In the dim past the peaks were people. Swy-loobs, one of the peaks, married a peak called Tacoma. The peaks grew up and were constantly extending and crowding one another. To give more room to the others, Tacoma offered to go away and settle where she could get more room, as she was growing bigger and bigger all the time. To the rising sun she said: "The people who live there have no mountains, I'll go there and live there and give my limbs a chance to expand." Tacoma had a hard time to get free and loose as the other peaks had grown so close together. To show her good intentions she took all the salmon and fish, which were very plentiful, along with her, so those who might live with her might have plenty of food. In her haste she forgot to take the tail of a fish, so Hood's Canal always had plenty, despite her intentions to take all the fish away.

The last story, or legend, or myth, is this: When Tacoma had settled where she now is, instead of being a benefit to the people who lived there as she said she would be, she turnd into a devouring monster. Any one going to the mountains to hunt or pick berries never returned. There was great consternation and wailing; then others volunteered to go out one after another. The first ones never returned, the last got back. The mountain sucked or drew in all who got near her into her inner parts and devour them. In those days there was an animal, the "mink" some say; others say it was the "fox," who was known as the "Changer" of such things. He was sent for and as a precaution had made lots of ropes from the hazel bushes, twisted and tied together. The "Changer" challenged the mountain to a duel of drawing or sucking on one another by their breaths. The mountain did its utmost and many rocks and boulders and trees and what not were hurled into her innermost.

The challenger, being tied securely, held his own; but the boulders and rocks rolling by him nearly killed him. Bruised and battered, the "Changer" finally won out. Tacoma's blood vessels broke open everywhere and that caused the many rivers and streams. She died and the "Changer" said hereafter the mountain shall be harmless and the rivers teem with fishes for the good or benefit of all the people.

I regret that my people have not left more reliable data of the past than legends. Yet that was their way of memorializing events. The broken fragments of a once happy, contented, liberty-loving, proud race all over this wide continent, can not but affect the present-time writer or keen observer with pity. Coming as a representative of that vanishing race, I hesitate to say, yet true, that all tangible memorials of my fallen race have vanished. Yet we have played our part, possibly not as forcibly as others, and when we are no more you, the conquerors, although with little sympathy or consideration in your treatment, will admit you have absorbed something of worth from us. One of the real joys of mankind is the American weed—the tobacco. May you always smoke it in peace. The lowly wapato, now dignified in an immense commerce, and the potato, with the Indian corn, are two of the greatest staples of food, which always grew in America. The great American bird, the turkey, has given us one of the finest customs of mankind. You had to come to America to learn to give thanks. The Great Spirit has always been recognized here. The red man without question was the original exponent of liberty. Of a native population of 300,000, 2 per cent, or 6,000 red blooded red men have enlisted to fight for your liberty. Many more, not accounted for, have gone over the boundary line and joined the British forces. Had 2 per cent of your 100,000,000 population enlisted you would have an army of 2,000,000 without drafting.

I want to thank you, you the American people, for having perpetuated aboriginal names. I shall not worry nor question your good judgment for justice, if left to your decision. Past records have shown that you have been very liberal in retaining original or aboriginal names. Twenty-four states of our great country of forty-eight states, and one territory, and countless villages, towns, cities, counties, rivers, lakes, and so forth, have aboriginal names and as time rolls by these names become so established in the English language that even now, and by and by, you will not know that the vanishing race had furnished them.

The 1918 Mazama Outing to Mt. Hood

By A. BOYD WILLIAMS

The annual Mazama Outing to Mt. Hood, which is given principally to qualify persons for membership in the Mazama Club, was conducted this year by the Local Walks Committee on August 10th and 11th. This was a later date than usual, giving those who usually make the climb earlier in the season an opportunity of viewing the mountain in its fall aspect when the glaciers are icy and almost devoid of snow.

About one hundred and thirty persons registered for the trip, and as fast as they gathered at the starting point on Saturday afternoon they were whisked away by the autos which were waiting for them. The ride to Government Camp was made through a steady drizzle, which increased as we proceeded. One by one the cars arrived, their occupants were served with hot soup and a hearty dinner, after which they started merrily on their tramp to Timber Line. Here an advance party had strung up a couple of canvas flys and these, with the U. S. Forestry Service cabin afforded us some shelter from the rain, which, by this time, was coming down so steadily that even the thinnest of us could not step between the drops. A blazing campfire helped somewhat to keep our knees from bumping together.

Although no one had visions of real slumber, we all turned in by ten o'clock and from this time on the true spirit of mountaineers was very evident. The girls were packed in about two layers on the floor of the cabin and most of them wanted to know why it had been built so high and narrow and short at both ends, since there was plenty of room from about one foot above the floor straight up. The men, under the tent flys, were only in one ply, but so arranged that some one's ear was reposing on the other fellow's boot, or a pair of knees abutting against some other unfortunate's chest, and all this time the gentle rain drops were oozing through the water proof canvass, playing tag with us, and it wasn't long before everyone was "it." Under these conditions all members of the party conducted themselves like end-men in a minstrel show. Jokes, repartee and quips were indulged in until about 2:30 A. M., when some one suggested that we end the miserable features of the night by getting up.

Coffee was soon brewing and this, with our fruit and sandwiches, served as our breakfast.

We began the ascent from camp at 3:15 A. M. in absolute darkness, with the exception of three flashlights, and as we left the timber, proceeding over the moraines, we were greeted by a storm sleet which froze to our clothes as fast as it struck. This was the final test of the nerve of the crowd. A few looked back at the faint gleam of a camp fire with a "why-did-we-leave-it" expression on their faces and decided that discretion was the better part of valor and unceremoniously disappeared. From this time on our orderly line was turned into a mass formation. Some were arguing for more speed ahead while from the rear came shouts of "What's the rush? we're all strung out for a mile." "The mountain will keep, this kind of weather wont?" etc.

About seven o'clock the clouds broke away and there before us lay old Mt. Hood in all its glory, with its ever present challenge of "Come on up." This gave fresh stimulus to everyone and by eight-thirty we were resting on Crater Rock, trying to encourage the sun to hand us a little more heat. In about an hour we had sufficiently thawed out and recuperated so as to make a final and supreme effort up the steep slope and "over the top" where the wonderful scenic award awaited us. On all sides, far below, in the foothills, great fleecy banks of sun-kissed clouds drifted slowly along and we realized that those who stayed below were still getting the rain while we basked in the sunshine and feasted our eyes on the wonderful view.

One hundred and seven out of the one hundred and twenty that left Timber Line reached the summit, which was a very creditable showing under the circumstances, for most of the climb was made under very disagreeable weather conditions. A great many had made their initial climb and, therefore, were deserving of much credit for the way each individual encouraged and helped the other fellow and plugged along with an intensity of purpose and determination which wins the title of Mazama.

After an hour or so on the summit the descent was made. Sliding was not up to par as the slopes were too icy at this season of the year. Upon returning to Government Camp we were all served with another hearty dinner and were well satisfied to climb into our machines, bound for home, tired but happy in the thought that we had accomplished the thing for which we had set out.

Jaunt of the Four Ex-Presidents

By R. L. GLISAN

The jaunt of the four ex-presidents of the Mazamas in 1916, when they visited the Mazama Camp at the Three Sisters, proved such a success that last summer the four decided to repeat, by visiting the Club at Wallowa Lake.

On the afternoon of July 10, 1918, C. H. Sholes, John A. Lee, R. L. Glisan and Jerry Bronaugh left Portland in Sholes' car. Bronaugh, who was also an ex-president, took the place of M. W. Gorman, who was prevented by illness from going.

We motored up the Columbia Highway as far as Cascade Locks. The road beyond being impassable, due to construction work, we ran the car on an improvised ferry, and, seated in the car, enjoyed the glorious sunset tints as the ferry took us up the grandest portion of that grand old river. The ferry landed us at Hood River, and we made camp in the Cottonwoods on the river bank. We were fully equipped to camp out. On the running board Sholes had constructed a long box to hold the commissary. On the opposite side a broad sheet iron, on hinges, held the dunnage bags and cooking utensils in place. The sheet iron in camp was lowered to form an indestructible table. We all had considerable experience in camp cooking. Each had a sleeping bag and we had two silkolene flys, 10x10, for shelter, if necessary.

From Hood River we motored over the Mosier Hill, a long, stiff climb, then over the Ortley hill, with its twenty-five per cent grades, where we caught a glimpse of Eastern Oregon. Lunched at The Dalles and went on over another grade on an open slope, where we could look down on the cascades and locks at Celilo. Beyond Celilo we dropped down grade to the Deschutes River, crossed on the toll bridge and followed the Columbia a few miles and then left the river, the road winding up a narrow defile through the mountain wall and out into the rolling wheat belt of Wasco County. We crossed the John Day River on a small ferry about dusk, while a chilly wind and no trees made camp prospects look forlorn. We followed its further bank to Rock Creek and up the creek several miles; made camp, stretching our flys over our sleeping bags as heat lightning and mutter-

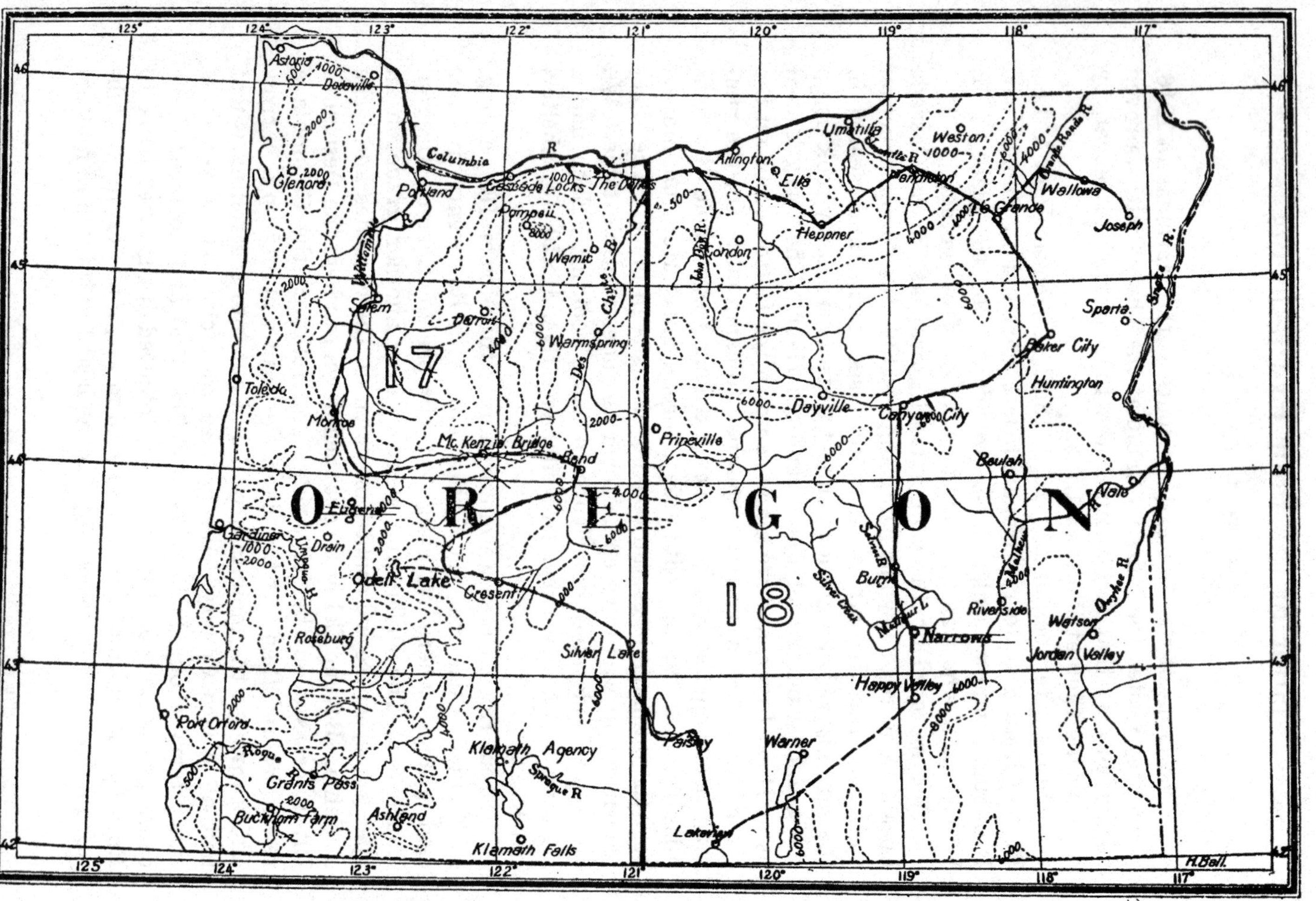

ITINERARY OF THE FOUR PRESIDENTS

ing thunder promised rain. A light shower followed, cooling and clarifying the air, and best of all, laying the dust for the day following. Doves, chats and quail called to us from the bushes along the creek and a rancher supplied us with hay for bedding, rich Jersey milk and clear, cold spring water. What had seemed cheerless proved just the contrary. The next day we crossed an open, rolling, treeless country, interspersed with golden yellow grain fields, intersected with deep cut ravines, where the washed-out, rock-strewn road made hard navigation for the skillfully piloted Hupmobile. Ground owls, weird and witch-like, came out of gopher holes and rolled their bleary eyes at us. Olex supplied us with gas. A recent cloudburst at Ione forced us to pick our way over mud plastered streets and at Heppner we lunched in the park, with flood and fire sufferers. We were warned not to use city water, the flood polluting the supply. Any one inclined to grumble should see the brave spirit manifested in that fated twice-stricken community.

Ten miles out from Heppner we toiled over two long grades, making Pendleton late in the afternoon, camping on a boulder-strewn flat in a park-like cluster of scattering trees and bushes. Jerry thought he could improve his gravel bed by using a discarded piece of sheet iron for a mattress.

Beyond Pendleton lay the limitless grain fields of the Umatilla Indian Reservation, golden yellow, waving and rippling in the breeze.

Harvest here required superhuman efforts in the way of a giant combination thresher which cut, gathered, gleaned, sacked and left the almost bursting grain bags in its path by heaps of threshed out straw.

Rising above the plains we made the crest of the Blue Mountains, looking back over and over again at the ever expanding view. We are inclined to think of Western Oregon and its timber and Eastern Oregon with its sagebrush. Here conditions were reversed. The open, treeless plains extended behind us to the west, while from the crest of the Blue Mountains down the eastern slope spread the finest yellow pine. Crossing the crest we started down the eastern slope through the forest. The road ran just as the early pioneers laid it out, the old Immigrant road straight up and straight down, heedless of contours and ignorant of grades. We dropped down grade to Meacham, only to rise again.

We made La Grande about mid-day and from there went on through Elgin to the rim of Wallowa Canyon. Here the road plunged down the steep wall in dizzy zig-zags, and then followed the river out into Wallowa Valley, well irrigated, one of the most fertile in the state.

While taking supper at Wallowa a brisk thunder shower cooled the air. We kept on enjoying the vivid sunset effects and camped on the banks of the Wallowa among huge cottonwood.

The next morning we passed Enterprise and Joseph and came to Wallowa Lake. White clouds, blue water and the setting of snow-tipped peaks made a wonderful composition.

We arrived at the main Mazama Camp about an hour ahead of the party who came by train to Joseph, where the autos of that town hospitably gathered to take them up to camp . We were camped on the West Fork of the Wallowa River just above the lake, near the junction with the East Fork. From this as a base camp we tramped and knapsacked, taking in Aneroid Lake, where we caught three pound Eastern Brook trout. Upper Lake Basin, Horseshoe Lake, where we caught gamey Rainbow trout, Mirror Lake and other lakes close to Eagle Cap. We climbed Eagle Cap, visited Ice Lake and returned to Wallowa Lake and the auto. It was a wonderful region, as others will more fully describe.

Leaving the lake we reversed our way to La Grande and camped that night by Catharine Creek, on the edge of Union town.

A hospitable miller turned his electric lighted barn over to us. Jerry and Sholes chose the hay for their beds, while John and I spread our sleeping bags on fragrant yielding corn shucks in the ample feed pen and slept under the stars. The next morning our friend renewed his hospitality and gathered some fresh hen fruit and handing me a small shotgun to slay a prize spring game chicken, a most welcome addition to our breakfast.

We then admired his thoroughbred live stock and under his guidance visited the State Experimental Station where the son of our Governor expanded on the methods employed to secure beardless barley and rye, and finished our inspection at the Shetland Pony farm, the best on the coast.

Taking a southerly course we passed North Powder and made Baker by noon, parking our auto in the shade, for the sun was tropical.

Beyond Baker we approached the Blue Mountains. Again crossing timbered slopes, we followed the valley down to Sumpter, of gold mining fame. The town was hardly there, fire having completely wiped out all of its business section a year ago. We secured fine water from a pipe rising from the blackened timbers, and fresh milk and advice from one of the very few still loyal citizens who had lost all but their pluck and cheerfulness. That night we gathered armfuls of tumbleweed for our bed, and advised others to do the same, if they wished to rest in bouyant comfort.

The following morning we rose with the sun and were soon on our way. Again we entered pine timber on the mountain crest and again we came down to another valley, the John Day. It was hot, Prairie City, where we lunched, being the hottest place on the trip. Following down the John Day we went up Canyon Creek, stopping to watch the huge gold dredger devour gravel and cast it out in mounds while the gold remained on the riffle bars within.

Canyon City, of placer mining fame, is now a peaceful village. Once again we crossed the Blue Mountains through pine timber and out into the Juniper Country. Beyond us lay Harney Valley. The setting sun lingered long enough to throw a crimson glow over the wide spreading valley, and we had just time before dark to locate camp on the sage brush fringed bank of Silvies River on the outskirts of Burns.

A broken clutch pin had prevented proper gear shifting and hitting high places had proved too severe on our rear springs. We spent a day at Burns imploring and brow-beating overworked garage and blacksmith workers to replace the broken spring.

Late afternoon we headed southerly again, crossed the Narrows between Malheur and Harney Lake at sunset, followed the southerly shore of Harney, the lake edge having receded three miles in the year past, the driest season known.

A biting wind made us hurry at Dietz Hot Springs to get camp shaped up for the night. Steam rose from a dozen or more hot pools. The place was deserted save for jackrabbits and coyotes. We were near the southwest edge of Harney Lake. We tried to gather sagebrush to lift our sleeping bags off the powdery alkali dust, but found the brush was chic, a bush covered with sharp thorns. At midnight it rained in a country that looked as if it never knew rain.

In the morning the sun tried to break through masses of white clouds.

We used different hot pools for cooking, washing and baking and the sun quickly dried out our scanty belongings.

We left the springs by the only road in sight and started across the desert.

The rain had cooled the air, laid the dust and fleecy clouds shut off the intense sun's rays, making what we had dreaded appear more attractive. For forty miles we traveled without a sign of human life, passing only one cabin, and that deserted. Finally we passed a rancher on his way out with his family for good and all, leaving behind the best ranch that lay out of doors, the only drawback being lack of water. We asked him where we were and his laconic reply "Jackass Hills" seemed especially appropriate.

About noon we arrived at Catlo, the entire population, consisting of the postmaster, being on the step of his adobe cabin, surveying the pools of water in the road, the first rain since last September.

From Catlo we went northwesterly over a sagebrush and rock strewn country, sage hens taking flight, jack rabbits bouncing up here and there, coyotes slinking stealthily away as we advanced. We were crossing where Finley had taken movies of antelopes, but we failed to see any.

Just at sunset the road dropped into a slight coolie or ravine, and then out in a most spectacular manner around the edge of a 2000 foot wall, rising abruptly from the Warner Lakes Basin. Below us lay a chain of lakes of bloody crimson tinted by the sun. The road was literally hewn out of the side wall for three miles, so narrow that we had to remove rocks from ruts we dared not dodge. We looked anxiously ahead as an approaching auto would have been a serious problem.

Long after dark we entered Plush at the lower end of the lakes, and asked a cowboy in chaps, leaning against the hotel bar, for the proprietor. It must have been a typical frontier place before the state went dry, the rendezvous of cow punchers from the big cattle ranches where they count their acres by the thous-

Top—McKenzie Pass—Timber Mound Surrounded by Lava Flow.
Center—Site of Prouty Monument.
Bottom—McKenzie River.

Top—Deschutes and Columbia Rivers.
Center—Diamond Peak from Odell Lake.
Bottom—Fort Rock.

ands. The cook had retired and refused to budge so the hotel keeper dished up some leftovers and we "fell to."

Camp was made with little ceremony, darkness and weariness intervening.

We were surprised the next day to see the stretch of pine timber we had to pass through on the way to Lakeview, which we reached about noon. The name implies proximity to a lake, but Goose Lake has shrunken and is five miles away.

Taking the road northerly we passed Abert Lake and camped at Summer Lake, recently sold by the state for the valuable deposits of potash and salts. We were surprised at the fertile ranches we passed, as we expected alkali and barren waste.

Silver Lake, which we passed the next day, was so nearly dry that they were cultivating what was formerly the lake bed. Beyond the lake we passed Fort Rock, a lava pile, standing alone in a flat country, looking like a shell broken fortress.

Some miles beyond we left the Highway and took a cross road westerly to Crescent and on to Odell Lake. A stiff breeze forced us down the outlet where we were protected by the trees and bank from the wind.

The next day we rowed up to the further and, securing several dozen fair sized trout and a view of Diamond Peak close by. It took two solid hours of steady rowing to return, proof of the size of the lake.

After late lunch, with trout, and more trout for the menu, we took the road to Davis Lake, and then on to the Wickiup Ranger's Station on the Deschutes, where we bathed and fished with fair success.

Rough roads and limited time prevented going in to Crane Prairie. Taking the road down stream through attractive yellow pine, we passed Pringle Falls and stopped to drink from the source of Spring River, where the river jumps full size from a dry hillside.

Crossing the Deschutes we camped on the further bank and again attempted, with poor results, to catch more trout while broiling some we had caught the previous day.

We easily made Bend the next morning, lunched at Sisters and camped at Frog Meadows on the McKenzie Pass. Crossing the Pass is always fascinating, no matter how many times you try it. You suddenly meet the lava flow as if only yesterday it had flowed down into the forest. Crossing its dark, undulating surface we saw the Three Sisters close on our left, Washington and Jefferson further away to the north, their snow mantles making them the more conspicuous in the long, slanting rays of the late afternoon sun.

At Frog Camp, without waiting to make camp, we took the trail to the base of the Middle Sister, and the four ex-presidents did homage at the bronze tablet marking the ashes of Ex-President Prouty. As we stood there the sun dropped below the horizon.

We returned late to camp, keeping the trail with some effort in the darkness.

We lunched next day near Blue River on the McKenzie, following down that beautiful rushing stream to its junction with the Willamette at Eugene.

Portland was reached the next day, having traveled close to 1500 miles and having traversed twenty-one counties. Not once had we slept under shelter during the three weeks outing.

I have detailed our itinerary as a guide to other fishermen and scenery seekers.

Give me Thy harmony, O Lord, that I
May understand the beauty of the sky,
The rhythm of the soft wind's lullaby,
The sun and shadow of the wood in spring
And Thy great Love that Dwells in everything!

—Alexander Pringle.

The Total Eclipse of the Sun

By BLAINE COLES

The United States was favored during the late Spring of 1918 by that most beautiful of all celestial phenomena, a total solar eclipse. The path of totality entered this country from the Pacific Ocean near Grays Harbor, Washington, and extended in a general southeasterly direction across the states, finally leaving the continent on the eastern coast of Florida. The Pacific Northwest was particularly fortunate in the matter of the eclipse in that totality occurred in the middle of the afternoon when the sun was high above the horizon, and further by the fact that the total phase was of longer duration than elsewhere in the United States.

Several expeditions were sent out from the observatories of the country, and in addition many amateur astronomers and laymen journeyed to the path of totality for the purpose of making scientific observations or simply to witness this unusual phenomenon, which perhaps most of them will never see again.

The Lick Observatory expedition located a station at Goldendale, Washington, the United States Naval Observatory at Baker, Oregon, and the Mt. Wilson Solar Observatory and Yerkes Observatory at Green River, Wyoming. The line of totality passed directly through the city of Denver, Colorado, and preparations to view the eclipse were made at the Chamberlain Observatory there, but clouds obscured the sun and no observations of totality were secured, the observers being only able to see the color effects of the eclipse as reflected by the clouds on the horizon.

The writer witnessed the eclipse at Goldendale, Washington, and as it happened, at the time of totality the sky was clear in the region of the sun and the "seeing" was excellent. As far as I have been able to ascertain conditions of observation were better at Goldendale than in any other station in the United States.

Dr. W. W. Campbell, Director of Lick Observatory, has kindly permitted the use of the eclipse pictures with this article. The original negatives are exceptionally fine, but it has been impossible to faithfully reproduce the delicate structure of the corona. Acknowledgment of the pictures is also made under each plate.—Editor.

It is my purpose in this article to treat of the eclipse in a non-technical manner, without burdening the reader with abstruse scientific figures and formulae, which, however interesting they may be to an astronomer, have no place in a work of this kind.

By virtue of the relative positions and motions of the sun, the earth and the moon, it sometimes happens that the moon interposes itself between the sun and the earth, thereby shutting off the light of the sun for a time and causing a total solar eclipse, and again, occasionally the earth is placed between the sun and the moon, which position results in an eclipse of the moon. Eclipses of the sun occur much more frequently in the main, than eclipses of the moon, but at any given point on the earth lunar eclipses take place a greater number of times than eclipses of the sun. This apparent anomaly is easily explained; the moon shines by reflected light only, the light coming, of course from the sun. Therefore when, as in a lunar eclipse, the earth cuts off the light from the sun the moon is darkened and so appears to that entire hemisphere of the earth which is at that time toward it. On the other hand the shadow of the moon as projected on the surface of the earth in a total solar eclipse is much smaller in area than the earth itself, being on the average 70 or 80 miles in diameter. Thus it will be seen that in a solar eclipse the path of totality is only 70 or 80 miles wide, and it is only in this path that the beauties of a total eclipse can be seen, the eclipse being partial outside of the path. The line of totality does not follow the same track at every eclipse, but may appear in any part of the world, and as a general thing we may say that a total solar eclipse will re-occur at a given locality only after an interval of some 250 years.

The writer was not a member of the Lick Observatory Expedition, but was accompanied to Goldendale by Professor E. B. Van Osdel, Director of the Observatory of McMinnville College; Charles Butterworth, of Portland, a photographer interested for many years in astronomical photography, and Howard J. Turner, an instrument maker, the builder of one of the telescopes used by the party.

Our purposes were to study the inner corona, to keep an accurate record of the temperature, and to note any deflection of the magnetic needle. Professor Sydney D. Townley, of Le-

By Permission of Lick Observatory.

Exposure 1-3 of a second, made at the third second of totality. The short exposure exhibits the prominences but does not show the corona to any great extent. The above photograph was taken with the 40 foot camera shown in another picture accompanying this article. The prominence on the southeast limit of the sun is about 40,000 miles high.

By Permission of Lick Observatory.

Exposure 1 m. 52 s.—This picture was taken with one of the 15 foot cameras and the exposure was purposely of long duration, in order that the full extent of the corona might be shown. Owing to the great length of the exposure the prominences are blotted out, and while the corona seemed to the eye to extend further, the above picture is a good representation of the eclipsed sun as the same appeared to the naked eye.

land Stanford Jr. University, established his station at Baker, Oregon, and we co-operated with him in the study of the inner corona.

We selected a site about a mile southwest of the City of Goldendale on the top of a hill, and perhaps 1000 feet west of the Lick Observatory Station. Our location gave us a clear view of the horizon in all directions, with an occasional panorama to the northwest, with Mt. Adams in the distance. In locating the station this was a controlling factor, because the shadow of the moon approached us from the northwest and we wished to see this feature, and as it happened our hopes were fully realized as we shall show hereafter.

A word or two about our instruments might not be out of place. We brought with us a well mounted refracting telescope of 4 inch aperture, equipped with proper eye-pieces for viewing the sun, and a diagonal attachment which enabled observations to be made in a comfortable position. We had also a refracting telescope of 9 inch aperture, built on the Newtonian plan by Mr. Turner. This instrument was completed only a few days before the eclipse and adjustments had to be made after we reached the ground, but these were satisfactorily attended to and the telescope gave excellent definition. For the temperature readings we used a centigrade thermometer, which read to a quarter of one degree, and in addition to the above we possessed a standard compass equipped with a hairline sight and vernier, graduated to one-half of one degree, but with care the same could be read to one-fourth of one degree. A temporary thermometer shelter was arranged and readings were commenced at noon and taken every fifteen minutes until 1:30 P. M., and from then on every five minutes until the eclipse was over. During the ten minutes before and after totality, readings were taken every two minutes.

Before proceeding to further discussion of the eclipse I feel it might be wise to acquaint the reader with the nature of the corona. The corona is virtually the outer atmosphere of the sun, and is composed mainly of a very finely attenuated gas; probably lighter than hydrogen. The gas is apparently an element unknown upon the earth and astronomers have given it the name of coronium. The corona completely surrounds the sun and extends many thousands of miles in all directions, but that part of the corona which originates near the equatorial regions of the sun extends into space a much greater distance than that which starts from the poles. The corona is only visible at the time of a total eclipse, as the light of the main body of the sun is so brilliant that no instrument has ever been devised with which to reduce this light and thus render the corona visible. It is for this reason that observatories send expeditions to all parts of the world to view the eclipses and to study the corona.

The members of our party and all persons in the region of Goldendale will long remember the day of the eclipse. At 2 o'clock on the morning of the 8th, the sky was clear, but later on clouds completely covered the sky; however, at 10 o'clock in the morning things began to look favorable and the sun broke through the clouds from time to time. That was all; and at 1 o'clock our record shows that the sun was covered with thin clouds and the possibility of seeing the eclipse seemed very, very slight, indeed. In fact the sky was practically overcast with the exception that a small break occurred in the clouds in the region of the sun about a minute before totality. Strange to say, the rift in the clouds that gave us the view of the eclipse closed within two minutes after totality was over. It was as if a celestial drama were being played. The curtain rose; gave us a glimpse of the actors and fell again after the climax—but it was sufficient. The small region of unclouded sky surrounding the eclipsed sun was absolutely clear; the air was steady and the "seeing" was perfect.

The second contact, or beginning of totality, was due at 2:57 Goldendale mean time, or 3:57 summer time, which is one hour in advance of the old standard time.

At noon the temperature stood at 33.2 degrees C., and at 2:48, the beginning of the eclipse or first contact, was 35 degrees C. At 3:50 the temperature had dropped to 27.7 degrees C., and at the beginning of totality was 26 degrees C, and so remained during totality. At 4:02, approximately two minutes after totality, the temperature was 25.5 degrees C., or in other words, at two minutes after totality the temperature was 1-2 degree C. lower than during totality, which is equivalent to .9 degrees on the Fahrenheit scale.

The instruments were all adjusted and arranged at noon, and the observers waited for the first contact, meanwhile speculating whether the eclipse could be seen at all, owing to the thick clouds. The first contact was not observed as the sun was behind a dark cloud, and when next seen the moon had already encroached upon the face of the sun and the eclipse had started.

At 3 o'clock it was manifestly growing cooler, and at 3:30 it seemed quite chilly in comparison with the high temperature an hour before. About this time the moon was well on the sun and soon the quality of the light from the sun was very much altered. The light from the calcium vapors of the sun became predominant, and was of a sickly yellow, and growing decidedly darker every moment. Everything in the landscape took on a most unearthly hue and even at that time it was clear to the most unobservant that something out of the ordinary was taking place.

During the morning and early afternoon the birds had been flittering about the station and singing joyously, but when the light became peculiar and as it grew darker their singing was strangely hushed and everything was quiet. The birds no longer flew about but sought the low lying brush and a few less fearful than the rest perched on the nearby telephone wires, remained quiet, and in their minds, no doubt, were preparing for the impending night. There was a farmhouse and a barn at the foot of the hill, and we noted that the chickens went into the chicken house and that several cows came up to the barn from the pasture. In general the effect was of approaching twilight, but such a twilight as can only come from a solar eclipse.

Gradually the darkness became more and more intense and the temperature continued to fall, a sort of phantom breeze stirred the air ever so slightly and above all a perfect stillness settled down on the earth. Even the observers were affected. One could not speak in the presence of this great demonstration of Nature.

Brilliant shades of red and purple illuminated the clouds on the northwest horizon towards Mt. Hood. The ever diminishing light of the sun competed with the gorgeous colors around the horizon, producing a most unearthly and supernatural effect, and yet this was merely the overture preceding that which was to come. The darkness became quite intense, so dark in fact that the observers could scarcely see to write on their charts. As we watched toward the northwest Mt. Adams was suddenly blotted from our sight. The advancing shadow of the moon effaced the mountain as completely as if it had never been there. Then with terrifying swiftness it advanced upon us and suddenly, in the twinkling of an eye, it had come, and totality had commenced.

The writer was at the eye-piece of the four-inch refractor, and suddenly the clouds cleared away, giving a beautiful field. The moon had nearly obscured the sun and the remaining crescent of light, through bright in the telescope, was vanishing rapidly. Soon two brilliant red prominences of hydrogen gas burst into view on the west limb of the sun from behind the dark moon. These prominences of glowing gas thrown out by the sun in tremendous eruption contrasted strangely with the black moon and the now feeble light of the crescent sun. Suddenly, it seemed that the moon covered the sun save for a row of lights on the east limb—"Bailey's Beads" as these are called—caused by the sunlight streaming between the mountains on the edge of the moon. Just as suddenly the glorious corona burst into view. All that had gone before was for the moment forgotten. At least in a perfect sky we had seen the corona. It completely surrounded the eclipsed sun but not in a uniform manner. Long streamers extended from the east and west limbs to a distance of nearly two solar diameters. There it shone with a soft pearly light, nearly lavender, nearly purple, and yet withal a beautiful green, through which the other colors played and strove for predominance. Nearer the eclipsed sun the flaming red prominences cast their glow through it all—indescribable, magnificent, superb. Can we blame the ancients for worshipping the sun, or can we ridicule their superstitious fears of an eclipse? In the presence of such a spectacle man can only watch. It is impossible to describe—words cannot properly tell the story of the eclipse.

But for us no time was to be lost. Drawings, measurements, readings and scientific notes had to be made. The program moved without hitch, the observers performed their duties faithfully, and matters were so arranged that during the one minute and fifty-seven seconds of totality every member of our party had the opportunity to look through each instrument and make measurements and at least two drawings. Before totality was over the writer was back at the four-inch refractor, and altogether twelve drawings were made and thirty-five measure-

Above—General view of Lick Observatory Station, showing the 40 foot camera the spectrographs and the 15 foot Vulcan and Enistein cameras.

Below—Portion of the writer's station, showing the 9 inch reflector in the background, left; the 4 inch reflector in the background, center; the 4 inch reflector in the foreground. The thermometer shelter and compass stand are not shown.

ments were taken during the two short minutes while the moon completely covered the sun.

All too soon the moon crept from the face of the sun and the crescent of the sun, ever so thin, reappeared, this time, however, on the west limb. The corona vanished, the total phase was over. For about two minutes the brighter prominences remained visible, but these, too, shortly faded away.

Directly clouds again covered the sun and as there seemed no chance of observing the fourth and last contact (the moment when the moon entirely leaves the sun) we set about our task of dismantling the instruments, preparatory to our return.

On the evening of the same day, while waiting at Maryhill, Washington, for a train, we chanced to set up the four-inch refractor and while sweeping the southern sky we observed the now famous new star in the constellation Aquilla.

The star was first seen by us at 11:04 P. M., new summer time, on the evening of June 8th. Its stellar character was at once apparent in the instrument, and although we had no star charts with us, we felt sure that the star was a new one, and so it proved upon examination of the charts next morning upon our return to Portland.

In conclusion let me say a word about the total solar eclipses in the near future. On May 29th, of next year, there will be a total eclipse of the sun in which the total phase will last six minutes and fifty seconds, as against the Goldendale eclipse of one minute and fifty-seven seconds. This eclipse will be visible in South America, Chile, Bolivia and Brazil.

A total eclipse on October 1st, 1921, will occur in the Antarctic regions and will probably not be observed.

The northern part of Australia will be favored with a total eclipse on September 21st, 1922.

A total eclipse will occur on September 10th, 1923, which will probably be widely observed. It was thought that the path of totality would cross the United States, entering the continent at about the center of the California coast line, but this appears to be incorrect and later calculations indicate that the shade of totality will be across the peninsula of lower California. In any event, however, the line of totality should be easily accessible and weather conditions at that time of the year should be excellent. The duration of this eclipse will be nearly four minutes.

On January 24th, 1925, a total solar eclipse will be visible in the northeast part of the United States during the early forenoon, but I am not now aware of the exact location of the path of the shadow.

Finally, the writer desires to acknowledge the privilege afforded him of the opportunity to recount the phenomena of the recent eclipse, and if he shall have been able by this means to excite in anyone an interest in astronomy and a desire to know more of the subject, he shall feel something more than grateful.

The Mazamas observed the total eclipse from many points of advantage. The Educational Committee arranged an excursion to Hood River, which was largely attended. The party went by train and were entertained by the Hood River people who furnished two hundred automobiles for a sight-seeing trip through the famous Hood River Valley. After luncheon the party viewed the eclipse from an elevation near the city. The weather conditions were about the same as those described in the preceeding article, although being nearer the edge of the shadow zone the duration of totality lasted only thirty-one seconds. All reported seeing the beautiful corona effects and described the phenomenae as wonderful.

Several motored to Cascade Locks and others climbed Mt. Defiance where only a momentary view was had, these places being on the edge of the shadow zone. A few went to Winlock, a small town south of Chehalis. This place was right in the center of the shadow zone and the weather conditions, usually uncertain, turned out to be nearly as good as at other places. Trips to the summits of St. Helens and Adams were considered, but owing to road conditions it was impossible to get into Spirit Lake at the base of the former and dangerous ice and snow conditions on the latter mountain led the Committee to advise against making the ascents.—Editor.

Address of the Retiring President

FELLOW MEMBERS:

Agreeable to custom and propriety as your presiding retiring, I give a brief review of the activities of the concluding year, with such suggestions as may grow out of my brief experiences as such.

The energies and activities of our country have been given over exclusively and universally to the prosecution of the world's war in which our own country has become the chief factor. Pleasure seeking and social affairs have given place to the stern realities of overcoming our country's enemies. Each month has seen new recruits for the army, the navy, or some other war service added from our membership. Already there are seventy-two stars in the Mazama Service Flag in evidence of our part in the war. No star as yet has been turned to gold. A token that every star represents a man who has offered, and may even yet give his life in freedom's cause.

In a small way our organization is participating to uphold the hands of our representatives in the field of honor by paying the dues of our absent members and keeping alive their memberships, even as we cherish their memory and appreciate their supreme offer of sacrifice for our homes and firesides. Would that we might do more. We could not do less.

The organization has yielded up its club rooms for the Ladies Contingency of the Liberty Loan drives, the Third and Fourth; an entertainment was given at the Little Theater during the year, all parts were taken by the Mazamas. The theater was "sold out" a week before the performance. The talent displayed equaled that of any camp fire performance. The receipts were donated to the Red Cross fund. The Club has also purchased five one hundred dollar Liberty Bonds.

The social life of our Club has suffered naturally and properly because of the sadness and gloom that prevades our land. Herbert Spencer has said: "A look through a powerful telescope will prove to anyone that his own heart beat shakes the building in which he stands." So the recoil of cannon, the bursting of shells, the cries of onslaught and of the wounded set vibrating loyal hearts throughout Christendom, and shall we find time or have inclination for social pleasure? Yet the Sunday local walks have been kept up with the usual regularity.

The two weeks annual outing in the Wallowa Mountains was thoroughly enjoyed by all and was a financial success which speaks well for those managing the finances of the outing.

To the three hundred and sixty-six members, we have added eighty-four new ones, one of these being a life member. While I regret to relate that some forty members have been dropped from the roll for non-payment of dues, while fifteen have withdrawn temporarily to renew their membership perhaps, again, when conditions change. These are abnormal times. The unusual is happening daily and we are not chagrined at this apparent lapse. When the war ends and the white-winged angel of peace hovers over stricken battlefields, then will our organization take on new and added life.

Our Library Committee has found time to do splendid work. They have catalogued and numbered the books and magazines of the Mazama headquarters and have added a number of good volumes. This was done as an approach to the fulfillment of the recommendation of the retiring president of last year. It has been thought not feasible to carry out recommendations then made as to securing a lodge "somewhere in the mountains," but recognizing the importance of that recommendation, I reincorporate the same—to be carried out as soon as conditions have become normal again after the war.

On the 19th of July, 1918, the organization reached its twenty-fifth birthday. A quarter of a century of life has been ours. Founded on a lofty purpose, with clean and wholesome aims, the organization lives on, and may it persist in opening up to the people of our state the latent beauties of mountain, forest and stream, otherwise hidden from view. Like the aerial squadron that serves as eyes to the contending army of the allies, so the mountain climbing contingent will long serve the state in revealing the hidden treasures of the frontier of discovery to a people who live in a land whose beauties, though transcendant, are unseen.

For the splendid aid of officers and members of the association I extend my thanks and my gratitude and bespeak for my successor your zealous intelligent and sympathetic co-operation.

Report of Local Walks Committee

By A. Boyd Williams

The local walks for the Mazama year 1917-1918, were very successful, when taking into consideration the war conditions which called so many of our active hikers away. The total attendance on the fifty-one trips was 2580, making an average of about 51.

The trips to Aschoff's Mountain Resort, of which there were three, proved the most popular. The annual Mt. Hood outing was also very successful, 107 out of 125 persons that started reaching the summit.

The annual boat picnic had to be called off on account of the prohibitive boat prices. The Labor Day trip also had to be changed at the last minute.

The following is a complete list of local walks taken during the year:

Date	Time	Places Visited	Leader	Attendance
1917				
Oct. 7	1 Day	Rocky Point—Logie Trail....	S. M. Fries	53
Oct. 14	1½ Days	Battleground Lake	(Sue MacCready ......) (Bertha Marshall .....)	30
Oct. 21	1 Day	Devil's Rest—Angel's Rest...	Anne Dillinger	42
Oct. 29	1 Day	Beacon Rock	V. L. Ketchum	66
Nov. 4	1 Day	Council Crest—Beaverton	Guy Thatcher	32
Nov. 11	½ Day	Palatine Hill—Terwilliger Blvd	Georgene Case	47
Nov. 18	½ Day	Alameda—Mt. Tabor Park....	Louise Vial	49
Nov. 25	½ Day	Willamette Blvd.— Columbia Blvd.	W. P. Hardesty	51
Dec. 2	½ Day	Barr Butte	Geraldine Coursen ...	23
Dec. 9	½ Day	Cornell Road—Kings Heights.	Vera E. Taylor	36
Dec. 16	1 Day	Dundee—Mistletoe Trip	Harry L. Wolbers	40
Dec. 23	½ Day	Shattuck Road—Canyon Road.	Reta Sammons	33
Dec. 29-30	1½ Days	Aschoff's Mountain Resort ...	Local Walks Com.	68
1918				
Jan. 6	½ Day	Fulton—Oswego ..	Geo. Meredith	25
Jan. 12-13	1½ Days	Larch Mountain ..	(Jay Bush .. ..) (Roy Ayer ....)	150
Jan. 20	½ Day	Skyline Trail—Meredian ..... Monument—Blasted Butte ...	John Lee	53
Jan. 27	½ Day	Sylvan—Humphrey Blvd.	Clarence Hogan	59
Feb. 3	½ Day	"Mystery Trip"	Committee of Five....	48
Feb. 10	1 Day	Columbia Highway	Kan Smith	62
Feb. 17	½ Day	Kelly Butte—Gates	Harold Babb	36
Feb. 24	½ Day	Council Crest	Dr. Wm. F. Amos ...	49
Mar. 3	1 Day	Oregon City—Coalca Pillar ...	A. F. Parker	29
Mar. 10	½ Day	Gilbert Station—St. Johns ...	S. M. Fries	7
Mar. 17	½ Day	Mt. View—Tualatin	Mrs. A. B. Williams..	10
Mar. 24	½ Day	Vancouver Barracks	Sue MacCready	43
Mar. 30	½ Day	Milwaukee Station	C. B. Woodworth	27
Apr. 7	1 Day	Oregon City—Clackamas River	Margaret Griffin	59
Apr. 14	1 Day	Vancouver Lake—Fruit Valley.	(Augustus High .....) (Mrs. Ira Harper ......)	35
Apr. 21	1 Day	Moffett Creek Falls	H. H. Riddell	30
Apr. 28	1½ Days	Aschoff's—Sandy River	Local Walks Com.	105
May 5	1 Day	Eagle Creek	A. F. Parker	137
May 12	1 Day	Parrott Mountain	(C. B. Woodworth ....) (J. A. Ormandy ......)	61
May 19	1 Day	Connell—Tunnel Spur	W. P. Hardesty	64

Date	Duration	Walk	Leader	Attendance
May 22	½ Day	Moonlight Walk, Washington.. Park—Arlington Heights	Mary Gene Smith	55
May 25-26	1½ Days	Scappoose—Chapman Station..	Geo. Meredith	46
June 1-2	1½ Days	Mt. Defiance	E. F. Peterson	36
June 8-9	1½ Days	North Plains—Rocky Point ...	Carl Sakrison	59
June 15-16	1½ Days	Gales Creek—Round Top Mt..	(Agnes G. Lawson ....) (Vera E. Taylor ......)	17
June 19	½ Day	Moonlight Walk, Marquam Hill	Mary L. Knapp	48
June 23	1 Day	Twin Mountain	Roy W. Ayer	15
June 30	1 Day	Hamilton Mountain	Rodney L. Glisan	47
July 4	1 Day	Dowling Farm	Mr. and Mrs. Dowling	30
July 7	1 Day	Cazadero—Eagle Creek	Frank Redman	57
Aug. 10-11	1½ Days	Annual Mt. Hood Outing	Local Walks Com. ..	130
Aug. 18	1 Day	Willamette Valley— Beaver Creek	E. H. Dowling	25
Aug. 24	½ Day	Annual Outing Reunion, Mountain View	Outing Committee	64
Aug. 25	1 Day	Sauvies Island	J. H. Clark .	15
Aug. 31 Sept. 2	2½ Days	Dowling Farm	Mr. and Mrs. Dowling.	30
Sept. 7-8	1½ Days	Vancouver—Mt. Pleasant	(August High) (Mrs. Ira Harper ....)	13
Sept. 14-15	1½ Days	Aschoff's Mountain Resort ...	Geo. Meredith	88
Sept. 22	½ Day	Whitwood Court—St. Johns ..	Cinita Nunan	48
Sept. 29	1 Day	Vancouver—Sifton	(Misses Sue and......) (Alda MacCready)	43
Oct. 6	1 Day	St. Johns—Union Stock Yards.	Lola Creighton	55

Local Walks Committee Financial Report

RECEIPTS

Amount collected on local walks	$133.99
Profit from Mt. Hood outing	93.00
Total	$226.99

EXPENDITURES

Expenditures	$151.13
Balance on hand	$ 75.86

Report of Certified Public Accountant who Examined the Financial Affairs of the Mazamas

INCOME AND PROFIT AND LOSS ACCOUNT

For the Period September 29, 1917 to October 7, 1918

INCOME:			
Members' Dues		$1,137.00	
Life Membership		50.00	1,187.00
MISCELLANEOUS:			
Interest on Liberty Bonds		$ 22.92	
Badges, Keys and Magazines Sold		9.15	
Chairs Sold		18.00	
Stationery		.80	50.87
			$1,237.87
LOSS IN COMMITTEES' TRANSACTIONS:			
LESS			
LOSSES:			
Magazine Publication		$ 366.66	
PROFITS:			
Mt. Hood Outing	$93.48		
Local Walks	73.26	$ 166.74	$ 199.92
Gross Income			$1,037.95
EXPENSES:			
Club Room Rent		420.00	
Telephone Rent and Tolls		64.50	
Printing and Stationery—General		359.45	
Entertainment Expenses		20.35	
Lecture Expense		5.75	
Associated Clubs Dues		15.00	
Insurance		7.15	
Sundries		96.84	989.04
NET INCOME			$ 48.91
ANNUAL OUTING OF 1918:			
Income			$1,183.28
Expenses			$1,183.28

Balance Sheet

As at October 7, 1918.

ASSETS

United States National Bank—General Fund	$1,092.26
United States Liberty Bonds	500.00
Bond Interest Receivable	6.67
Club Room Furniture and Camp Equipment	900.00
	$2,498.93

LIABILITIES

Surplus	$2,498.93
	$2,498.93

Portland, Oregon, October 15, 1918.

To the Members of
THE MAZAMA COUNCIL,
Portland, Oregon.

DEAR SIRS:

In accordance with your instructions I have audited the accounts of the Mazamas for the fiscal year ended October 7, 1918, and present herewith my report. The accompanying Income and Profit & Loss Account shows the transactions for the period under review, which resulted in a profit of $48.91, and the Balance Sheet reflects the financial condition of the Club as at October 7, 1918.

I reconciled the funds in the custody of the Treasurer with the bank statement, and received a certificate from the bank verifying the balance.

The various invoices, cancelled checks and statements and accounts of the Treasurer and the several committees were examined and found to be in order.

Yours truly,

ROBERT F. RISELING,
Certified Public Accountant.

Roll of Honor

CHARLES E. ATLAS,

Second Lieutenant, U. S. Air Service,
279 Aero Squadron,
American Expeditionary Forces.

FRANK S. BAGLEY,

66th Aero Squadron,
American Expeditionary Forces.

BYRON J. BEATTIE,

Naval Operating Base, Reserve Officers' School
Hampton Roads, Virginia.

LEE BENEDICT,

Co. F, 12th Infantry,
Camp Fremont, California.

ROSCOE G. BIGGS,

Receiving Ship, Puget Sound, Washington.

CLEM E. BLAKNEY,

659th Aero Squadron,
American Expeditionary Forces.

W. E. BODWAY,

Yeoman, 55 Vittoris Emmanuel III,
care Postmaster, New York, N. Y.

GUSTAVE BROCKMAN,

Co. G, 13th Infantry,
Camp Fremont, California.

ALBERT S. BROWN,

Camp Taylor, Kentucky.

DAVID CAMPBELL,

Camp Hancock, Ga.

GEORGENE M. CASE,

Base Hospital, Camp Lewis, Washington.

K. P. CECIL,

First Lieutenant, 65th Artillery, C. A. C.,
American Expeditionary Forces.

WALTER E. CHURCH,

Corporal, Hdqrs. 2nd Bat., 65th Artillery, C. A. C
American Expeditionary Forces.

ARTHUR M. CHURCHILL,

Camp Taylor, Kentucky.

WILLIAM D. CLARK,

M. S. E. 404th Squadron, A. S. S. C.,
Vancouver Barracks, Washington.

ARTHUR COOK,

Sargeant, Headquarters 37th Engineers,
Fort Myer, Virginia.

D. J. CONWAY,

153d Aero Squadron, American Air Service,
American Expeditionary Forces.

RAYMOND T. CONWAY,

Corporal, Co. L, 157 Infantry, 40th Division,
American Expeditionary Forces.

NELSON ENGLISH,

37th Engineers, 1st Battalion Headquarters,
American Expeditionary Forces.

W. W. EVANS,

Corporal, Q. M. Warehouse No. 1, American P. O. 727,
American Expeditionary Forces.

J. N. FLESHER,

2nd Division, First Battalion,
166th Depot Brigade,
Camp Lewis, Washington.

W. P. FORMAN,

Corporal, Co. B, 158th Inf., American P. O. 729,
American Expeditionary Forces.

FORREST L. FOSTER,

Sergeant, M. P. E. S., American P. O. 729,
American Expeditionary Forces.

HERBERT J. FOSTER

Co. D, 117th Engineers,
American Expeditionary Forces.

GEORGE GARRETT

A. C. S. Engineers' Section, American P. O. 714,
American Expeditionary Forces.

A. H. S. HAFFENDEN,

Second Lieutenant, F. A. N. A., American P. O. 710,
American Expeditionary Forces.

E. R. HAWKINS,

Lieutenant, Ordnance Supply Depot,
Springfield, Mass.

CHAS. S. HELFRICH,

105 Spruce Squadron, Portland, Oregon.

E. G. HENRY,

2nd Lieut., Barracks 2, Park Field, Memphis, Tenn.

CLARENCE HOGAN,

Co. B, 2d A. A. P.,
Fort MacArthur, California.

RALPH S. IVEY,

5th Co., Casualty Detachment, 41st Division,
Camp Hill, Newport-News, Virginia.

HENRY G. JOHNSON,

Lieutenant, care Balloon School, Fort Omaha, Neb.

F. G. KASH,

Sergeant, 8th Company, Oregon Coast Artillery,
Fort Stevens, Oregon.

EMMA B. KERN,

R. N. A. N. C., U. S. A., Base Hospital No. 46
American Expeditionary Forces.

HENRY A. LADD,

U. S. A. Base Hospital, Unit No. 46
American Expeditionary Forces.

ELMER LERDELL,

Vancouver Barracks, Washington.

F. P. LUETTERS,

Co. C, 147th Infantry, American P. O. 763,
American Expeditionary Forces.

W. T. LUND,

116th Engineers' Detachment, U. S. P. S. 701,
American Expeditionary Forces.

DR. J. W. McCOLLOM,

Lieut., 1st U. S. Infantry, Camp Lewis, Wash.

F. H. McNEILL,

War Risk Section, American P. O. 717,
American Expeditionary Forces.

J. D. MEREDITH,

U. S. A. Base Hospital, Unit No. 46,
American Expeditionary Forces.

S. MILES,

437th Engineers' Depot Detachment,
Washington, D. C.

JESSE MILLER,

U. S. S. South Dakota, care Postmaster, New York N. Y.

ALAN BROOKS MORKILL,

Lieutenant, 7th Battalion,
Canadian Expeditionary Forces.

JOHN F. NALLY,

Corporal 19th Spruce Squadron, B. A. T.,
Vancouver Barracks, Washington.

B. W. NEWELL,

Sergeant, 65th Field Artillery, C. A. C., care Supply Co.,
American Expeditionary Forces.

W. D. PEASLEE,

Captain, 537th Engineers,
Camp Travis, San Antonio, Texas.

ALFRED F. PARKER,

Receiving Ship, Puget Sound, Washington.

P. G. PAYTON,

338 Aero Service Squadron,
American Expeditionary Forces.

ARTHUR D. PLATT,

First Student Co., S. O. T. C.,
Camp Mead, Maryland.

GEORGE X. RIDDELL,

Captain, Co. 7, E. O. T. C.,
American Expeditionary Forces.

CECIL V. REDDEN,

Sergeant, 8th Spruce Squadron,
Vancouver Barracks, Washington.

STANLEY C. RICHMOND,

8th Squadron, S. P. D., B. A. P.,
Vancouver Barracks, Washington.

C. H. SAKRISON,

Co. H, 76th Infantry, Camp Lewis, Washington.

E. C. SAMMONS,

Lieut.-Colonel, 137th Infantry,
American Expeditionary Forces.

ALFRED C. SHELTON,

Base Hospital, Camp Lewis, Washington.

DR. J. DUNCAN SPAETH,

Camp Jackson.

GEO. A. STUDER,

Co. F, 12th Infantry,
Camp Fremont, California.

J. MONROE THORINGTON,

American Ambulance Corps,
2031 Chestnut St., Philadelphia, Pa.

F. P. UPSHAW,

Ensign, U. S. Naval Training Camp, Seattle, Wash.

D. VAN ZANDT,

Corporal 157th Aero Squadron.

WM. S. WALTER,

Lieut., American Expeditionary Forces.

B. F. WENNER,

Vancouver Barracks, Washington.

RONALD M. WILSON,

Second Lieut., care Rev. Wilson, Dover, Mass.

MARY WING,

General Delivery, Bremerton, Washington.

HARRY L. WOLBERS,

Corporal, Co. C, 322d Field Signal Battalion,
American Expeditionary Forces.

JOHN S. WARD,

Co. 1, S. O. T. S., Camp Meigs, Washington, D. C.,
S. M. C.

Y. M. C. A. WORKERS

R. H. ATKINSON,

care Y. M. C. A., Camp Lewis, Wash.

ROBERT D. SEARCY,

Paris, France.

LINN L. REIST,

1322 Bryant Ave., North,
Minneapolis, Minnesota.

Y. M. C. A. CANTEEN WORKERS

ENID C. ALLEN,

12 Rue d'Agguesseau, Paris, France.

EVA BRUNELL,

LAURA HATCH,

12 Rue d'Agguesseau, Paris, France.

MARY C. HENTHORNE,

12 Rue d'Agguesseau, Paris, France.

MARY C. JACOBS,

RED CROSS WORKERS

JESSIE RAY NOTTINGHAM,

Barnard Unit American Red Cross,
No. 4 Rue De l'Elysee, Paris, France.

MRS. M. R. PARSONS,

Hotel de Richelieu,
Mont de Marsam, Landef, France.

Book Reviews

Edited By MINNIE R. HEATH.

"SUNSET CANADA." Early historical events, including the life and customs of the aborigines and possessions of a vast domain, by the Hudson Bay Company, are described in an interesting style by Archie Bell, author of "Sunset Canada."

Several chapters are devoted to the principal industries, such as mining, lumbering, fishing, agriculture and a description of the larger industrials centers, giving an idea of the potential wealth and future possibilities of Western Canada.

The volume is valuable for its general information, but its strongest appeal will be to the prospective tourist and all lovers of the "out of doors."

Many of the illustrations could well be classed as works of art, and the descriptions of the numerous scenic points, of which British Columbia and the Canadian Rockies are justly famous, were written by a true lover of the mountains.

Taken throughout the book is well worth while and will prove a valuable addition to the "See America First Series."

W. W. ROSS.

BELL, ARCHIE. Sunset Canada, British Columbia and Beyond. 1918. The Page Company, Boston. $3.50 net.

"ON THE HEADWATERS OF PEACE RIVER" Paul Leland Haworth is a man whom it has been impossible to extinguish by years of super-civilization, the instinctive lure of the primitive and desire to penetrate the unknown. A student, author and teacher by training and environment, he has chosen for a time to penetrate with one companion, "a region beyond the farthest campfire and the last tin can." He has given us, in his book, the story of this trip, by canoe and packsack, into one of the few remaining unexplored regions of our continent.

The titling of the book, "On the Headwaters of Peace River," indicates the scope of his trip. The Peace River is formed by the junction, in northern British Columbia, of the Parsnip River and the Finlay. The Parsnip is a well known stream, having been used as a highway for years by trader, trapper and prospector. The course of the Finlay, too, is fairly well known on its lower reaches. The Upper river, however, as well as some of its branches, notably the Quadacha and the Fox, and the territory drained by these streams, is practically unknown. It is this unknown region that the author intended to explore when he set out by canoe from the head of the Parsnip River.

Down the Parsnip they progressed with little effort, paddling leisurely along in quiet reaches and running swiftly down its "wagon roads" as the cleared channels in the rapids were locally termed. At Finlay Forks, the juncture of the Parsnip and the Finlay, the character of their trip changed.

Up the Finlay their progress was necessarily slow. Poling and lining the canoe up the stream, which was too swift for the use of paddles, jeopardizing both canoe and contents in many dangerous traverses, they fought their way steadily up to the mouth of the Quadacha.

Here canoe and outfit were cached, and burdened with packs and armed with rifle and camera, they set out afoot to explore the Quadacha country. The extent of this trip was limited by their failure to find game on which they had depended for a part of their subsistence. For this reason they were turned back before reaching the three-peaked mountain and the great glacier in which the Quadacha originates. Later the mountain and glacier were observed from a peak they ascended on their return to the cache. Thus was settled between the two pioneers the often discussed question, "What makes the Quadacha white?"

Returning to the cache they proceeded somewhat farther up the Finlay and again left the canoe and set out with heavy packs, following the range that forms one wall of the Fox River valley. After following this course for some distance they crossed the range and were again in the Valley of the Finlay in a district in which they hoped to find game. In this hope they were not disappointed and spent several days in hunting and exploration. The return to their canoe was made by raft down the long canyon, as this portion of the Finlay is known.

The final stage of their journey is here begun in the descent of the Finlay and Peace Rivers. At Hudson's Hope, on the Peace, they disposed of the canoe and completed the trip by gasoline launch to the railroad.

The trip thus briefly outlined, was rich in the variety and extent of its experiences. The book is equally rich in the material it offers the reader.

To the angler, the big game hunter and the canoeist it will equally appeal as well as to those interested from a technical standpoint in the distribution of animal life and origin of species. The native Indians or "Siwash" were everywhere observed and their customs and habits frequently commented upon. It is full of well chosen anecdotes, illustrating the quality of the pioneers in the less remote regions. The possibilities of the country for future development and settlement is given thorough consideration. In common with every new country this region has witnessed the rise and fall of its boom town. This is especially true along the newly completed Grand Trunk Pacific, by which the author reached the starting point of his trip. These towns and the various factors which controlled their rise and fall, are vividly pictured by the author.

For us, interested as we are in the mountaineering possibilities afforded by the region, the book presents a view of unclimbed peaks, unexplored glaciers and vast ranges of mountains, that make us long for the opportunity to follow the trail of the author into this wonderful country.

HAROLD S. BABB.

HAWORTH, PAUL LELAND, "On the Headwaters of Peace River." Charles Scribner's Sons., N. Y. 1917.

"THE MELODY OF EARTH" An Anthology of Garden and Nature Poems from present day poets, selected and arranged by Mrs. Waldo Richards, contains poems by 182 singers, who have listened and caught the harmony of nature. The poems are arranged in twelve groups, under appropriate titles, beginning with "Within Garden Walls," and ending with "The Garden of Life," and among the poets represented are Bliss Carman, Josephine Peabody, Yeats, Angela Morgan, Edwin Markham, Clinton Scollard, Richard Burton, James Whitcomb Riley, Ramindranath Tagore and James Wheelock, who sings in "Earth."

"All we say and all we sing
Is but as the murmuring
Of that drowsy heart of hers
When from her deep dream she stirs:
If we sorrow, or rejoice,
You and I are but her voice."

CHRISTINE N. MORGAN.

RICHARDS, MRS. WALDO. "The Melody of Earth." Houghton-Mifflin Co.

"THE CRUISE OF THE CORWIN" The author's own journal of the Arctic Expedition sent out by the U. S. Government to seek the Jeannette and missing whalers lost in the ice off Point Barrow. There is an introduction by William Frederic Bade, and an appendix giving the Glaciation of the Arctic and Subarctic Regions, and Botanical Notes.

In "The Cruise of the Corwin," Mr. Muir tells of the difficulties encountered in their attempt to reach Wrangell Land, where it was expected they would find some record left by the Jeannette. The book not only tells of the glaciation and tundra and fauna of the Northland, in which Mr. Muir was particularly interested, but also of the natives, their life, habits and manner of abode. It is full of personal interest to those who are familiar with the Arctic, as well as to those who do not know it.

MARY L. KNAPP.

MUIR, JOHN. "The Cruise of the Corwin." 1917. Houghton-Mifflin Co. $2.75 net.

"FINDING THE NORTH WHILE IN THE SOUTHWEST" The author of this new book places the reader in intimate association with the ins and outs of the Southwest and imparts the why and wherefore of all the interesting corners of this district. It is historically valuable.

L. E. ANDERSON.

SAUNDERS, CHARLES FRANCIS. "Finding the North While in the Southwest." 1918. Robert McBride & Co. $1.25 net.

"OUR NATIONAL FORESTS" The author uses in the introduction to his subject, "Our National Forests," this sentence—"Its grandeur makes you love it; its vastness makes you fear it; yet there is an irresistible charm, a magic lure, an indescribable something that stamps an indelible impression upon the mind and that makes you want to go back there after you have sworn an oath never to return."

He is describing our west and the appeal is strong, for Mazama hikers know many of the national forests from actual contact.

The book deals with the natural resources of the forests, comprising 155,000,000 acres, in a comprehensive manner. Statistics are presented in interesting, readable form and the government disposition and management of these resources is attractively presented.

The author has condensed, under one cover, the various phases of forestry work. In his introduction, discussing forestry as a National problem, he treats of such subjects as—our consumption of wood, the lumber industry, forests and stream flow, how the government obtained the National Forest lands, their extent and character, the romance of the National Forest region, famous scenic wonders near the forests, the topography and climate of the National Forest region, why the forests were created, how the National Forest policy has benefitted the people, financial returns and the new eastern National Forests.

The contents of the book are such as to interest the casual reader or satisfy one more intent upon research work. It is illustrated by many half-tone cuts made from original photographs.

Our National Forests are attractively presented, with sufficient data to make the volume valuable as a book of reference.

MINNIE R. HEATH.

BOERKER, RICHARD H. "Our National Forests." 1918. The MacMillan Co. $2.50.

"TENTING TONIGHT" "Tenting Tonight," by Mary Roberts Rinehart, is a very entertaining account by that well known writer of two expeditions into some of the more inaccessible spots of the northwest. It is first a recital of her experiences in taking herself, her family, some eight packers, guides, cooks and thirty-one pack horses—not to forget two boats—through the western and practically unknown side of Glacier Park, up to the Canadian border. In the course of this trip they run the rapids of the North Fork of the Flathead River, according to the author, an unheard of feat, and to her "a prolonged four days' gasp."

In the second expedition they crossed the Cascades from the Chelan country to Puget Sound, by way of Cascade Pass, over an absolutely virgin trail—a trip abounding in thrills. Characteristic and mirth-provoking dissertations on the relative vices and virtues of bough beds, the terror caused by the untimely defection of the cook and a plaint as to the positive tortures of down hill going enliven the account.

MARY GENE SMITH.

RINEHART, MARY ROBERTS. "Tenting Tonight." 1918. Houghton-Mifflin Company.

"THE HUMAN SIDE OF ANIMALS" Portrays in an interesting, readable manner the human-like qualities of animals. The intelligence, the reason, the method of protection, the government, the industries, the musical ability, the play and the help to man are all compared favorably with the faculties possessed by civilized man along these lines. The author finds that animals are possessed of love, hate, joy, grief, courage, pain, revenge, pleasure, want and satisfaction. The last chapter deals entirely with proofs showing that the author is convinced that animals have as much hope of a future life as man. The illustrations in color and the photographs are well in keeping with the text in their effort to portray the animals from a friendly point of view. On the whole the book is one to attract an interest in and even to create lovers of animals.

OLGA HOLLINGSBY.

DIXON, ROYAL. "The Human Side of Animals." 1918. Frederick A. Stokes Co. $1.75 net.

"SIGN TALK" A very complete and well written dictionary of the Sign Language, which Mr. Seton claims is the oldest, most universal, and most easily understood language ever introduced. It is still used to a great extent by the Indians when talking to members of other tribes.

"Seeing is believing," is the fundamental basis of "Sign Talk," for by signs the same idea is expressed, no matter what words are spoken. If a person be in Berlin or Timbucktoo, no matter what language he speaks, he will use the same signs to express the same thoughts as another person from a different part of the globe who speaks an entirely different language.

The book is profusely illustrated and a study of it should prove interesting and very helpful to all travelers and lovers of Indian lore.

GEORGE MEREDITH.

SETON, ERNEST THOMPSON. "Sign Talk." 1918. Doubleday, Page & Co. $3.00.

"CAMPING AND WOODCRAFT" This convenient sized book is one which every hiker should read whether he intends to leave the beaten tracks of the forests or not. It is intensely interesting while being practical and instructive.

Volume One was reviewed in the 1916 Mazama Magazine, and the present volume (Number Two) is a confirmation of the anticipated pleasure which the previous number aroused.

Traveling with the author, one learns to extricate himself from the many difficulties encountered when lost in a strange forest, or when one's camping equipment has been destroyed by some disaster. Under his advice one learns to know how to find and prepare food when remote from civilization. With him, one builds a shelter, and, in fact, "lives off the country" quite happily. It is a book to own as well as to read.

M. R. H.

KEPHART, HORACE. "Camping and Woodcraft. Vol. II. 1917. Outing Publishing Co., New York. Cloth $1.50 net. Leather $2.00 net.

"STEEP TRAILS" This, the last of John Muir's posthumous book, edited by William Frederic Bade, is a collection of letters and articles, written in the field, covering a period of nearly thirty years—from 1874 to 1903, during which time they appeared mostly in local publications. The papers are arranged in chronological order, making a book of nearly 400 pages, divided into twenty-four chapters, and with a dozen illustrations.

The volume opens with "Wild Wool," a delightful bit, giving us a glimpse of the bighorn sheep of the Sierra Nevada, and describing the "love work of nature" in caring for her many bairns of clothing them with "smoothly imbricated feathers, shining jackets and shaggy furs. The squirrel has socks and mittens, and a tail broad enough for a blanket; the grouse is densely feathered down to the end of his toes; and the wild sheep, besides his undergarment of fine wool, has a thick overcoat of hair that sheds off both the snow and the rain."

The chapters on Mt. Shasta tell of rambles in that region, describing the woods and the wild life; also the lava beds, made famous by the Modoc war, with a vivid account of an ascent of the mountain for barometrical observations, when the party was overtaken by a storm and spent the night on top "midst the hissing, spluttering fumaroles."

In "Steep Trails" we wander through Utah and the Nevada dead towns and the pine forests where the Indians of that region harvest the nuts for their winter food; on to the great Northwest, with observations on the Puget Sound country, with its lumber industries, interminable forests; once more wending upwards on the rise of Mt. Rainier; from thence into the Oregon country, with descriptions of its picturesque coast, through beautiful woodlands, filled with the singing of innumerable streams. Here we find an interesting account of David Douglas in the Umpqua Hills—the Douglas for whom the noble Douglas spruce is named—gathering specimens and exciting the curiosity of the Indians as he wandered through the woods.

The volume closes with a wonderful description of the Grand Canyon, depicting it in one place as a "collection of stone books covering thousands of miles of shelving—myriad forms of successive floras and faunas, lavishly illustrated with colored drawings, carrying us back into the midst of the life of a past infinitely remote."

All lovers of Nature's work will be charmed by reading these accounts of the life out-of-doors—those days that enrich one's life and of which Muir had so many; the thrilling experiences one has when Nature frowns; but the "night will wear away and tomorrow we go a-Maying, and what campfires we will make, and what sun-baths we will take."

CHRISTINE N. MORGAN.

MUIR, JOHN. "Steep Trails." Houghton-Mifflin Co., Boston. $3.00.

"A GUIDE TO THE NATIONAL PARKS OF AMERICA" A comprehensive compilation of detailed information, useful and necessary to the traveler visiting the national parks. Viz: Yellowstone, Crater Lake, Mt. Rainier, Glacier, Yosemite, Sequoia, Mesa Verde, Rocky Mountain, as well as the Grand Canyon of Arizona and the Hot Springs of Arkansas. Also notes on the Hawaiian, Mt. McKinley, Lassen and other national parks and monuments and on Canadian national parks. Contains accurate data on transportation to and in the parks, accommodations, regulations and privileges, distances and elevations, gives definite touring directions and describes features of interest. "Our national parks are the greatest and most individual recreation grounds that any American may visit, whether he cross the water or not."

FRANK JONES.

ALLEN, EDWARD FRANK, Editor of "Travel;" revised and enlarged edition with illustrations and four maps. 1918. Robert M. McBride & Co., Union Square, New York. $1.25 net.

"A GUIDE TO THE WHITE MOUNTAINS" This revised edition of Sweetzer's White Mountain Guide is a regional handbook of the White Mountains, which furnishes sufficient data to make clear the details needed by an enthusiastic hiker or climber.

It is of convenient pocket size with accompanying maps and will prove interesting and companionable, as well as decidedly useful to any one visiting the White Mountains.

MINNIE R. HEATH.

SWEETZER, M. F. "A Guide to the White Mountains." 1918. Edited and revised by John Nelson. Houghton-Mifflin Company. $2.75 net.

"THE BIRD STUDY BOOK" Every year more people become interested in out-of-door things. Bird study has become popular because children are taught in the schools to love birds, not only for their beauty, but because of their economic value to the state and nation.

It is not difficult for those versed in ornithology to find books for the classification of birds, but so often the inquiry is made for a good beginner's book. This book meets the requirement for both old and young. It tells in simple language what to look for and where to find it. It is an introduction to the feathered people of the field and forest. It is a guide to the traits and character of birds, which, if followed, will lead to those bird companionships which will make life happier.

Mr. Pearson has long been a lover of bird life. For years, he has been the chief executive of the National Association of Audubon Societies, the organization that has done more than any other to bring birds into our homes by systematic teaching among the schools of the nation. He was more fitted to write this book than any one else. The book is of great value in bird study because it is a part of his life.

WILLIAM L. FINLEY.

PEARSON, T. GILBERT. The Bird Study Book. 1918. Doubleday, Page & Company.

"VOYAGES ON THE YUKON AND ITS TRIBUTARIES" It covers the field thoroughly from every point, as only a writer could who has spent a great part of his life there, living intimately with the people and knowing them and the whole country which is treated in this late book.

L. E. ANDERSON.

STUCK, HUDSON. "Voyages On the Yukon and Its Tributaries." 1917. Scribners, New York.

"DESCRIPTION..OF AND GUIDE TO JASPER PARK" In the booklet of the above title we are shown how the Canadian government makes known to the nature-lover the attractions of its national parks. The book is issued by the Department of the Interior, at Ottawa, Canada.

Jasper Park is in Alberta Province, the central part of it (with the little town of Jasper as the hub) being located about 200 miles west of Edmonton. The park covers about 4400 square miles, and it is reached by two transcontinental railways that enter by way of the valley of the Athabasca River.

Before compiling the booklet (Guide) the department had a photographic survey made of the central part of the park, and the results are shown in a splendid topographic map consisting of six sheets which accompanies the Guide. The treatise is divided into several chapters, including the location and early history of the park, a general description of the park, with special chapters devoted to outlines of possible trips in every direction from the principal towns. There is even included an appendix giving a condensed schedule of all possible trips requiring from only an hour to several days. The numerous mountains in the park are done full justice with illustrations, representations on the maps, and are arranged in a special alphabetical list, giving location, elevation, etc.

It will be seen from the foregoing that the Canadians are most up-to-date in providing the intending tourist with succinct and easily referred-to descriptions and data covering their national parks. A feature that must appeal strongly to the genuine seeker of knowledge is the early history of the park region. We read in this guide of the explorers, fur-traders, naturalists and others who traversed the present park long ago, as early as 1810. Noted characters in the early history of the provinces are mentioned, and extracts of narratives written by them are quoted freely. It may be explained that the Hudson's Bay Company and other fur-trading companies then operating in the Canadas and the Pacific Northwest had to establish lines of communication between Hudson Bay and the Pacific Ocean. The discovery of Athabasca Pass in 1811 opened a new route to the trading posts along the Columbia River, and that of Yellowhead Pass in 1826 opened a route afterwards followed by transcontinental railways. The use of both of the passes involved passage through the present park.

We read in the Guide of how, a century ago, the trading brigades, twice a year, started from Edmonton, passed through the park and on to the headwaters of the Columbia, thence down the latter to our present Vancouver, near Portland. The voyage along the Columbia and a portion of the Athabasca was made in boats, but the rest of the way by pack-horses. Twice a year, in each direction, this voyage was made, each occupying about three and one-half months and involving an amount of hardship and toil not appreciated by the present generation.

We also read of the founding of Astoria at the mouth of the Columbia in 1811 by John Jacob Astor, mentioned by Washington Irving in his book "Astoria" as causing such a sensation in this vast wilderness at that time. The miscarriage of Astor's plans, by which the new post of Astoria had to be virtually surrendered to the British in 1813, is related. The next year, the annual spring brigade which passed up the Columbia, bound for the Hudson Bay region (via our park), consisted of men, women and children of many nationalities—described as "an extraordinary collection of human beings."

All in all, the book under review is an admirable exponent of the right method to follow in presenting to intelligent persons the features of the scenic regions that they may contemplate visiting. The thorough methods used and the manner of presentation may largely account for the partiality shown by men of science, scholars and nature-lovers generally for Canadian scenery.

W. P. HARDESTY.

"Description of and Guide to Jasper Park." Department of the Interior, Ottawa. 1917.

"PRACTICAL BAIT CASTING" A book intended for the beginner and one to induce the inexperienced to become a proficient bait casting fisherman. It will also prove of much value to those of more experience.

The different kinds and qualities of rods, reels and lines and how and when to use each is well discussed. Live and artificial baits are covered in detail.

The closing chapters are devoted to all kinds of waters, the fish to be found therein and the best method and time to be successful in obtaining a good catch under varying conditions.

ED. PETERSON.

ST. JOHN, LARRY. "Practical Bait Casting." 1918. The MacMillan Company, $1.00. Illustrated.

"CAMPING OUT" This book carries information as to every possible phase of camp life. It is the tested result of thirty years practical experience in camping out and is intended to meet the needs of the out-of-doors enthusiast under any and all circumstances. Whether it be a hunting, fishing or canoe trip, a horseback jaunt or auto trip, a beach hike or a pack into the mountains, one will find a wealth of valuable information which if heeded would enable any one of these to be done with the maximum of comfort and the minimum of effort.

It is designed to meet all the exigencies of weather, heat, rain, snow or cold, insect life, cooking and shelter. It contains details as to every kind of equipment, packs, clothing, tents, etc. A chapter on camp cooking and one on wilderness guideposts contain many valuable hints. No possible detail is neglected from camping de luxe to the lone hike with the barest necessaries. Not the least important—to quote the author—is "how to sleep warm without loading oneself like a furniture van."

Best of all it is not a commercial product but a very reliable account of pleasure trips taken by the author in company with Joan, the Kid, Micky, Doc or some other of his particular pals.

MARY GENE SMITH.

MILLER, WARREN H. "Camping Out." 1918. Geo. H. Doran Company, New York.

A Book

He ate and drank the precious words,
His spirit grew robust;
He knew no more that he was poor,
Nor that his frame was dust.

He danced along the dingy days;
And this bequest of wings
Was but a book. What liberty
A loosen'd spirit brings!

—Emily Dickinson.

Mazama Membership List, November 1, 1918

ACTON, HARRY W., 519 W. 121st St., New York, N. Y.

ACTON, MRS. HARRY W., 519 W. 121st St., New York, N. Y.

ADAMS, DR. W. CLAUDE, 1010 E. 28th St. N., Portland, Oregon.

AITCHISON, CLYDE B., 504 Real Estate Trust Bldg., Washington, D. C.

AKIN, DR. OTIS F., 919 Corbett Bldg., Portland, Oregon.

ALLARD, NAN F., Foot of Miles St., Portland, Oregon.

ALLEN, ARTHUR A., care Portland Rowing Club, Portland, Oregon.

ALLEN, ENID C., 917 Andrus Bldg., Minneapolis, Minn.

ALMY, LOUISE A., Box 426, Dillon, Montana.

AMOS, DR. WM. F., 1016 Selling Bldg., Portland, Oregon.

†ANDERSEN, DR. FREDERICK, Merwyn Bldg., Astoria, Oregon.

ANDERSON, LeROY E., 213 Northwestern Bank Bldg., Portland, Ore.

ANDERSON, WM. H., 4464 Fremont Ave., Seattle, Washington.

APPLEGATE, ELMER I, Klamath Falls, Oregon.

ASCHOFF, ADOLF, Marmot, Ore.

ASCHOFF, OTTO, Marmot, Ore.

ATKINSON, R. H., 1028 E. Washington St., Portland, Oregon.

ATLAS, CHAS. E., care Hurley Mason Co., Tacoma, Wash.

AVERILL, MARTHA M., 1144 Hawthorne Ave., Portland, Oregon.

AYER, ROY W., 689 Everett St., Portland, Oregon.

AYER, LEROY, JR., Chawfordsville, Oregon.

BABB, HAROLD S., 578 Miller St., Portland, Oregon.

BACKUS, LOUISE, 122 E. 16th St., Portland, Oregon.

BACKUS, MINNA, 122 E. 16th St., Portland, Oregon.

BAGLEY, FRANK S., Portland, Ore.

BAILEY, A. A., JR., 644 East Ash St., Portland, Oregon.

BAILEY, VERNON, 1834 Kalorama Ave., Washington, D. C.

BALLOU, O. B., 80 Broadway, Portland, Oregon.

BALMANNO, JACK H., 5508 50th Ave. S. E., Portland, Oregon.

BANFIELD, ALICE, 570 East Ash St., Portland, Oregon.

BARCK, DR. C., 205-207 Humboldt Bldg., St. Louis, Mo.

BARNES, M. H., 657 Schuyler St., Portland, Oregon.

BARRINGER, MAUDE, 5th and M Sts., Fredonia, Kansas.

BATES, MYRTLE, 448 E. 7th St., Portland, Oregon.

BEATTIE, BYRON J., 830 Rodney Ave., Portland, Oregon.

BELL, MISS HALLIE, 330 E. 9th St. N., Portland, Oregon.

BENEDICT, LEE, 185 E. 87th St. N., Portland, Oregon.

BENEDICT, MAE, 185 E. 87th St., N., Portland, Oregon.

BENTALL, MAURICE, General Delivery, Hathaway, Montana.

BERG, MRS. G. ALBERT, 304 So. 6th St., Marshalltown, Iowa.

BIGGS, ROSCOE G., 547 E. Pine St., Portland, Oregon.

BISSELL, GEO. W., 223 W. Emerson St., Portland, Oregon.

BLACKINGTON, PAULINE, 32 Jeffersonian Apts, Portland, Ore.

BLAKNEY, C. E., R. F. D. No. 2, Box 151, Milwaukie, Ore.

BLUE, WALTER 1306 E. 32d St. N., Portland, Oregon.

Bodway, W. P., General Delivery, Portland, Oregon.

BORNT,, LULU ADELE, 641 E. 13th St. S., Portland, Oregon.

BOWERS, NATHAN A., 501 Rialto Bldg., San Francisco, Cal.

BOWIE, ANNA, 297 E. 35th St., Portland, Oregon.

BOYCE, EDWARD, 207 St. Clair St., Portland, Oregon.

BOYCHUK, WALTER, 174 Meade St., Portland, Oregon.

BRENNAN, THERESA, 380 Tenth St., Portland, Oregon.

BROCKMAN, GUS, 15 E. 15th St., N., Portland, Oregon.

BRONAUGH, JERRY E., Gasco Bldg., Portland, Oregon.

BRONAUGH, GEORGE, 350 North 32d St., Portland, Oregon.

BROWN, ALBERT S., Portland, Ore.

BROWN, G. T., 500 E. Morrison St., Portland, Oregon.

BRUNELL, EVA, 18th and Abbott St., Woscester, Mass.

BULLIVANT, ANNA, 269 13th St., Portland, Oregon.

BUSH, J. C., 683 E. Morrison St., Portland, Oregon.

CALHOUN, MRS. HARRIET S., 38 Deleware Ave., Detroit, Mich.

CALLWELL, CHARLOTTE, 309 San Rafael St., Portland, Oregon.

CAMPBELL, GRACE, 415 Tenth St., Portland, Oregon.

CAMPBELL, DAVID, Monmouth, Oregon.

CAMPBELL, P. L., 1170 13th Ave., E., Eugene, Oregon.

CARL, MRS. BEULAH MILLER, 629 Ash St., Portland, Oregon.

CASE, GEORGENE M., Foot of Miles St., Portland, Oregon.

CECIL, K. P., U. S. Forest Service, Portland, Oregon.

CHAMBERS, MARY H., 729 11th Ave., E., Eugene, Oregon.

CHASE, J. WESTON, 216 Lumbermen's Bldg, Portland, Oregon.

CHENOWETH, MAY, 104 E. 24th St., N., Portland, Oregon.

CHRISTIANSON, WM. D., 134 Colburn St., Brantford, Ontario.

CLARK, WM. D., Vancouver Barracks, Washington.

CHURCH, WALTER E., 1170 13th Ave., E., Eugene, Oregon.

CHURCHILL, ARTHUR M., 1229 N. W. Bank Bldg., Portland, Oregon.

CLARK, HOMER J., 706 Glisan St., Portland, Oregon.

COLBORN, MRS. AVIS EDWARDS, 355 Prospect Ave., Buffalo, N. Y.

COLBY, F. H., 14th and Jefferson Sts., Portland, Oregon.

COLLINS, W. G., 510 32d Ave., S., Seattle, Washington.

*COLVILLE, PROF. F. V., care Dept. of Agriculture, Washington, D. C.

CONNELL, DR. E. DeWITT, 628 Salmon St., Portland, Oregon.

CONWAY, D. J., 4705 60th St., S. E., Portland, Oregon.

CONWAY, RAYMOND T., 4705 60th S. E., Portland, Oregon.

COOK, ARTHUR, 243 W. Park St., Portland, Oregon.

COOK, F. R., 430 E. 40th St., N. Portland, Oregon.

COURSEN, EDGAR E., 658 Lovejoy St., Portland, Oregon.

COURSEN, GERALDINE R., 658 Lovejoy St., Portland, Oregon.

COWPERTHWAITE, JULIA, Station E., Portland, Oregon.

COWIE, LILLIAN G., Wellesley Court, Portland, Oregon.

CRANER, HENRY C., P. O. Box 672, Los Angeles, Cal.

CREIGHTON, LOLA I., 920 E. Everett St., Portland, Oregon.

CROUT, NELLIE C., 1326 E. Tillamook St., Portland, Oregon.

CURRIER, GEO. H., Leona, Ore.

CURTIS, EDWARD S., 614 Second Ave., Seattle, Washington.

CUTTING, RUTH M., 4603 59th St., S. E., Portland, Oregon.

*DAVIDSON, PROF. G., San Francisco, Cal.

DAVIDSON, GRACE M., 10th and Montgomery Sts., Portland, Ore.

DAVIDSON, R. J., 458 E. 49th St., N., Portland, Oregon.

DAVIDSON, MRS. R. J., 458 E. 49th St., N., Portland, Oregon.

DAY, BESSIE, 690 Olive St., Eugene, Oregon.

*DILLER, PROF. JOS. S., U. S. G. S., Washington, D. C.

DILLINGER, ANNA C., 692 E. Ash St., Portland, Oregon.

DILLINGER, MRS. C. E., 692 E. Ash St., Portland, Oregon.

DOWLING, EUGENE H., 742 Belmont St., Portland, Oregon.

DOWLING, MRS. COLISTA M., 742 Belmont St., Portland, Oregon.

DUFFY, MARGARET C., 467 E. 12th St., Portland, Oregon.

ELLIS, PEARL, 454 E. 22d St., N., Portland, Oregon.

ENGLISH, NELSON, 267 Hazel Fern St., Portland, Oregon.

ERREN, H. W., 285 Ross St., Portland, Oregon.

ESTES, MARGARET P., 692 E. 43d St., N., Portland, Oregon.

EVANS, WM. W., 246 20th St., N., Portland, Oregon.

EWELL, ELAINE, 608 E. Taylor St., Portland, Oregon.

FAGSTAD, THOR, Cathlamet, Wash.

FARRELL, THOMAS G., 328 E. 25th St., Portland, Oregon.

FARRELLY, JANE, 1072 E. 29th St., N., Portland, Oregon.

FINLEY, WM. L., 651 E. Madison St., Portland, Oregon.

FISH, ELMA, 259 E. 46th St., Portland, Oregon.

FLEMING, MISS M. A., 214 Post Office Bldg., Portland, Oregon.

FLESHER, J. N., Carson, Wash.

FORD, G. L., 309 Stark St., Portland, Oregon.

FORMAN, W. P., Portland, Oregon.

FOSTER, FORREST L., 354 49th St., S. E., Portland, Oregon.

FOSTER, HERBERT J., 323 E. 162d St., New York, N. Y.

FRANING, ELEANOR, 549 N. Broad St., Galesburg, Ills.

FRIES, SAMUEL M., 691 Flanders St., Portland, Oregon.

FULLER, MARGARET E., 64 E. 67th St., Portland, Oregon.

GARDNER, BERNICE J., 2611 62d St., S. E., Portland, Oregon.

GARRETT, GEORGE, 646 Cypress St., Portland, Oregon.

GASCH, MARTHA M., 9 E. 15th St., N., Portland, Oregon.

GEBALLE, PAULINE, 689 Northrup St., Portland, Oregon.

GEORGE, MELVIN C., 616 Market Drive, Portland, Oregon.

GEORGE, LUCIE M., Rexford Apts., Portland, Oregon.

GILE, ELEANOR, 622 Kearney St., Portland, Oregon.

GILMOUR, W. A., Title & Trust Bldg., Portland, Oregon.

GIRSBERGER, MABEL R., care Modoc Lbr. Co., Chiloquin, Ore.

GLISAN, RODNEY L., 612 Spalding Bldg., Portland, Oregon.

GOLDAPP, MARTHA OLGA, 455 E. 12th St., Portland, Oregon.

GOLDSTEIN, MAX, 575 Third St., Portland, Oregon.

**GORMAN, MARTIN W., Forestry Bldg., Portland, Oregon.

GRAVES, HENRY S., U. S. Forestry Service, Washington, D. C.

*GREELEY, GENERAL A. W., Washington, D. C.

GREY, HARRY C., Shermerville, Ills.

GRIFFIN, MARGARET A., 303 Title & Trust Bldg., Portland, Ore.

GRIFFITH, B. W., 1736 Kane St., Los Angeles, Cal.

HAFFENDEN, A. H. S., 4236 49th Ave., S. E., Portland, Oregon.

HALLINGBY, OLGA, 767 E. Flanders St., Portland, Oregon.

HARDESTY, WM. P., 60 E. 31st St., N., Portland, Oregon.

HARDINGHAUS, EVELYN, 1026 Beulah Vista St., Portland, Ore.

HARNOIS, PEARLE E., 1278 Williams Ave., Portland, Oregon.

HARPER, IRA H., 2801 H Street, Vancouver, Wash.

HARPER, MRS. IRA H., 2801 H St., Vancouver, Wash.

HARRIS, CHARLOTTE M., 1195 E. 29th St., N., Portland, Oregon.

HARZA, L. F., Box 345, Jacksonville, Florida.

HAZARD, JOSEPH T., 4050 First Ave., N. E., Seattle, Washington.

HATCH, LAURA, 36 Bedford Terrace, Northampton, Mass.

HATHAWAY, WARREN G., Manette, Washington.

HAWKINS, E. R., 655 Everett St., Portland, Oregon.

HEATH, MINNIE R., 665 Everett St., Portland, Oregon.

HEDEN, PAUL F., 720 E. 22d St., N., Portland, Oregon.

HELFRICH, CHARLES S., Allentown, Penn.

HENDERSON, G. P., 1155 E. Yamhill St., Portland, Oregon.

HENRY, E. G., 488 N. 24th St., Portland, Oregon.

HENTHORNE, MARY C., Cor. E. 71st and Morrison Sts., Portland, Ore.

HERMANN, HELEN M., 965 Kerby St., Portland, Ore.

HEYER, A. L., JR., care Foundation Co., Tacoma, Wash.

HIGH, AUGUSTUS, 300 W. 13th St., Vancouver, Wash.

HILTON, FRANK H., 504 Fenton Bldg., Portland, Oregon.

HIMES, GEO. H., Auditorium, Portland, Oregon.

HINE, A. R., 955 E. Taylor St., Portland, Oregon.

HITCH, ROBERT E., 503 Fenton Bldg., Portland, Oregon.

HODGSON, CASPAR W., Rockland Ave., Park Hill, Yonkers, N. Y.

HOGAN, CLARENCE, 591 Borthwick St., Portland, Oregon.

HOLDEN, JAS. E., 1652 Alameda Drive, Portland, Oregon.

HOLMAN, F. C., 558 Lincoln Ave., Palo Alto, California.

HOLT, DR. C. R., 217 Failing Bldg., Portland, Oregon.

HORN, C. L., The Wheeldon Annex, Portland, Oregon.

HOWARD, ERNEST E., 1012 Baltimore Ave., Kansas City, Mo.

IVANAKEFF, PASHO, 246 Clackamas St., Portland, Oregon.

IVEY, RALPH S., R. F. D., Milwaukie, Oregon.

JACOBS, MARY B., 315 11th St., Portland, Oregon.

JAEGER, J. P., 131 6th St., Portland, Oregon.

JEPPESEN, ALICE, 891 Albina Ave., Portland, Oregon.

JOHNSON, FRED J., 275 Pine St., Portland, Oregon.

JOHNSON, H. C., 618 Nicollet, Minneapolis, Minn.

JOHNSTON, AMY, 545 E. 23rd St., N., Portland, Oregon.

JONES, F. I., 307 Davis St., Portland, Oregon.

JOYCE, ALICE V., O. A. C., Corvallis, Oregon.

KACH, F. G., Portland, Oregon.

KERN, EMMA B., 335 14th St., Portland, Oregon.

KERR, DR. D. T., 556 Morgan Bldg., Portland, Oregon.

KNAPP, MARY L., 656 Flanders St., Portland, Oregon.

KOERNER, BERTHA, 481 E. 45th St., N., Portland, Oregon.

KOOL, JAN, 1309 Yeon Bldg., Portland, Oregon.

KREBS, H. M., 245 E. Broadway St., Portland, Oregon.

KRESS, CHARLOTTE, 1026 Beulah Vista St., Portland, Oregon.

KRUSE, JOHANNA, Route A, Portland, Oregon.

KUENEKE, ALMA R., 869 Clinton St., Portland, Oregon.

KUNKEL, HARRIET, 405 Larch St., Portland, Oregon.

KUNKEL, KATHARINE, 857 Garfield Ave., Portland, Ore.

LADD, HENRY A., care Ladd & Tilton Bank, Portland, Oregon.

LADD, W. M., care Ladd & Tilton Bank, Portland, Oregon.

LANDIS, MARTHA, 2019 E. Main St., Portland, Oregon.

LANE, JOHN L., 410 Harrison St., Portland, Oregon.

LANE, MRS. JOHN L., 410 Harrison St., Portland, Oregon.

LAWFFER, G. A., 309 Stark St., Portland, Oregon.

LAWSON, AGNES G., 767 Montgomery Drive, Portland, Oregon.

LEADBETTER, F. W., 795 Park Ave., Portland, Oregon.

LEE, ALTHEA E., 4828 32d Ave. S. E., Portland, Oregon.

LEE, JOHN A., 505-6 Concord Bldg., Portland, Oregon.

LERDALL, ELMER, Vancouver Barracks, Wash.

LEPPICH, ELSA L., 733 Washington St., Portland, Oregon.

LETZ, JACQUES, State Bank of Portland, Portland, Oregon.

LIBBY, HARRY C., 1278 E. Taylor St., Portland, Oregon.

LIND, ARTHUR, care U. S. National Bank, Portland, Oregon.

LOUCKS, ETHEL MAE, 466 E. 8th St., N., Portland, Oregon.

LUETTERS, F. P., 689 Everett St., Portland, Oregon.

LUND, WALTER, 191 Grand Ave., N., Portland, Oregon.

LUTHER, DR. C. V., E. 34th & Belmont St., Portland, Oregon.

McARTHUR, LEWIS A., 561 Hawthorne Terrace, Portland, Ore.

McBRIDE, AGNES, 1764 E. Yamhill St., Portland, Oregon.

McCLELLAND, ELIZABETH, 267 Shawnee Path, Akron, Ohio.

McCOLLOM, DR. J. W., 553-7 Morgan Bldg., Portland, Oregon.

McCORKLE, J. F., Second and Oak Sts., Portland, Oregon.

McCOY, SALLIE E., 211 Lumbermen's Bldg., Portland, Oregon.

McCREADY, SUE O., Box 147 Etna Logging Co., Vancouver, Wash.

McCULLOCH, CHARLES E., 1410 Yeon Bldg., Portland, Oregon.

McDONALD, LAURA, 354 E. 49th St., S., Portland, Oregon.

McISAAC, R. J., Parkdale, Oregon.

McNEIL, FLORENCE, 607 Orange St., Portland, Oregon.

McNEIL, FRED H., care Journal, Portland, Oregon.

MacDOUGALL, CHARLOTTE, 661 Monroe St., Portland, Oregon.

MACKENZIE, WM. R., 1002 Wilcox Bldg., Portland, Oregon.

MAHONEY, HELENA C., 1238 Commonwealth Ave., Boston, Mass.

MAHONEY, PAUL, 1838 Commonwealth Ave., Boston, Mass.

MARBLE, W. B., 3147 Indiana Ave., Chicago, Ill.

MARCOTTE, HENRY D. D., 218 E. 56th St., Kansas City, Mo.

MARKHAM, B. C., 343½ Washington St., Portland, Oregon.

MARSH, J. W., 528 S. Ivanhoe St., Portland, Oregon.

MARSHALL, BERTHA, 1445 B St., San Diego, Cal.

MEARS, HENRY T., 494 Northrup St., Portland, Oregon.

MEARS, S. M., 721 Flanders St., Portland, Oregon.

MEREDITH, GEORGE, 8½ N. 11th St., Portland, Oregon.

MEREDITH, JOHN D., 329 Washington St., Portland, Oregon.

*MERRIAM, DR. C. HART, 1919 16th St., N. W., Washington, D. C.

MERTEN, CHAS. J., 307 Davis St., Portland, Oregon.

MILES, S., 32 E. State St., Albion, N. Y.

MILLER, ARCHIE J., Box 22, Enterprise, Ore.

MILLER, JESSE, 726 E. 20th St., Portland, Oregon.

MILLS, ENOS A., Longs Peak, Estes Park, Colo.

MONROE, HARRIET E., 1431 E. Salmon St., Portland, Oregon.

MONTAGUE, JACK R., 1310 Yeon Bldg., Portland, Oregon.

MONTAGUE, RICHARD W., 1310 Yeon Bldg., Portland, Oregon.

MORGAN, MRS. CHRISTINE N., 320 E. 11th St., N., Portland, Oregon.

MORKILL, ALAN BROOKS, 1971 Oak Bay Ave., Victoria, B. C.

NALLY, JOHN S., Washington, D. C.

NELSON, L. A., 410 Beck Bldg., Portland, Oregon.

NEWELL, BEN W., care Ladd & Tilton Bank, Portland, Oregon.

NEWTON, JOSEPHINE, 1350 Pine St., Philadelphia, Pa.

NIEHANS, MARGARET, 353 Harrison St., Portland, Oregon.

NICKELL, Anna, 410 Stanley Apts, Seattle, Washington.

NILSSON, MARTHA E., 320 E. 11th St., N., Portland, Oregon.

NISSEN, IRENE, 969 E. 23d St., N., Portland, Oregon.

NORDEEN, EDITH, 361 Graham St., Portland, Oregon.

NORMAN, M. OSCAR, 698 E. 62d St., N., Portland, Oregon.

NOTTINGHAM, JESSIE RAY, 271 E. 16th St., N., Portland, Oregon.

NUNAN, CINITA, 489 W. Park St., Portland, Oregon.

O'BRYAN, HARVEY, 602 McKay Bldg., Portland, Oregon.

**O'NEILL, MARK, Worcester Block, Portland, Oregon.

OGLESBY, ETTA M., Baron Apt., 14th and Columbia, Portland, Ore.

ORMANDY, HARRY M., 501 Weidler St., Portland, Oregon.

ORMANDY, JAMES A., 501 Weidler St., Portland, Oregon.

PARKER, ALFRED F., 374 E. 51st St., Portland, Oregon.

PARKER, ROSE F., Milwaukie, Oregon.

PARSONS, MRS. M. R., Mosswood Road, University Hill, Berkeley, California.

PATTULLO, A. S., 500 Concord Bldg., Portland, Oregon.

PAUER, JOHN, 485 E. 20th St., N., Portland, Oregon.

PAYTON, PERLEE G., 3916 64th St., S. E., Portland, Oregon.

PEARCE, MRS. LLEWELLYN C., 1016 Clackamas St., Portland, Ore.

PEASLEE, W. D., 125 E. 11th St., Portland, Oregon.

PENDLETON, CECIL, M., 285½ First St., Portland, Oregon.

PENLAND, JOHN R., Box 345, Albany, Oregon.

PENWELL, ESTHER, 95 E. 74th St., Portland, Oregon.

PERNOT, DOROTHY, 242 5th St., Corvallis, Oregon.

PETERSEN, AUGUST, Y. M. C. A., Portland, Oregon.

PETERSON, ARTHUR S., 780 Williams Ave., Portland, Oregon.

PETERSON, E. F., 780 Williams Ave., Portland, Oregon.

PETERSON, H. C., M. A. A. C., Portland, Oregon.

PETERSON, LAURA H., 395½ Clifton St., Portland, Oregon.

PILKINGTON, THOMAS J., Sebastopol, California.

**PITTOCK, H. L., Imperial Heights, Portland, Oregon.

PHILLIPS, MABEL F., 335 14th St., Portland, Oregon.

PLATT, ARTHUR D., Cambridge, Mass.

PLUMMER, AGNES, Third and Madison Sts., Portland, Oregon.

PRATT, JULIA E., 1200 E. Taylor St., Portland, Oregon.

PRENTYS, R. P., Congress Hotel, Portland, Oregon.

PREVOST, FLORENCE, Highland Court Apts., Portland, Oregon.

REA, R. W., Prineville, Oregon.

REDDEN, CECIL V., Vancouver Barracks, Wash.

REDMAN, FRANK M., 1014 Northwestern Bank Bldg., Portland, Ore.

REED, MRS. ROSE COURSEN, 308 Eilers Bldg., Portland, Oregon.

*REID, PROF. HARRY FIELDING, Baltimore, Md.

REIST, LINN L., 600 Chamber Commerce Bldg., Portland, Oregon.

RHODES, EDITH G., 5005 42d Ave., S. E., Portland, Oregon.

RICE, EDWIN L., 1191 E. Yamhill St., Portland, Oregon.

RICHARDSON, EDWARD L., 1217 Lee St., Evanston, Ills.

RICHARDSON, JEAN, 131 E. 19th St., Portland, Oregon.

RICHMOND, STANLEY C., Vancouver Barracks, Wash.

RIDDELL, GEO. X., Portland, Ore.

RIDDELL, H. H., 415 E. 19th St., N., Portland, Oregon.

RILEY, FRANK BRANCH, Chamber Commerce Bldg., Portland, Ore.

RISELING, ROBERT F., 1427 Northwestern Bank Bldg., Portland, Oregon.

ROBERTS, ELLA PRISCELLA, 109 E. 48th St., Portland, Oregon.

ROBINSON, DR. EARL C., 660 Morgan Bldg., Portland, Ore.

ROEMER, LOWELL, 4405 E. 89th St., S. E., Portland, Oregon.

*ROOSEVELT, THEODORE, Oyster Bay, N. Y.

ROOT, BELLE OTT, 328 Mill St., Portland, Oregon.

ROSENKRANS, F. A., 335 E. 21st St., N., Portland, Oregon.

ROSS, RHODA, 1516 E. Oak St., Portland, Oregon.

ROSS, WILLIS W., 494 Yamhill St., Portland, Oregon.

SAKRISON, C. H., 356 Fargo St., Portland, Oregon.

SAMMONS, E. C., 69 E. 18th St., Portland, Oregon.

SCHNEIDER, MARION, 260 Hamilton Ave., Portland, Oregon.

SCHNEIDER, KATHERINE, 260 Hamilton Ave., Portland, Oregon.

SEARCY, ROBERT D., 616 Chamber of Commerce Bldg., Portland, Ore.

SELF, NORA, Camas, Wash.

SHARP, J. C., Prineville, Oregon.

SHELTON, ALFRED C., 1390 Emerald St., Eugene, Oregon.

SHEPARD, F. E., 490 E. 33d St., Portland, Oregon.

SHERMAN, MINET R., 774 Everett St., Portland, Oregon.

SHIPLEY, J. W., Underwood, Wash.

SHOLES, CHAS. H., 1530 Hawthorne Ave., Portland, Oregon.

SHOLES, MRS. CHAS. H., 1530 Hawthorne Ave., Portland, Oregon.

SIEBERTS, CONRAD J., 683 E. Stark St., Portland, Oregon.

SIEBERTS, MRS. CONRAD J., 683 E. Stark St., Portland, Oregon.

SILL, J. G., 539 Vancouver Ave., Portland, Oregon.

SILVER, ELSIE M., 100 Sixth St., Portland, Oregon.

SMEDLEY, GEORGIAN E., 262 E. 16th St., Portland, Oregon.

SMITH, MARY GENE, Campbell-Hill Hotel, Portland, Oregon.

SMITH W. E., 589 E. 12th St., N., Portland, Oregon.

SMITH, KAN., U. S. Forest Service, Portland, Oregon.

SMITH, LEOTTA, Palatine Hill, Oswego, Oregon.

SMITH, PROF. WARREN D., 941 E. 19th St., Eugene, Oregon.

SNEAD, J. S. L. 572 E. Broadway St., Portland, Oregon.

SPAETH, J. DUNCAN, Princeton, N. J.

†SPURCK, NELL I., Seward Hotel, Portland, Oregon.

STARKWEATHER, H. G., R. F. D. No. 1, Milwaukie, Oregon.

STARR, NELLIE S., 6926 45th Ave., S. E., Portland, Oregon.

STEARNS, LULU, Ladd & Tilton Bank, Portland, Oregon.
STEVENTON, JOSEPHINE, 720 Oberlin St., Portland, Oregon.
STONE, DR. W. E., Purdue University, Lafayette, Ind.
STONE, MRS. W. E., 146 North Grant St., Lafayette, Ind.
STRINGER, A. R., Jr., 179 Bancroft Ave., Portland, Oregon.
STUDER, GEORGE A., 608 Schuyler St., Portland, Oregon.
STURGES, DANIELA R., 540 Elizabeth St., Portland, Oregon.

TAYLOR, VERA E., 604 Spalding Bldg., Portland, Oregon.
TENNESON, ALICE M., R. F. D. No. 4, Yakima, Washington.
THATCHER, GUY W., 302 Sacramento St., Portland, Oregon.
THAXTER, B. A., 391 E. 24th St., Portland, Oregon.
THOMAS, EMMA M., Marmot, Ore.
THORINGTON, DR. J. M., 2031 Chestnut St., Philadelphia, Pa.
THORNE, H. J., 452 E. Tenth St., N., Portland, Oregon.
TOMPKINS, MARGARET, Foot of Miles St., Portland, Oregon.
TREICHEL, CHESTER H., 535 Mall St., Portland, Oregon.
TREICHEL GERTRUDE, 535 Mall St., Portland, Oregon.
TUCKER, RALPH J., 389½ 16th St., Portland, Oregon.

UPSHAW, F. B., 401 Concord Bldg., Portland, Oregon.

VAN BEBBER, L., 503 Fenton Bldg., Portland, Oregon.
VAN ZANDT, DEAN, 6119 87th St., S. E., Portland, Oregon.
VANDER SLUIS, HELEN, 769 E. Broadway St., Portland, Oregon.
VEAZIE, A. L., 695 Hoyt St., Portland, Oregon.
VERNON, HOWARD L., 22 Reade St., New York, N. Y.
VESSEY, ETHYLE, 1207 W. 18th St., Vancouver, Wash.
VIAL, LOUISE ONA, 580 E. Main St., Portland, Oregon.

WALDORF, LOUIS W., Western, Nebraska.
WALTER, WILLIAM S., 55 North 21st St., Portland, Oregon.
WARD, JOHN S., New York, N. Y.
WARD, W. W., 53 E. 11th St., Portland Oregon.
WARD, MRS. W. W., 53 E. 11th St., Portland, Oregon.
WARNER, CHAS. E., 454 Taylor St., Portland, Oregon.
WEER, J. H., P. O. 1563, Tacoma, Washington.
WEISTER, G. M., 653 E. 15th St., N., Portland, Oregon.
WENNER, B. F., 675 Glisan St., Portland, Oregon.
WHITE, WM., 1214 Commonwealth Bldg., Philadelphia, Pa.
WILBURN, VESTA, care Richmond Paper Co., Seattle, Washington.
WILDER, GEORGE W., 226 14th St., Portland, Oregon.
WILLARD, CLARA, 801 Franklin St., Vancouver, Wash.
WILLIAMS, A. BOYD, King Davis Apts., Portland, Oregon.
WILLIAMS, MRS. A. BOYD, King Davis Apts., Portland, Oregon.
WILLIAMS, GEO. M., 713 F St., Centralia, Wash.
**WILLIAMS, JOHN H., 2671 Filbert St., San Francisco, California.
WILLIAMS, THOMAS H., 945 Weidler St., Portland, Oregon.
WILSON, CHARLES W., Bellevue, Idaho.
WILSON, RONALD M., care Rev. Wilson, Dover,Mass.
WING, MARY, 1124 Macadam Road, Portland, Oregon.
WOLBERS, HARRY L., 577 Kerby St., Portland, Oregon.
WOODWORTH, C. B., care Ladd & Tilton Bank, Portland, Oregon.
WYNN, DR. FRANK B., 421 Hume Mansur Bldg., Indianapolis, Ind.
YORAN, W. C., 912 Lawrence St., Eugene, Oregon.
YOUNG, CRISSIE C., 547½ Sixth St., Portland, Oregon.
ZANDERS, RUTH, Milwaukie, Oregon.
ZIEGLER, MAE, 89 Mason St., Portland, Oregon.

Active	399
*Honorary	8
**Life	4
Total	411

†Deceased.

Printed by YORAN PRINTING HOUSE, EUGENE, OREGON

Record of Mountaineering
in the Pacific Northwest

Mazama Organization for the Year 1919-1920

OFFICERS

E. C. SAMMONS (U. S. National Bank) .... President
GEO. X. RIDDELL (689 Everett St.) .... Vice-President
ALFRED F. PARKER (330 Northwestern Bank Bldg.) ...... Corresponding Sec'y
A. BOYD WILLIAMS (King-Davis Apts.) .... Recording Secretary
MARTHA E. NILSSON (320 E. 11th St. North) .... Financial Secretary
MARION SCHNEIDER (260 Hamilton Avenue) .... Treasurer
LOLA I. CREIGHTON (920 East Everett St.) .... Historian
ROY W. AYER (689 Everett St.) .... Chairman of Outing Committee
EUGENE H. DOWLING (742 Belmont St.) .... Chairman Local Walks Com.

COMMITTEES

OUTING COMMITTEE—Roy W. Ayer, Chairman; Charles J. Merten, Miss Martha E. Nilsson.

LOCAL WALKS COMMITTEE—Eugene H. Dowling, Chairman; W. J. Paeth, Ben Newell, Miss Crissie C. Young, Miss Minna Backus.

HOUSE COMMITTEE—Mrs. Laura McDonald, Chairman; C. M. Pendleton, Dr. D. T. Kerr, Miss Sallie E. McCoy, Miss Bernice Gardner.

ENTERTAINMENT COMMITTEE—Geo. Meredith, Chairman; Jamieson Parker, Harry L. Wolbers, Miss Olga Hallingby, Miss Evelyn Hardinghaus.

PUBLICATION COMMITTEE—Geo. W. Wilder, Chairman; Miss Crissie C. Young, Miss Agnes G. Lawson, Alfred F. Parker, Miss Minnie R. Heath.

EDUCATIONAL COMMITTEE—Frank I. Jones, Chairman; Miss Laura Peterson, Miss Katherine Schneider, John A. Lee, Rodney L. Glisan.

LIBRARY COMMITTEE—Miss Lola I. Creighton, Chairman; Harry C. Libby, O. B. Ballou, Mrs. B. M. Carl, Miss Florence McNeil.

MEMBERSHIP PROMOTION COMMITTEE—Miss Harriett E. Monroe, Chairman; Frank M. Redman, T. Raymond Conway.

AUDITING COMMITTEE—Robert F. Riseling, Chairman; Adrian Smith, B. C. Nelson.

PUBLICITY COMMITTEE—Fred H. McNeil, Chairman.

MAZAMA

A Record of Mountaineering in the Pacific Northwest

Alfred F. Parker
Agnes G. Lawson

Publication Committee
G. W. WILDER, Editor

Minnie R. Heath
Crissie C. Young

VOLUME V PORTLAND, OREGON, DECEMBER, 1919 NUMBER 4

Contents

the weird feeling that has taken possession of us. Not a murmur of complaint is heard. Everyone is eager to get something to eat, and find a shelter for the night, which is already beginning to close in upon us.

The quarters for the women are on a ridge about a quarter of a mile away from the cook tent. The trip across that long, long snow-field seems never-ending, this first drizzly, foggy night. The men carry the dunnage across and set to work putting up shelter tents. At one spot a large fly is strung out from a log and propped up with alpenstocks to shelter some twenty girls. What a lot of amusement the "circus tent" brings to us! Everyone shares with everyone else to make up for the lack of dunnage. Miss Nilsson does her best to make us all comfortable for the night.

Monday morning arrives, dark and drizzly. Few fail to answer the call for breakfast. As the crowd is gathered about, laughing and chattering, the fog lifts for a few moments. There, right in front of us, is the majestic mountain peak. We gaze with awe and reverence at that towering mass of rock and snow. Suddenly the curtain of fog falls once more, and we are alone on the mountain.

Monday is a busy day. Camp sites must be located, tents put up properly, wood gathered for the needed fires, dunnage sorted, and shoes dried. Were you ever so busy in your life?

As if to compensate us for the experiences of Sunday and Monday, Tuesday brings us perfect weather. Who can keep down the smiles? Everyone radiates sheer joy. Mt. Rainier is there in all of its majesty. The Tatoosh range looms up across from us, luring us on to adventure. Our president, Mr. Coursen, plans a trip up Pinnacle Peak. Ninety respond with enthusiasm. At nine o'clock we are off across the snow-field beyond the cook tent and along the Reflection Lake trail. We know this is "Paradise" Valley. We look down at our feet; flowers, myriads of them, in wild masses of brilliant alpine coloring, border the trail. We look back; Mt. Rainier towers above. We look beyond; near us we find little gems of snow lakes, and away beyond are the jagged pinnacles of the Tatoosh range. At Reflection Lake we pause to marvel at the perfect image of the mountain. We zig-zag up the ridge, crossing and recrossing long and steep snow-fields. Just below a series of falls we find a rocky ledge carpeted with abundant heather, where we rest for lunch. Now we begin the real climb. Alpenstocks are not needed on the rock work, so hand over hand we pull ourselves up the steep slope. Long stretches recall most vividly the pinnacle of Mt. Jefferson.

The panorama spread out before us is fitting reward for the energy expended in getting to the top. The whole of Paradise Valley lies at our feet. In every direction mountain peaks loom up on the horizon. There, nestled on Mazama Ridge, is our camp. Down below is Paradise Inn. How near to camp it is; and we had thought it was so far away. We gaze and gaze and try to impress the loveliness, the peace, the beauty of it all upon our minds. We are loath to climb down; but who wants to miss dinner or be late for a camp-fire session?

We really have a camp-fire that night—one that persists in lingering in our minds, for it is the first real one of the outing. Everyone is fairly bubbling over with joy at the change in the weather and at the appearance of the full moon. "Bill" Yoran stands there in the center of the natural amphitheatre and directs us in our fun. The sessions are of nightly occurrence. Memory fails to make one more attractive than another. Each one is a success. There is abundant talent in camp. Selections on violin, mandolin and flute, each in turn delight us. Recitations, solos, scientific lectures, story-telling, mock trials, reminiscences of Stevenson, readings from the daily Courzama, limericks, jokes, minstrel shows, dances—a conglomerate mass of entertainment—give us fun and laughter.

On each day we are up for a seven-thirty breakfast, eager for the pleasures that offer. There are so many attractive side trips; to Stevens Ridge, Bench Lake, Indian Henry's, Narada Falls, Louise Lake, Martha Falls, Camp Muir and Unicorn Peak. Our days are filled with social events; swimming parties to the nearby lakes, tea-parties, birthday parties, special "owl" sessions, and dinners and dances at Paradise Inn.

Before many days have passed, groups drift up to the Finleys' quarters on "Married Hill", where there is an unusual experience in store for most of us. Here the little wild chipmunks eat out of our hands, walk the wire and pull up the peanuts suspended from the string in a queer paw-over-paw sailor fashion. Here Peter is guard and barks joyous greetings to his many friends. Dog-like, he has adopted us all. Many a laugh he gives us at the camp-fire, chasing his tail or baying at the moon.

It is Sunday morning. At ten o'clock the party gathers in a flower-covered amphitheatre in the shade of the trees. The day is perfect. In every direction a marvelous picture of loveliness meets the eye. Dr. Marcotte, who has just recently climbed to the top of the mountain, delivers a never-to-be-forgotten sermon to the assembled group, who will start for the summit in the afternoon. Surely we know that "God's in His Heaven, all's right with the world" as we listen to the singing and strains of the violin out there in the open.

Another Monday is here. In camp, groups gather about, chattering and laughing. Unconsciously they watch the mountain, eager for a sight of those who are up there putting their strength to a test. Eighty-three, under the leadership of Mr. George Riddell, left camp at two o'clock on Sunday afternoon. The lonely stay-in-camps cannot keep them from their minds. If said lonely ones could have glimpsed the Forest Service lookout station at six o'clock on Sunday evening, they would have seen the group thoroughly enjoying the hot tea served so generously by Mr. Brunn of the station. A glimpse of the party at Camp Muir would show them spending a comfortable and cosy night. What matter the over-zealous pilferings of the rats about the cabin, except to those who lost their lunch, when the northern lights are on display in the far horizon, or when one can enjoy a perfect setting of the moon and a radiant sunrise?

Gibraltar, never in better condition for climbing, is conquered, and after hours—or is it years?—of upward striving, sixty happy and successful climbers gather on the summit (14,408 feet). Eagerly the well-known peaks are pointed out: Hood, St. Helens, Adams, Jefferson, Three Sisters, Baker and Glacier Peak. Such a scene is surely worth the effort of the climb. There below lie mountains, ridges, valleys and lakes, as far as the eye can see. How quickly, under the spell of the mountain, the tired ones recuperate as they linger over the last crumbs of lunch! Mid-day seems too early for the homeward start, but the precipitous wall of Gibraltar must be left behind by two o'clock, for it is at about this hour that the rocks begin to loosen and fall.

At interesting stretches in the downward climb, as on the ascent, the "movie" man is ever present. Somehow his mere presence, with no doubt the extra rest he calls for, seems to lessen the distance to Camp Muir. Again packs are shouldered

ON COWLITZ GLACIER ON RAINIER

Photo by Boychuk.

SUNRISE OVER THE COWLITZ GLACIER ON RAINIER

Photo by Boychuk.

and old landmarks sighted. At six o'clock everyone is safely back in camp.

In happy companionship and recreation one day quickly follows another. Each has its own thrill, its own joy—yes, and sorrow too, for the last days of camp are saddened by the loss of one of our beloved companions, Jack Meredith.

Again we are in the dinner line; again we are eating crackers and soup as we circle up the hill above the cook tent. Now we are seated at the long tables, laughing and chattering with our neighbors. Some of these neighbors are our guests for the first time, others have been with us before. Some are members of other outdoor clubs: Sierra, Colorado, Mountaineers, Forest and Field. Some are known nationally in the fields of politics, science and education. All have contributed much to the success of the outing. We want them to be with us on other trips.

The two weeks are over and we are at home. It is a difficult task to put down into words what the outing means to us—what it will mean to us in the days that follow. We know that time cannot rob us of our happy memories of days spent at camp Coursen.

List of Members of the Mt. Rainier Outing Who Reached the Summit of Mt. Rainier

Willard Allphin, Portland, Oregon
Gertrude Andrae, Portland, Oregon
Harold Babb, Portland, Oregon
Alice Banfield, Portland, Oregon
Mae Benedict, Portland, Oregon
Lee Benedict, Portland, Oregon
Clement Blakney, Portland, Oregon
E. Boehme, Portland, Oregon
Lulu Bornt, Portland, Oregon
Dorothy Brownell, Portland, Oregon
Walter Boychuk, Portland, Oregon
P. L. Campbell, Eugene, Oregon
Randolph S. Carroll, Portland, Ore.
Herbert I. Corning, Portland, Ore.
Edgar E. Coursen, Portland. Oregon
George H. Currier, Leona, Oregon
Balfour Daniels, Princeton, N. J.
Arthur J. Emmrich, Portland, Ore.
William W. Evans, Portland, Oregon
Fred Everson, Portland, Oregon
Th. Fagstad, Cathlamet, Wash.
Wm. L. Finley, Portland, Oregon
Mrs. Wm. L. Finley, Portland, Ore.
W. C. Foster, Portland, Oregon
F. G. Franklin, Albany, Oregon
Margaret E. Fuller, Portland, Ore.
Bernice Gardner, Portland, Oregon
Martha Gasch, Portland, Oregon
E. G. Gearhart, Astoria, Oregon
Mrs. E. G. Gearhart, Astoria, Oregon
F. Giesecke, Portland, Oregon
Mabel R. Girsberger, Chiloquin, Ore.
Olga Hallingby, Portland, Oregon
George H. Harvey, Denver, Col.
Clarence A. Hogan, Portland, Ore.
Evelyn Hardinghaus, Portland, Ore.
George Hartness, Portland, Ore.
Hazel Howard, Portland, Oregon
Pasho Ivanakeff, Portland, Oregon
Amy Johnston, Portland, Oregon
Marie Koennecke, Portland, Oregon
Agnes G. Lawson, Portland, Oregon

Martha Landis, Portland, Oregon
John A. Lee, Portland, Oregon.
Mary Knapp Lee, Portland, Oregon
Jacques Letz, Portland, Oregon
Harry Libby, Portland, Oregon
Ethel M. Loucks, Portland, Oregon
Georgia Lyon, Chicopee Falls, Mass.
Sallie McCoy, Portland, Oregon
Henry Marcotte, Kansas City, Mo.
Sabina E. Mason, Portland, Ore.
Duncan Moore, Chicago, Ill.
J. D. Meredith, Portland, Oregon
O. W. T. Muellhaupt, Portland, Ore.
Arthur H. Marshall, Vancouver, Wn.
Ruth Olson, Portland, Oregon
Katherine Ogilbe, Portland, Oregon
Emily F. Otis, Portland, Oregon
Rose F. Parker, Milwaukie, Oregon
Alfred F. Parker, Portland, Oregon
Jamieson Parker, Portland, Oregon
Edward L. Patzelt, Portland, Oregon
P. G. Payton, Portland, Oregon
Cecil Pendleton, Portland. Oregon
R. A. Perry, Portland, Oregon
E. F. Peterson, Portland, Oregon
R. P. Prentys, Portland, Oregon
Edwin L. Rice, Portland, Oregon
Louis Rice, Portland, Oregon
Cecil Redden, Vancouver, Wash.
Bessie M. Renfro, Portland, Oregon
Joe H. Renfro, Portland, Oregon
Henrik Renstrom, Squantum, Mass.
George X. Riddell, Portland, Oregon
Rose E. Rothe, Bridgeport, Conn.
Minet Sherman, Portland, Oregon
Mary Gene Smith, Portland, Oregon
Gretta Smith, Portland, Oregon
Leotta Smith, Portland, Oregon
Alice M. Tenneson, Yakima, Wash.
Margaret E. Tompkins, Portland, Ore.
Harry Beal Torrey, Portland, Oregon
Elizabeth Torrey, Portland, Oregon
Lyle Turner, Portland, Oregon
George W. Wilder, Portland, Oregon
Robert P. Walsh, St. Louis, Mo.
A. Boyd Williams, Portland, Oregon
Mrs. A. Boyd Williams, Portland, Ore.
L. L. Wilson, Tracy, Calif.
Crissie Young, Portland, Oregon
Ruth Zanders, Portland, Oregon
Oneita Webb, Portland, Oregon

Thin, thin the pleasant human noises grow,
And faint the city gleams;
Rare the lone pastoral huts—marvel not thou!
The solemn peaks but to the stars are known
But to the stars and the cold lunar beams;
Alone the sun arises, and alone
Spring the great streams.—*Arnold.*

Around the Great West Side

JOHN A. LEE

In 1914 the Mazamas had their summer's outing camp at Mystic Lake, on the north side of Mt. Rainier. At the conclusion of that outing R. L. Glisan, J. Wheelock Marsh and the writer made a knapsack trip from Mystic Lake to Paradise Park, going around to the east and following what is known as the "high trail." An account of this trip by Mr. Glisan, entitled "Mystic Lake to Paradise Park," appeared in the "Mazama" of that year. During the course of the outing the writer had also skirted the mountain to the westward as far as Spray Park.

Inasmuch as the Mazama camp this year was to be at Paradise Park, I decided that the occasion was opportune to connect up the termini of my previous excursions around the mountain by tramping westerly from Spray Park to Paradise Park. Since the outing of 1914, a fair "nannie" from the Mazama flock had signed up to travel with me on life's journey, so that she was my companion on this trip. A game little pal she proved to be.

On the evening of Friday, August 1, Mrs. Lee and I left Portland, arriving in Tacoma the morning following and at Fairfax by noon, where the hike was to begin. We were equipped for all contingencies, for well did I recall having been marooned in a fog above Cowlitz Canyon on the previous trip, with huckleberries the party's only rations for the last day. Almost everything but the "kitchen range" had gone into our packs and come fog, come rain, come snow, what recked we? When the heavier of the two packs tipped the scales at 85 pounds at Fairfax, with full 15 pounds in the lighter pack, it was a bit disconcerting, we will confess, but pride would not permit the removal of a single ounce.

The first leg of our journey was to be a hike of 21 miles over the Grindstone trail to Crater Lake, which lake is situate somewhat below Spray Park, at an elevation of 5000 feet. The name of this lake is inappropriate and confusing; the Indian name of Mowich Lake should receive official recognition. This first interval we negotiated in two days, camping the first night at the last crossing of Evans Creek, five miles or so from Fairfax, and the second in Grindstone cabin, 12 miles out, with fog and drizzle changing to a downpour of rain ere we reached the haven of the cabin at Crater Lake. The trail is an old one and excellent for the most part, though a number of the bridges have fallen from decay. Except for the first mile out of Fairfax the trail leads through virgin forest, some of it of splendid stand. The frowning of the elements we did not mind, as we knew the days to follow would be the clearer, with the smoke and haze dispelled; and then the cool air helped us to carry our heavy packs up the long slope.

Crater Lake, where we bivouacked for two nights, is most picturesque and a delightful place to camp. The lake is circular in shape, a half-mile or so in diameter, with one-half of its circumference girt about with beetling crags and the other half with alpine groves and heath-clad meadows. In spots where the heather had not claimed full possession, erythroniums were blossoming in great profusion.

The air was clear and crisp as we emerged from the cabin the first morning of our stay at Crater Lake, and there to the southeast, so near it seemed we might reach out and touch it, loomed the great white dome of Rainier, with the two Mowich ice streams leading down. What a different world from that of the previous day! The tang in the air impelled to action and the heights above were calling, in a way that could not be resisted, "Come up! Come up!" With breakfast over and lunch and camera in pack, we set out for Spray Park, four miles distant by trail and a thousand feet higher on the mountain's slope.*

Spray Park is in every way one of the most beautiful of the many alpine parks on Rainier, but its crowning glory is Spray Falls. The many little streams that flow down through the park converge into one of considerable size at the westerly margin of the park, there to plunge over a precipice of several hundred feet to form Spray Falls. The water spreads out fan-like as it descends and the effect is most beautiful and impressive.

From the high vantage point of Spray Park we gazed intently at the lofty buttresses bordering the Mowich glaciers, debating whether on the morrow we should brave those bastions,

---

*As we neared the park the shrill whistle of a marmot, uttered close to the trail, brought us to a sudden halt. The cry of this animal, however often heard, is always a little startling, and deceptive too, if one is not familiar with it, resembling much the whistling of a man. A quick glance enabled us to catch a glimpse of this interesting rodent with his shaggy coat and grizzled collar as he scampered into his hole.

the most formidable on the mountain, or follow the horse trail leading to Sunset Park. But with heavy packs to carry we decided that discretion was the better part of valor. Pinning to a tree near Spray Falls a message for the Mountaineers, who were to pass that way in a few days in their circuit of the mountain (a message we heard from later), we returned to Crater Lake, and in the late forenoon of the next day set out over the horse trail for Sunset Park.

A new trail has been constructed from Crater Lake to the crossing of the two Mowich streams, which descends by many switchbacks the north slope of the porth Mowich canyon, some three miles down to the stream, and saves the long ten mile detour around by Meadow Creek. In the exhilaration of the descent after so much of "upping," we fairly loped over the springy turf of the newly-made trail. Yes, we would have to ascend on the other side of the canyon, and on that same day too, every one of those more than two thousand feet of descent, but that was something for the future—after luncheon; now we were going down, "on high."

As we started, the music of the river, far down below, was wafted up to us in subdued undertone; swelling gradually as we descended, it burst into a mighty diapason as we came out upon the boulder-strewn bed of this mad Mowich stream. The trail crosses the two forks only a few rods above their confluence and, thanks to new bridges just completed by the National Park Service, we soon were across both streams and at luncheon by a crystal clear rivulet in the cool of the canyon. An eight-inch trout darted down from under my pail as I was dipping water for our meal.

After a hearty luncheon and a good rest, for the day was warm, we started on the toilsome climb up to Sunset Park. We had lunched on the 2700-foot contour and the highest point in Sunset Park, on the trail, is 5200 feet. So 2500 feet of elevation and seven miles of distance were to be our stint for the afternoon—as we thought, not reckoning with that bane of the mountaineer, the festive mosquito. But later as to this pestiferous little imp of Satan.

What pleasure is there, say you, in lifting oneself, plus a heavy pack, foot by foot up a steep mountain trail? Even more pertinent is your query if the trail leads, as did this one, through an unbroken forest, with no outlook save an occasional glimpse of the tree tops across the canyon. The answer is simple. It is

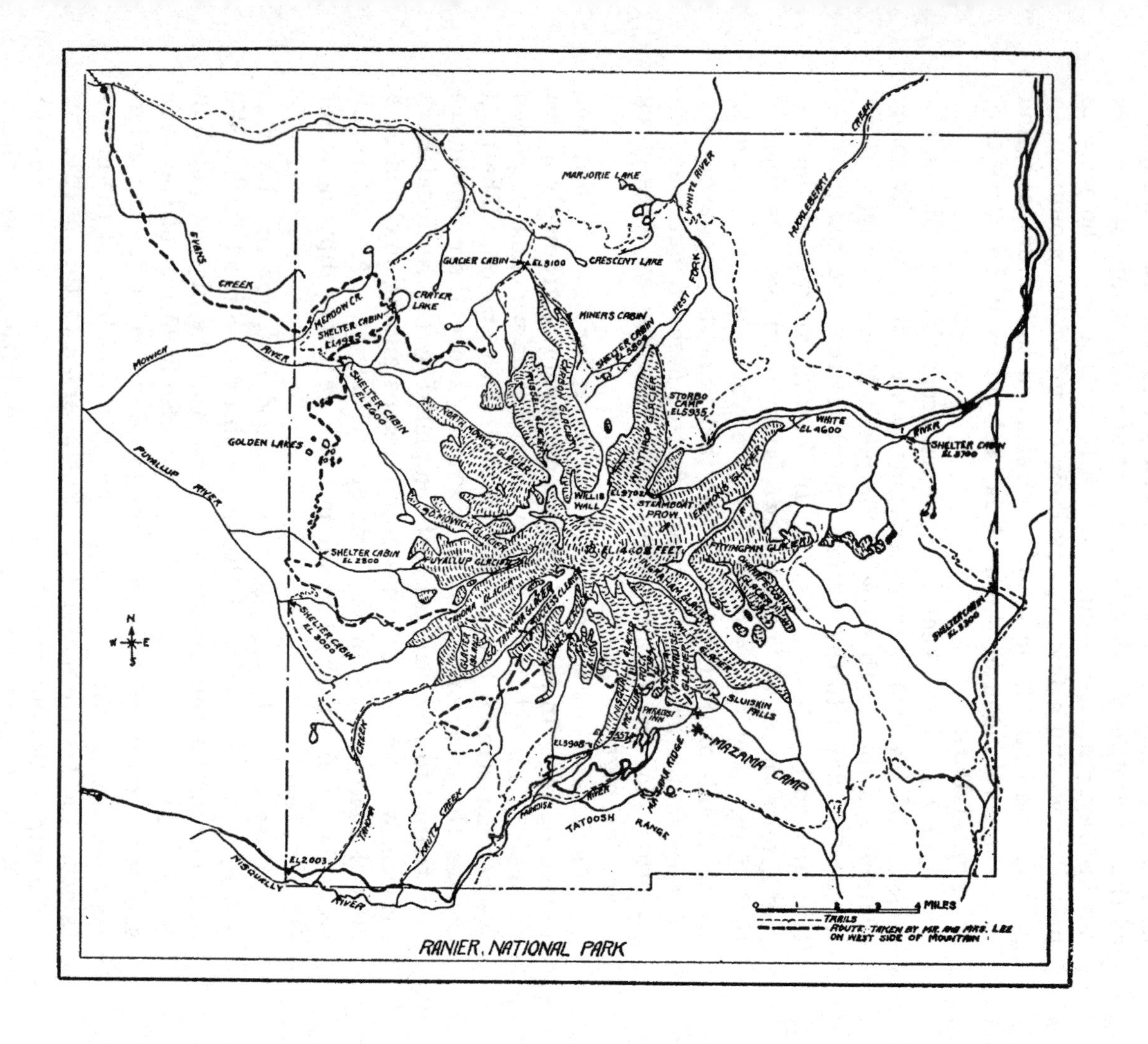
EVANS CREEK
MARJORIE LAKE
WHITE RIVER
HUCKLEBERRY CREEK
GLACIER CABIN
EL 5100
CRESCENT LAKE
WEST FORK
CRATER LAKE
MEADOW CR.
SHELTER CABIN EL 4985
MINERS CABIN
MOWICH RIVER
SHELTER CABIN EL 2600
STORBO CAMP EL 5935
WHITE EL 4600
RIVER
SHELTER CABIN EL 3700
GOLDEN LAKES
NORTH MOWICH GLACIER
CARBON GLACIER
RUSSELL GLACIER
WINTHROP GLACIER
PUYALLUP RIVER
SO. MOWICH GLACIER
WILLIS WALL
EL 9702
STEAMBOAT PROW
EMMONS GLACIER
SHELTER CABIN EL 2800
EL 14408 FEET
FRYINGPAN GLACIER
PUYALLUP GLACIER
INGRAHAM GLACIER
TAHOMA GLACIER
SO. TAHOMA GLACIER
SUCCESS CLEAVER
SHELTER CABIN EL 3300
SHELTER CABIN EL 3000
N
W
E
S
NISQUALLY GLACIER
PARADISE GLACIER
PARADISE INN
SLUISKIN FALLS
EL 5908
MAZAMA RIDGE
MAZAMA CAMP
KAUTZ CREEK
TAHOMA
PARADISE RIVER
TATOOSH RANGE
EL 2003
NISQUALLY RIVER
0 1 2 3 4 MILES
TRAILS
ROUTE TAKEN BY MR. AND MRS. LEE ON WEST SIDE OF MOUNTAIN
RANIER. NATIONAL PARK

then that you feel to the fullest the joy of rugged health, the satisfaction in the assurance that you are equal to the game. You drink deep of the cool air redolent with the incense of the forest, forcing it down with every breath into the very last lung cell. You must stop frequently to cool off and rest, and this affords opportunity for reflection. A sense of freedom, of independence, steals over you, almost the completest that one can feel in this world of complex human relationships and interdependencies. For with your bed and grub and house upon your back, you can camp, if you choose, where night overtakes you. And if night travel suits your fancy (and that of your partner in the game), what is there to say you nay? And then if you have been taught to observe, and particularly if you have given a little attention to botany and tree study, there is much to be seen in the forest to add to the joy of living.

On this particular climb we started in the belt of alder, vine maple, Douglas fir, white fir (*abies grandus*), Western red cedar and Western hemlock. Soon these species were displaced, as we ascended, by noble fir, lovely fir, and Western white pine, though the change was not abrupt but gradual. And then as we came out into the open park land on the summit of the ridge, these species in turn gave way to alpine fir, mountain hemlock, white-bark pine and Alaska cedar. The fruitage of all these conifers had been unusually bountiful the previous year and everywhere along the trail the seeds had taken root and were sending up their delicate cotyledons to struggle for existence alongside the parent trees.

With the steep ridge surmounted we soon were among the Golden Lakes. These were a disappointment. In the first place they proved to be merely shallow little snow lakes, formed in depressions of the ground from the melting of the snow and gradually receding. But worst of all, they were breeding ponds of myriads of mosquitoes, and hungry varmints they were too. Lingering not, we hurried on to Sunset Park proper, hoping that there we might find refuge from their vigorous onslaughts and a well earned rest. But it was not to be; the little demons pursued us still. And beautiful as was the verdure of the park, tired though we were, and with the shades of night fast falling, we could find no running water close about, so after a hurried consultation (still under fire), we continued on the trail down another long slope toward the North Puyallup River, intending to camp at the first stream. Two lusty rivulets were encountered, but both flowing in sharp ravines down the steep slope, so aided by a friendly moon and our electric "bug," we paused not until we had made the complete drop of 2200 feet down to the North Puyallup river. A neat lttle shelter cabin with plenty of new moss in the bunks was there to welcome us, and tired as we were and late the hour, we were not loath to accept of its hospitality. We had made fifteen miles with our packs since 10 a. m. that day, including 2500 feet of climb and 5500 of descent. A hurried fire and some hot soup and then the sweet slumber of the weary hiker.

We were awakened on the morrow by the sound of voices and two curious faces peering into our cabin door. These early morning visitors proved to be two gentlemen from the Co-operative Campers at Indian Henry's Hunting Ground, Francken and Bjorklund by name, who were on their way to Sunset Park and had camped that night a short distance from our cabin. They were the first wayfarers we had met since leaving Fairfax. After a few friendly exchanges they proceeded on their way and we to prepare a substantial breakfast and survey the scene about us. Dense forest met our gaze on every side save a small open space about the cabin. The elevation was 3000 feet.

Weary of so much up and down work, we determined that shortly we would leave the horse trail and, for the remainder of the distance to our destination, would attempt the "high trail" across the glaciers. Our best route, however, was to follow the horse trail until we had rounded Klapatche Ridge and then take the "snake trail" up to St. Andrew's Park, which we expected to reach that day. Our start was late, for we were loath to leave this restful little nook in the forest.

An hour's climb up Klapatche Ridge brought us to Raeburn Point, from which we could survey the Puyallup Glacier with its spectacular ice falls and could catch a glimpse too of St. Andrew's Park. The Puyallup Glacier is the source of the North Puyallup river, on the banks of which stream we had spent the night. Having encircled Klapatche Ridge, we stopped on the trail for our afternoon tea. Here our friends of the morning overtook us and were persuaded to accompany us to St. Andrew's Park. Right clever chaps they showed themselves to be.

At the first crossing of St. Andrew's Creek, after having left the horse trail, we came upon the camp of a crew of men who were engaged in building a new trail up to St. Andrew's Park and were hurrying construction so as to have the trail

1—Spray Falls
2—On the Kautz Glacier.
3—Mother Ptarmigan and Chick.
4—Mrs. Lee on the North Tahoma Glacier.
5—On Neva of North Tahoma Glacier.
6—North Tahoma Glacier and "The Island"

1—Indian Henry's Hunting Ground. 2—Upper portion of North Tahoma Glacier.
3—Giberalter from Ranger's Cabin.
4. Seracs and Neve of the North Tahoma Glacier. 5. Seracs of the Kautz Glacier.

ready for the coming of the Mountaineers. The small portion then completed was of no avail to us, so we struck out on the old "snake trail," which the blazes enabled us to follow. The climb was stiff, but steady plugging brought us to a delightful camp site on the first bench of the park, just at nightfall.

St. Andrew's Park is little known to the tourist because of its relatively secluded situation, but it will easily rank as one of the three or four most beautiful of the many beauty spots on Rainier. The park consists of three distinct benches, rising one above the other at intervals of perhaps five hundred feet, the topmost bench having an elevation of about 6500 feet. The outlook to the north is across Puyallup Glacier to the frowning battlements of the Colonnade Cleaver, which separates this glacier from the South Mowich Glacier, still farther to the north; to the south, across the great Tahoma glaciers to Pyramid Peak and Indian Henry's Hunting Ground. Glacier Island, a high spectacular crag, is set right in the middle of the broad ice stream of Tahoma Glacier as it descends from the mountain's summit, dividing the glacier into two parts, designated as the North Tahoma Glacier and the South Tahoma Glacier.

Our friends, after some wavering, decided they would not attempt the trip with us across the Tahoma glaciers the next morning, so after a few small exchanges of provisions and hearty western hand-clasps, we parted company; they to descend to the horse trail and follow it back to Indian Henry's, and we to continue on our selected course, up through the park and thence across the glaciers to Pyramid Peak and Indian Henry's. Climbing leisurely from bench to bench and enjoying the many beauties of the park, the noon hour found us at the extreme limit of trees, where we stopped for lunch. From this high lookout station we were able to map out our course in detail and it was decided that the most feasible route, as well as the most interesting would be to cross the glacier above Glacier Island, though this meant traversing a full two mile stretch of ice, all of which was absolutely new to us. The north wall of the canyon below Glacier Island rose high above the glacier and appeared to be an abrupt precipice that could not be descended except by aeroplane; an appearance which was borne out by closer inspection later. It was plain also that it would be foolhardy to attempt a crossing before the morrow, with the day already so far spent. So dropping down to a fringe of timber on the brink of the canyon we made an early camp.

We were off in good season the next morning. Skirting along the slope of Puyallup Cleaver to reach the point selected for our descent upon the glacier, a big billygoat all at once became silhouetted against the sky, upon the very crest of the Cleaver. He was not far distant, but too far to show up in a picture. His appearance was not entirely unexpected, for numerous tracks and occasional tufts of wool clinging to the stunted trees had shown that we were now in "goat country." Almost at the same instant a mother ptarmigan with quite young chicks appeared in our path. They sauntered off leisurely as we drew near, all being garbed in their summer's coat of brown.

The approach to the glacier proved even better than we had anticipated and the glacier itself, up as far as the crest behind Glacier Island, was less steep and less badly crevassed than we had supposed it would be. We made good time up to this point, though stopping frequently to get our breath and to take an occasional picture. The southerly half of the glacier proved not quite so easy to negotiate. By making a slight drop from the crest behind the island we had thought to push straight across on contour, but great seracs and bergschrunds were encountered, blocking our way.

It was at this point that the writer was treated to the most thrilling half-moment experienced on the whole trip. At the head of a steep slope leading down to where the glacier dropped off for a good five hundred feet, my wife sat down to change the film in her camera. Setting the camera down beside her on the ice, it took a start and went gliding down the slope. Quick as a flash she gave herself a shove and went sliding down in the wake of her cherished possession, making rapid little hitches so as to accelerate her speed and overtake the fast-moving object. I felt a chill come over me as I realized the situation. As soon as voice could be found to speak, I cried out, "Let the camera go." But those little hitches still went on with clock-like regularity. Now thoroughly alarmed, I repeated, in the loudest and most commanding tone I could assume, "Let that camera go!" Shortly her hand seized the camera, and, rolling over on her face and digging in her toes in approved mountaineering form, she came to a stop within a few yards of the brink of the precipice. As I met her on the slope and she noted the still anxious, and perhaps severe, expression on my face, she remarked, with the utmost sang-froid and with a smile and a jaunty toss of her head, "Well, I got the camera!" No more was said, but somehow the words of the old saw thrust themselves into my consciousness, "Where ignorance is bliss, 'tis folly to be wise," to be displaced shortly by that more cheerful, not to say more considerate philosophy, "All's well that ends well."

Climbing back to the crest of the glacier behind the island, we now pursued a diagonal course up and across the broad neve, threading our way among numerous crevasses, some of them of huge proportions, watching closely to avoid the slightest sign of "blind" ones, picking our steps carefully when the ice tongue betwen them was narrow, until finally we came out upon Success Cleaver, the southerly rim of the glacier, at about the 8200 foot contour and just at dusk. The cleaver is very narrow at this point and protrudes only slightly above the two glaciers which it separates, the Tahoma on the north, which we had crossed, and the Pyramid Glacier on the south. Crossing the cleaver and glissading rapidly down the steep though unbroken surface of Pyramid Glacier, we made welcome camp in Pyramid Park.

We were now, at this camp, only a short distance off the most direct "high line" route from Indian Henry's to Paradise Park. So, without breaking camp, we set out the next morning, which was Sunday and the eighth day since leaving Fairfax, to have a look at the much vaunted beauties of Indian Henry's; also to pay our respects to the Co-operative Campers, of whom Francken and Bjorklund had told us much. On the way over we followed the south route around Puyallup Peak; we took the north route on our return, finding the latter the better and more direct.

Though expecting much of Indian Henry's, we were not disappointed; except as to one circumstance, and fortunately one not always present; the winged pests from which we had fled in Sunset Park were here in countless hordes. The setting and verdure of the park are exquisite; and the view of the mountain that one gets from Indian Henry's is, to my mind, unsurpassed. At the Co-operative Campers' we found our coming had been expected and we were most hospitably received by Mr. and Mrs. Danilson, the managers of the camp. We regretted though, not to meet again our two friends of the trail, who chanced to be absent for the day.

Spending a second night in our cosy little camp in Pyramid Park, we set out early next morning on the last leg of our journey, realizing it might mean a long day to complete it. We planned to cross high up on the Krautz Glacier, as we had done on the Tahoma, so as to avoid the sharp walls of the canyon farther down. This would necessitate leaving Van Trump Park some distance below us, a matter of regret, as we had hoped to take in every feature of interest along our route. But the trip had been prolonged to one of greater duration than we had planned and we were anxious now to get in to Mazama Camp, in time at least to make the ascent of the mountain before the camp should disband.

Selecting the point of junction of the Success and Kautz glaciers as the most likely crossing of the Kautz, we headed straight for it. This course took us over an arm of Pyramid Glacier and at one resting place near the glacier some fair pictures of ptarmigan were obtained. Passing just below the end of the cleaver separating the Pyramid and Success glaciers, we stepped out upon the Kautz at the precise point chosen for the crossing. The prospect was not altogether inviting. The distance to be traversed to gain the other side was not great and the opposite wall not especially declivitous, but everywhere, in front of us and up and down, the glacier presented a tumbled, jumbled mass of seracs, piled high in endless confusion. Then too, the occasional moans and groans that would be wafted up from the ice depths below created a sort of uncanny feeling, a sense that this might, indeed, be the Esquimaux's inferno. Finally, by picking our way a few hundred feet farther up along the broken surface of the glacier, a way through was discovered and we had lunch on Wapowety Cleaver, at about the 8500-foot contour.

Striking out across the cleaver on to Van Trump Glacier, we soon were enveloped in a dense fog. This was by no means encouraging, as our larder was now almost nil and would scarcely have yielded one scanty meal. After a half-hour's chilly wait the fog lifted so that we could see something of what lay before us and then we made rapid progress down and across the generally smooth surface of this glacier. The fog closed down again while we were still on the glacier, but we kept on, watching carefully to keep our course, until finally we bumped into an abrupt rock wall at the glacier's southerly rim. After some search for a pass this obstacle was surmounted and there below us, shimmering in the bright sunlight, was the welcome Nisqually Glacier. Far across we could make out Sluskin Falls, pouring in feathery foam over the high east rim of Paradise Valley; and a little way to the right, on Mazama Ridge, could be discerned the white tents of Mazama Camp. Having just emerged from a wilderness of fog, we could well appreciate how Moses of old must have felt as he looked out upon the promised land.

To have dropped down to the Nisqually Glacier immediately in front of where we stood would have brought us to a difficult and perhaps impossible crossing, and besides would have meant the loss of much elevation that would have had to be regained on the other side. But some distance above and to our left and almost opposite McClure Rock on the further side could be observed what appeared to be a most excellent way across, with an approach, though steep, that seemed entirely feasible.

Having feasted our eyes to the full on the wonderful panorama spread out before us, which included, of course, the Tatoosh peaks and St. Helens and Adams beyond, we hastened on. Shortly we were attracted by a scene below that caused us again to pause. There on a grassy slope and not more than a quarter of a mile away a band of fourteen goats, five adults and the rest kids and yearlings, were grazing as peacefully as if in a farmer's meadow. After gazing for some minutes at this quiet pastoral scene we attempted to approach to within photographic range, but without success. Soon the nearest of the adults was observed to look intently for an instant in our direction and then to walk quietly over into the midst of the herd, when presto, as if by one impulse, the whole herd began to form in file and move smoothly toward the glacier; and then, as a little ridge was gained, they broke into a rapid run and soon were gone. Much disappointed, we continued on our course and, skirting the slope, had just reached a point where we expected to drop down to the selected point of crossing, when there again came our goats, this time apparently oblivious of our presence though much nearer and moving in long calvacade up and across the steep snow field. It was now too dark for a picture but we stood and watched them until they disappeared from view among the seracs of Wilson Glacier, which here joins the Nisqually.

The snow slope leading down to the glacier was long and for a distance very steep. To zigzag down it would have required a half-hour at least and the evening shadows were fast lengthening. By glissading we could do it with much more ease and in a small fraction of the time. Turning to my companion, who had been ever ready on the trip to tackle any venture that was suggested, I remarked, "Mary, are you game?" "Yes, if you think it's safe," came the answer promptly. On the instant we were off and in scarcely more time than it takes to tell it were on the glacier. My wife had take the steep slope like a veteran, not once losing her feet.

Once out upon the glacier we found the going excellent, as safe and easy as any boulevard, and soon were across. In crossing we had diagonalled downward slightly so as to strike the opposite canon wall where it was the least high and steep and this brought us quite near to Panorama Point. A long and tedious climb out of the canon, a rapid traverse of the moraines and snow fields below McClure Rock, then of Paradise Glacier, and we were in the Mazama Camp. The hour was 9:30 and darkness had come.

The first group of tents that we encountered chanced to be those of the Montagues, Finleys, Brewsters, Torreys and Dr. Marcotte. Intercepted, as we were passing, by these worthy folk, we were made to partake of the good things in the way of provender that they had in private stock, to all of which, as may be guessed, we did full justice. Then with a fanfare of trumpets heralding our approach (in reality the clarion voices of Messrs. Marcotte, Finley and Montague in well assumed burlesque) we were ushered blushingly before the Mazama camp-fire. After listening to a brief recital of our trip, the camp-fire session, which had been about to conclude its program, adjourned; and we to rest and sleep.

We had been nine and one-half days on the trip from Fairfax, had lived out of our knapsacks all that time without replenishment (except a pound of sugar and a half-pound of butter obtained at Indian Henry's), had crossed seven glaciers, had made in the aggregate 20,000 feet of elevation, and had hiked a total distance of 80 miles. A strenuous trip, say you? Yes, it was; but it was worth while.

With the Birds and Animals of Rainier

By WILLIAM L. FINLEY
State Biologist of Oregon

The fun of studying wild birds and animals with a note book or camera is not so much in odd time chances of observation, but in continued periods of leisure, so you can spend your entire time about the bird homes and in the haunts of wild animals, just as one takes a vacation at the seashore. The joy of nature comes to the amateur, not to the professional. But to get good photographs of shy creatures, one has fairly to make a business of lying in wait for his subjects hour after hour, or maybe week after week.

We were camped at Indian Henry's Hunting Ground on the west slope of Mt. Rainier early in July. The snow still covered the ground everywhere except in a few patches under tree clumps. On July 10, Flett, Hungate and Jewett were scouting for mountain goats up over Pyramid Peak and by chance they nearly stepped on a white-tailed ptarmigan sitting on her eggs. Next morning we loaded the moving picture camera on a pack horse and started for the nest. It was over the snow the entire way, but not a difficult climb except in one rather steep place when we got well up on the side of Pyramid Peak. When we approached the nest, the bird was not at home. The nest was in a patch of heather at the base of a big rock on the steep slope. We went on above to a level place, unloaded and sat down to wait. In half an hour I looked over the edge of the rock and the mother had returned to her eggs.

We moved around cautiously a little distance away, snapping the camera at nothing in particular and pretending we did not know anything about the ptarmigan's nest. I don't know whether the old lady on the eggs was deceived or not, but we approached nearer and nearer, until I sat down within eight feet.

Birds differ a good deal in individuality. Sometimes one will stand for a good deal, but at other times we find a fussy individual that is too particular about her home affairs. This is why a camera man has to go very cautiously and use a good deal of patience. After we had tried several exposures, I moved the motion picture camera up within shooting distance. It is an ungainly cyclops-eyed looking monster that any ordinary fowl might be afraid of, but I soon found Mother Ptarmigan thought her first duty was to hold the fort regardless of all invaders.

She sat flattened in the heather and although her gray mottled plumage was no match for green in color, yet with the lights and shadows in the heather and backed with a big gray rock, I would have defied any person to see her if he had not known the spot. Her sharp eyes watched every move we made as camera after camera—there were only four—moved nearer on her home.

It is fortunate for a camera hunter that some birds and animals live far up on the mountain sides or in deep woods where they really do not know how dangerous a man is. We put the lady of the party forward for a closer interview. She knelt down at four feet and put her hand slowly forward in friendly greeting. Inside of ten minutes the hand was within four inches of Mrs. Ptarmigan's bill, with the moving picture camera clicking off a footage that seemed rather wasteful. But in taking moving pictures one has to keep on grinding, for you never know just when something good is going to happen.

As the hand moved nearer, the bird on the nest began to cluck softly, almost inaudibly, showing that she did not like to be bothered. But when her feathers were stroked gently she was much more docile than an ordinary setting hen, for she did not show fight.

The main enemies of the snow grouse or ptarmigan here on the high mountain side are not men with guns, but perhaps hawks and eagles that occasionally fly over. The ptarmigan is best known because of its changing plumage, which is a good example of protective coloration. In the winter when the whole country is snow covered, it is dressed in pure white to match the snow fields, but in summer its body is clothed in a speckled dress of brown and white. The under part of the tail and the wings remain in pure white even in summer. When it crouches among the rocks it is difficult to see. As a rule the eye catches a bird only when it is moving.

Then Mrs. Finley reached under and took out one of the eggs, but the bird mother did not seem to care. When it was laid down in front of her she reached out her bill and pulled it back under her breast with an expression that plainly said, "Any woman ought to know that you can't hatch out a chick without keeping the egg warm."

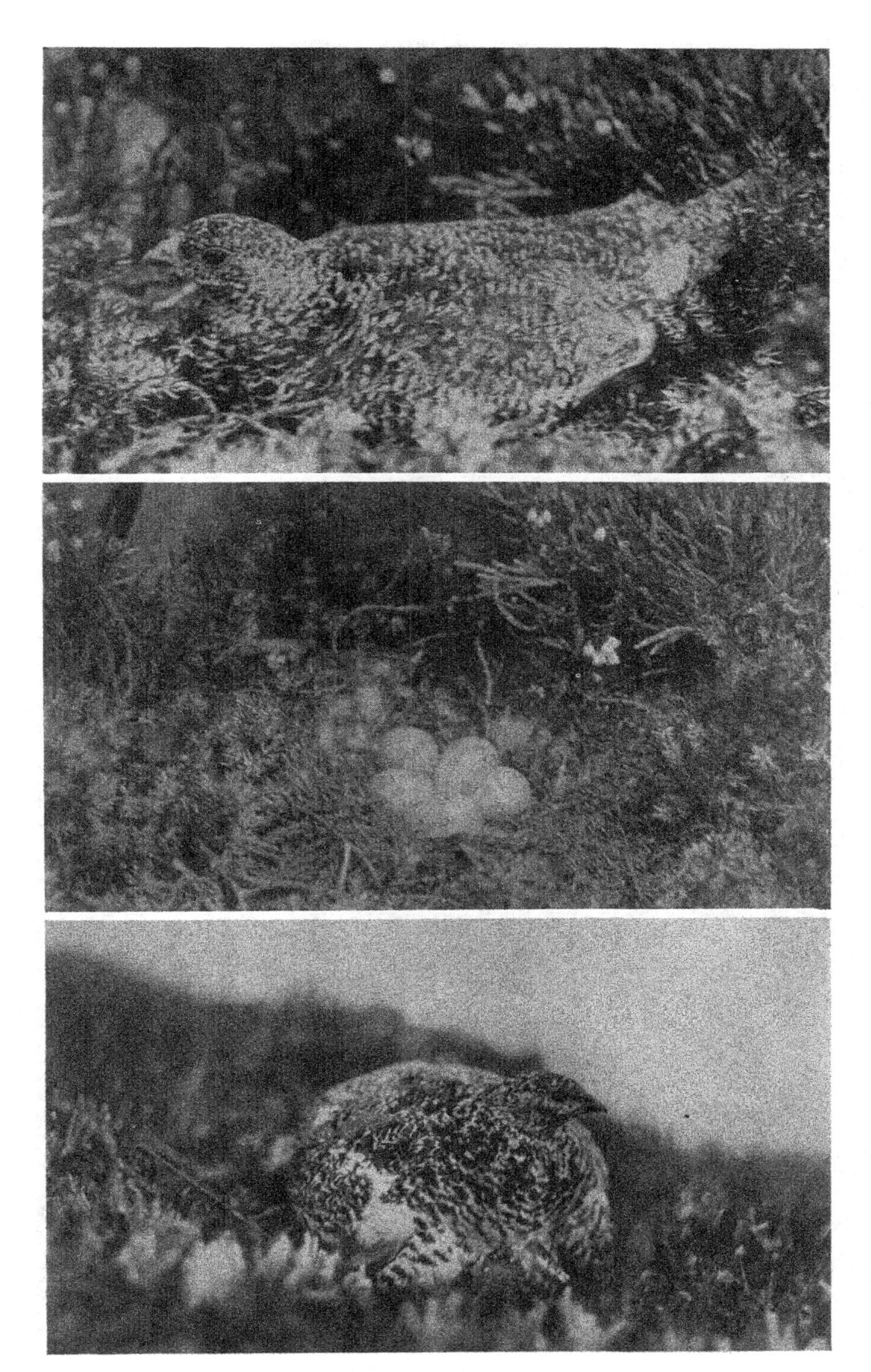

1—Mother Ptarmigan on nest. Pyramid Peak. above Indian Henry's.
2—Nest and Eggs of White-tailed Ptarmigan on Rainier.
3—Mother Ptarmigan hovering young, Mazama Ridge.
Photos by Wm. L. and Irene Finley.

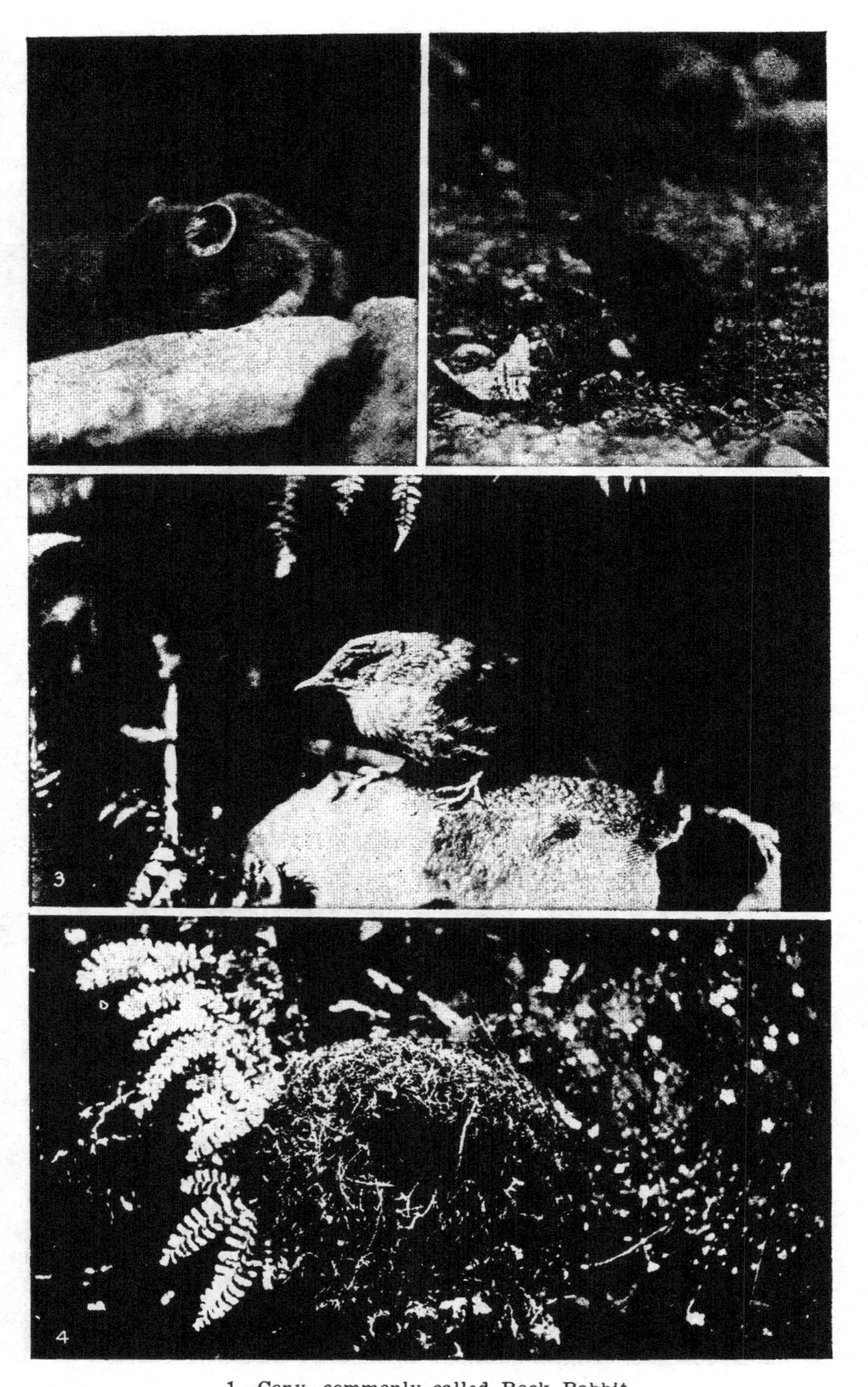

1—Cony, commonly called Rock Rabbit.
2—Varying Hare or Snowshoe Rabbit at Longmire.
3—Young Water Ouzel or Dipper.
4—Moss Nest of the Water Ouzel or Dipper.
Photos by Wm. L. and Irene Finley.

Then to show us that she knew her business, although surrounded by a battery of cameras and four people, she reached under and turned her eggs, for this, too, is an important part of bringing off a family of chicks.

A few minutes later when the moving picture camera was within three feet of her it was being reloaded and one of the magazines slipped with a loud click. She must have thought it was going to explode, for she jumped to one side and went walking off among the rocks. She seemed rather glad to have a little rest from household affairs, for she fell to preening her feathers and then started eating the blossoms and buds of the white heather. I judged by the way she snapped them off that they were one of the main items on her bill of fare. This gave us good additional material for photographs.

Fifteen minutes was long enough for the eggs to be uncovered, but her home was surrounded by cameras and people. She executed a flank movement, walked around from above and slipped in on her nest, giving the eggs another good turning over as she fluffed out her feathers and settled down again to business.

A few weeks later while we were camped on Mazama Ridge, late one foggy afternoon Mrs. McClain rushed up to our tent and said: "Come on with your camera. I want to show you a ptarmigan and her chicks." She led the way up the trail about a hundred yards and there was an old mother ptarmigan hovering over a family of chicks. She, too, seemed unafraid of people, as we edged toward her very slowly. I could see she did not want to move and let her children get out in the could. She permitted us to get within four or five feet of her and so finish our series of ptarmigan pictures with not only a mother on the nest, but also with her family of chicks.

We were camped at Indian Henry's for five days, from July 9 to July 13, and had a very good chance to study bird life in that locality. The Biological Survey of the Department of Agriculture is co-operating with the National Park Service of the Department of the Interior to publish a popular report on the wild birds and animals of Rainier National Park. Our party at Indian Henry's was working up the field material for this report. The plan which was carried out was to spend the entire summer circling the mountain, camping and making cross section surveys from lower to higher altitudes. The expedition was managed by Dr. Walter P. Taylor, Assistant Biologist to the Biological Survey of Washington, D. C. Another member of the party at Indian Henry's was J. B. Flett, who has been Park Ranger for seven years. He was formerly with the Department of Botany of the Tacoma High School. He is an expert on the flora of the park and has published a very interesting pamphlet printed by the Department of the Interior, entitled "Features of the Flora of Mt. Rainier National Park." William T. Shaw is Professor of Zoology at the State College at Washington, former instructor in zoology at Oregon Agricultural College. J. W. Hungate is in charge of the Department of Biology of the State Normal School at Cheney, Washington. Stanley G. Jewett, Predatory Animal Inspector of the Biological Survey, George Cantwell, Field Assistant of the Biological Survey, Mr. and Mrs. William L. Finley, William L. Jr. and Phoebe Katherine Finley comprised the rest of the party.

After a short stay at Longmire, the next regular camp was made at Reflection Lake. Paradise Valley and the nearby points of interest are visited by almost every person who enters the park. So perhaps more attention was paid this region than any other part of the park, as people in general will want to know especially the wild creatures of this section.

Another creature that, like the ptarmigan, has a change of coat for the seasons is the varying hare or snowshoe rabbit. He has the brown fur in summer, but changes to the white in winter. However, it is interesting to note that in some of the localities of western Washington and Oregon the snowshoe rabbit does not change to the white coat of winter. This is especially true of regions where there is little snow.

Ordinarily the snowshoe rabbit is very shy in the woods, but where protected he soon becomes acquainted with man. At Longmire, I counted twelve or fifteen snowshoe rabbits around the hotel and some of the other buildings. Some of these seem to have a home under the tent platforms. At the ranger's cabin, Mr. Flett fed two during last winter when they stayed most of the time under the cabin, and they were quite tame. I saw one of these several times and once or twice got up within about twelve feet of him. As he went hopping off leisurely he reminded me very much of a toy hobby horse, as his big padded hind feet rolled up as he jumped.

We of course expected to find the American dipper or water ouzel along these mountain cascades. One day Mrs. Finley and I were standing at the Third Crossing bridge over Paradise River when we saw a water ouzel carrying food. Mrs. Finley went on

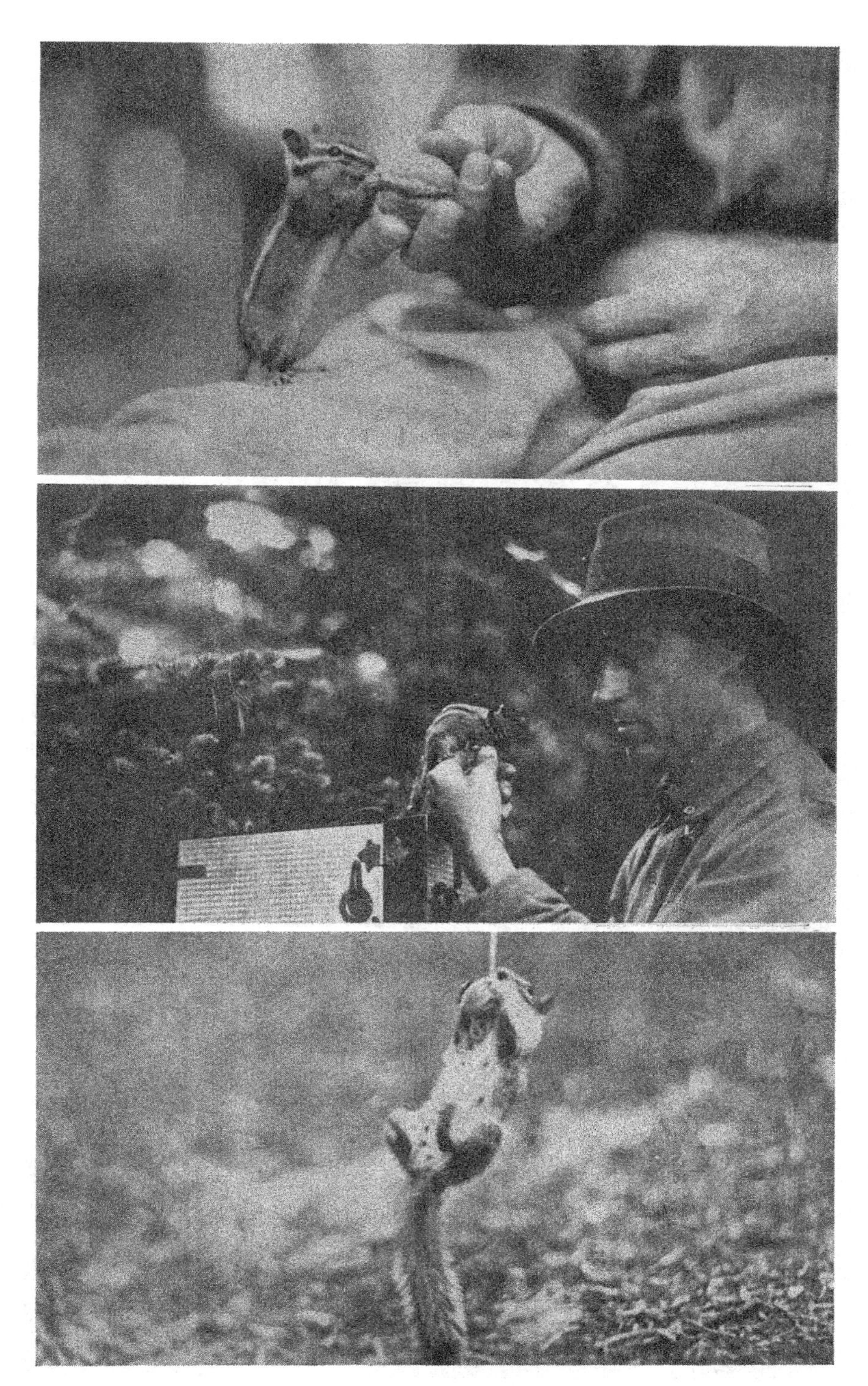

1—Taming a Chipmunk.
2—"Gimme that nut."
3—"Chippy" getting a nut that has been tied a little too high.
Photos by Wm. L. and Irene Finley.

1—View of Mt. Rainier from Indian Henry's.
2—Mt. Rainier from Indian Henry's.
3—Mt. Rainier from Indian Henry's.

one side of the stream and I on the other up to the falls just above the bridge. I discovered a spot on the ooposite side under a rock at the side of the falls where I saw a nest. It was a big mossy ball with a hole in the side, a typical home of the bird. We soon discovered that this was either an earlier nest of the present season or a nest from last year.

The main stream about fifty yards above the bridge drops over a rock wall perhaps twenty-five feet high. On the right side there is a narrow space of some four inches between the wall and the water. To our surprise, we discovered that the birds were carrying food, lighting on a jutting point in the rock wall and then fluttering along through this narrow space in under the main volume of water. It seemed impossible that there could be any place for a nest with such a great amount of water pouring over the rock.

The conditions showed very clearly that the birds had built the nest earlier in the season when the volume of water was smaller. It was impossible to get a sight of it from any angle, but it was very likely on a little shelf. With the melting snows, the water had increased and the old birds were nearly shut off from an entrance. The nest must have been in the dripping water, but the young birds were perhaps nearly grown so they could survive the flood. With the main current of the falling waters just missing the nest, the young birds would have to wait until they were expert on the wing to ever get out from under the heavy falls. By tying the camera to a small tree just at the edge of the cliff and pointing it down, we got some pictures of the old birds going in and out.

One of the impressions a person gets in the higher mountains is a lack of bird life. Yet if he keeps his eyes and ears open he is likely to discover birds that are shy and not seen by the ordinary observer. Of the hundreds or thousands of people passing over this Third Crossing bridge, perhaps not a single one during the summer season discovered the nest of a hermit thrush in the branches of a scrub fir that hung down from the top of a rock wall a few feet above the rushing waters and not more than twenty feet from the railing of the bridge. Standing on the bridge, we could watch the male and female feeding the young birds.

By climbing the rock six or eight feet I could get a foothold on a flat stone but could not stand up on account of the overhanging rocks. Both the parent thrushes were feeding young birds that were about a week old. They were shy at first and we had to sit crouched down for quite a while to get the birds used to our being so near the nest. They perched in the branches above and did not dare to go to the nest. After staying around the vicinity for three days and making frequent visits, we were on intimate terms with both birds. We found the shyness in the thrushes was an indication of their finer nature. We enjoyed visiting them and trying to get some good pictures.

While we were camped at the west end of Reflection Lake, we had a good chance to study conies and marmots in a big rock slide a hundred yards or so up the trail to the west. Here the boulders and slabs have shaled off a rocky cliff, making a typical home for these animals. Just west of the slide in the woods, we found a fair sized colony of mountain beaver.

The sewellel, commonly called the mountain beaver, is not related to and does not resemble an ordinary beaver except perhaps in its gnawing capacity. It does not inhabit the water, although it lives in moist places often near the water. A typical place where I saw a large colony of these animals is on the north side of Wahtum Lake near the upper end. The mountain beaver is something like the marmot or woodchuck, but it has a very short tail, small round eyes and long whiskers. It hunts and feeds at night time and as a rule stores up its supply of food for the winter by cutting the leaves and green twigs and letting them dry or cure.

The rock slide was inhabited by an old female hoary marmot and three young. The animal is well named because of its silver-gray back. It is rather sluggish in habit, but lopes about with ease over the rocks. The movement looks very much like that of a bear. This mother had a loud whistle that reminded me very much of that used by one boy to call another. As we approached, the mother used this as a signal to warn her children to get down under the rocks out of danger.

The habit of this old marmot was to feed during the early morning, but as soon as the sun was well up her greatest pleasure was to stretch out on a flat rock and take a sun bath. Her favorite food seemed to be bunch grass. Sometimes she would go around and clear up to the top of the cliff; at other times I saw her feeding down below the trail in the woods. The mother was not very wild, but cautious and always extremely anxious about her young when we were near the slide. One morning she was on the big flat rock, her favorite sunning place. One of the little marmots climbed up where she was. I was about forty feet away. She was uneasy. She pushed him over by putting her nose under one of his front legs, then picked him up in her mouth, just like an old cat picks up a kitten, except she had him by the side of the body. As she raised him up he curled in a ball and I saw her carry him ten or twelve feet over the rocks and down into a crevice. I think she was just weaning her young, for occasionally I saw her carrying a mouthful of grass from up the hillside down over the rocks.

There were a number of conies, commonly called rock rabbits, living in the big slide. The first day I saw one about thirty feet above me. He was coming down the rock with a big mouthful of grass. The grass was four or five inches long, carefully placed and held in his mouth like a small bundle. He had the root ends in his mouth; it stuck out on one side like a long moustache.

The cony looks like a little gray guinea pig. An interesting part of his life is his habit of making hay during the summer. He cuts grass and tender twigs, which are dried in the sun, and then stored away under a slab of rock for the winter. I have often seen his hay smoothly stored in little cocks very much resembling the hay in a farmer's field, except for size.

We had no sooner pitched our tent on Mazama Ridge than a little chipmunk ran up the tree where I was standing and sat looking down curiously as if to enquire what we had come for. He watched me as I nailed up an old box against the tree for a cupboard. As soon as we had stacked in our provisions, he called in all the neighbors and they began gnawing open and carrying off whatever was available. Fortunately, we discovered the raid made on the nuts in our dunnage bags before they were all gone. We decided after that they would work for any nuts they got.

I tied a string to a peanut and soon Chippy was running along after it as a cat tries to catch a spool. At first he would sit and open up the nut and store the kernels away in his cheeks. After a nut was pulled away from him a few times, his first idea was to cut the string and carry the whole thing away.

Chippy was ambitious to get a nut no matter where it was. He soon learned to walk ropes and a slack wire as easily as he could go from branch to branch. I next hung some nuts on a long string attached to the slack wire. They were within a few inches of the ground. I thought perhaps he would climb down the string to get the nuts. He tried to reach them from the ground, but finding it impossible, Chippy ascended the tree and walked the slack wire. When he came to the long string with the nut dangling at the end he did the simplest thing possible—hauled the string up hand over hand to get the nut. He soon learned that nuts grew on strings and he went around camp hauling up every string that he came to.

All of which furnished a fair amount of work for Chippy and some amusement for the Mazamas who gathered around.

MOUNT RAINIER

Something untrodden in the routine dust
Of unconcerned humanity, something
Unclaimed, some spot yet sacred, undefiled,
Above, beyond the daily round of form,
Still native, free, and pure—such seekest thou,
O idle dreamer? Yonder turn thy gaze
To that intrepid peak which fills the sky;
To human eyes still changeful, whether in
The hueless lights of cold and unsunned dawn,
Or in the warmer tints of brilliant sunsets;
Yet endlessly the same, uplifted—aye,
Unmoved, most strong, unmindful of the storms
Of human fate and human destiny.
Fact visible of God invisible,
And mile-post of His way, perpetual
And snowy tabernacle of the land;
While purples at thy base this peaceful sea,
And all thy higher slopes in evening bathe,
I hear soft twilight voices calling down
From all thy summits unto prayer and love.

—*Francis Brooks.*

Views of Tatoosh Range from Mazama Ridge, with Mt. Adams in the distance.
—By Boychuk.

THE MEIJE FROM THE WEST.
Photo by Munroe Thorington.

Dauphine Days

J. MUNROE THORINGTON
American Ambulance, France

More than half a century has passed since Edward Whymper and a few others brought us pen pictures of the little known Alpine district of the Dauphine. The region has continued to be one which is comparatively rarely visited and the descriptive literature is restricted to a few tourist booklets and to the more accurate though less entertaining Alpine guide books. It has therefore seemed not entirely out of place to record these recollections of a war-time holiday.

While the writer professes to be one of those whose chief joy consists in finding the "wrong way up" to some airy spire, he also unashamedly acknowledges the charm of a perfect mountain landscape as seen from the valley. From this less exalted point of view, contrast is essential, and the memory picture of a Dauphine valley with harvesters on the hillside, a lake with cattle, a little age-old stone chapel and snowy peaks that tower ever upward will last when the afterglow of many another mountaineering experience has faded beyond remembrance.

We had served in a base hospital during the Champagne drive (1917) and our good fortune had brought us into Paris on that day of days when the first American troops paraded. In a week of little medical work, we had even sneaked away to Chamonix for some days on Mont Blanc and the aiguilles. It was these events which formed the background for a last few days in the mountains, and the Dauphine was the region of our choice.

Leaving Paris on the night express, the next morning found us in Grenoble on our way to La Grave. Our party of three, including a U. S. lieutenant, my brother and myself, was increased by the addition of a Scotch captain just down from Arras and bound for Turin. Our lieutenant held the distinction of bringing the first American uniform to this part of France, his broad-brimmed campaign hat conspicuously in contrast with the blue-tam-o'-shanters of the "blue devils" of this district. My brother and I were in the uniform of the American Ambulance, and, with our Scotch reinforcement, it is not surprising that certain small boys constantly followed us about.

It was late in the season and we were quite alone in the P.L.M. motor as we rolled out of the old university city built high up on the hillsides of the Isere valley. Ahead of us in the distance the snow peaks of the Belladonna group rose blue grey in the morning mist of a clear September day. Taking the Briancon road, we soon reached the sleepy little town of Vizille with its picturesque chateau, known historically as the cradle of the French revolution, and in another hour arrived in Bourg d'Oisans. In the main street our progress was held up for a number of reasons, including a flock of geese, a balky mule and a little puffing train loaded with logs. The mixup looked hopeless, so we dismounted to walk about while the debris was being cleared. German prisoners engaged in construction work along the road eyed us curiously as we passed and noted the American uniform with much surprise. We paraded by several times for their benefit. The road once clear, we continued onward through the gloomy gorge of the Romanche and out into a broad sunny valley beyond.

The main valleys of the Dauphine have a climate much like that of the Riviera, favorable for fruit and grain cultivation, while every gap in the hills frames a soaring peak or a gleaming snow field. On the other hand, many of the side valleys are so narrow and deep that they are always in shadow, the boldness of their precipices, the rushing waterfalls and absence of vegetation giving them an unsurpassed wildness of character.

Here every available bit of ground was under cultivation, even steep slopes up beyond the smoke blackened villages on the hillside. Our Scotch friend facetiously but aptly described it as "farming of the highest type," and indeed we frequently saw the peasants making use of the forces of gravity in bringing their loads down the mountain side and in some cases even using a natural "take-off" from which to load their wagons.

As we passed along, peasant boys climbed on the machine to sell lavender flowers and we bought huge fragrant bunches for a few pennies. The soaring peak of the Meije came into view, the tumbling ice falls from the great Dauphine glacier of Mont de Lans apeared on the cliffs south of the road and in a few minutes we rolled into the quaint old hillside town of La Grave, which we had selected as our headquarters during the next few days.

After lunch we strolled up the mountain path past the village of Les Terasses with is twelfth century church and passed on

N. face of Meije and Breche de la Meije from village of Les Terosses.
Photo by Munroe Thorington.

Upper—Alpland on the Col du Loutant.
Center—Summit of the Tete de Toura Meige (left), Pic de la gave (right.)
Lower—Peaks of the Central Dauphine, looking south from the Col de la lange.
Photos by Munroe Thorington.

toward the Col d'Infernet road to the little chapel of Le Chazelet. Across the valley rose the majestic peak of the Meije with its rocky buttresses gleaming with new snow. Green valleys, dotted with villages and checkered with yellow terraced fields constantly came into view as our panorama broadened and the meadows over which we walked were a white carpet of edelweiss. We were happy and carefree and one quite understood how a sober army man could lose his dignity enough to frisk over the meadow in a vain attempt to capture a stray billygoat which appeared much startled at the eccentricities of the genus homo in uniform.

Continuing along the ridge, we reached the grassy summit of the Signal de la Grave, commanding a magnificent view, including the Grand Rousses and the Aiguille du Goleon toward the north. Southward, we overlooked village-dotted valleys to the majestic Monte des Agneaux at the sources of the Romanche, while the Pelvoux, Grand Ruine, the black crags of the Ecrins and the icy wall of the Meije and Rateau ending in the glacier du Mont de Lans closed in our view. We stopped for nearly an hour in this wonderful spot and walked down to the valley as the sun was setting over the westward snow fields.

The next morning we were content to remain down in the village, sprawled out on a grassy meadow with an unobstructed view of the Meije and its hanging glaciers. We were anxious to make at least one good ascent in the district and attempted to find a guide, but all the first class men were away on army service and we had to content ourselves with a porter (Emile Pic). In the afternoon we set out for the Refuge Evariste Chancel with the idea of reaching the Col de la Lauze (11,625 ft.) on the next day. This pass, lying between the Pic de la Grave and the eastern end of the Mont de Lans glacier, offers some interesting snow work and is famous as one of the great view points in the Dauphine. A beautiful walk through the woods on the south side of the valley brought us to the little lake of Puy Vachier, hidden away in a glacial cirque north of the Rateau, the refuge perching high up on the rocks to one side. Another half-hour found us at the cabin, looking down at the reflection of the Meije in the blue water below.

Alpine club houses are seldom visited in war time and we effected entrance by way of a rear window; the doors were soon open and shortly afterwards appetizing odors began to float out from the direction of the kitchen. Outside, the sunset was magnificent, the great peaks fiery red and taking on heliotrope shades in the afterglow. We had our evening meal by the light of a flickering candle, the villainous greasy face of our guide and the lieutenant's sombrero giving the place an appearance of the typical wild west den, seen only in the moving pictures.

After a comfortable night, it was no easy matter to make an early start, but despite the difficulties we were well on our way and making fast time over the hard snow slopes before daybreak. The black rocks of the ranges in the east were sharply outlined against a reddening sky, the slender Pic Central of the Meije seemed turned to silver, while we, the only human beings in this Alpine world, cast gigantic stalking shadows out over the snow as the sunlight reached us. We roped before climbing out onto the glacier, but there were very few crevasses and these we crossed by firm snow bridges, traversing the long slopes and cutting steps toward the glistening pass which lay ahead. We were soon in the snow saddle and climbed up to a resting place in the rocks to one side.

The panorama was marvellous. Northward, almost ninety miles away, the stupendous southwestern face of the Mont Blanc massif dominates the view and we could easily pick out the individual peaks lying between the Aiguille du Bionassy and the needlelike spire of the Geant. Nearer are the ranges of the Tarentaise and far below in the valley we could just make out the houses of La Grave as dots in the patchwork of the terraced mountain side. Eastward, we overlooked the valley of the Romanche toward the Col du Lautaret and far beyond to the peaks of Italy, then came the nearer summits of the Dauphine, the jagged ranges in the south seen across the trench of the valley des Etancons and silhouetted against a cloudless sky, while to the westward, across the tremendous expanse of the glacier du Mont de Lans, our view was of the plains of France stretching away in the distance. We fancied we could make out the misty ridges of the eastern Pyrenees. Below in the valley of the Veneon it was possible to see the chalets of La Berarde and across the ridges beyond we caught a glimpse of the Val Gaudemar.

We spent several hours following chamois tracks across the great snow fields of the glacier du Mont de Lans and climbed the little rocky peaks of the Jandri and the Tete de Toura to look over the precipices of La Berarde in the valley below. Then back to our pass where the remains of our bread and cheese disappeared as if by magic. Our guide smoked contentedly and blew great rings which floated lazily away, while we lay in a sunny corner of the rocks watching Mont Blanc disappear in the blue glint of the noon haze. And less than three days before we had been in the wards of a great hospital with its endless rows of beds filled with brave men with maimed bodies; it seemed so very far away, absurd and impossible.

We were aroused by Emile, who had begun to yodel in an ungodly fashion as a sign that his pipe was finished and that we must be on our way. With rope in place and axe well in hand we glissaded helter-skelter down below the Pic de la Grave and out to the crevassed fields below. The slopes up which we had laboriously cut steps in the early morning had softened and we glissaded down through snow almost knee deep with little rainbows of glittering spray following us.

Our afternoon was spent in wandering lazily through the woods ti La Grave wth no more exciting adventure than the waylaying of an unwary cow; be it known, however, that we discovered a new use for a lieutenant's sombrero.

Next morning we walked nine kilometers to the Col de Lautaret, which lies on the old Napoleonic road from Italy. To the north is the Col de Galibier, the highest road pass in Europe, while southward the road toward Briancon ends in the hazy distance of the plains. The Lautaret is a bit of grassy alpland with the old hospice in its midst. Goats graze contentedly by a little sparkling stream and like a setting for a beautiful jewel, this garden spot is hemmed in on all sides by gorgeous peaks of bold and fantastic outline.

And here I would leave you, in the midst of beauty that mortal pen will never perfectly describe. To know the charm of these isolated valleys you must seek them out for yourself with the love of the Old World out-of-doors in your soul. The Dauphine is the land of heart's desire for the lover of contrast and completeness of setting in Alpine scenery. Perchance one day you too may walk this road to the Lautaret and beyond,

"And, pausing, look forth on the sundown world,
Scan the wide reaches of the wondrous plain,
The hamlet sites where settling smoke lay curled,
The poplar-bordered roads, and, far away,
Fair snow-peaks colored with the sun's last ray."
(Alan Seeger.)

Mountain Sickness

HARRY BEAL TORREY

One gathers a wealth of impressions in camp with the Mazamas on Mt. Rainier; of the glories of massive mountains and crevassed glaciers, of alpine lakes and hillside tapestries, of rain and fog and crystal air and wonder-working clouds, of summer sun brilliant on winter snows; of the rare Mazama fellowship that throughout warmed our hearts; of vivid appeal to sense and spirit in almost every moment of the outing. And all of these dwell in one's memory, heightening the color of everyday life.

Other impressions one gathers also on Mt. Rainier, that belong to the athletic business of climbing. Of some of these I have been asked to speak.

Let me first of all give you a few figures.

To reach the summit from Camp Coursen on Mazama Ridge one must ascend 8700 feet in a little over five miles. The ascent falls conveniently into three parts: (1) from Camp Coursen to Anvil Rock; (2) from Anvil Rock to Gibraltar; and (3) from Gibraltar to the summit. In the first, one ascends 3900 feet in a little less than three miles; in the second, 3100 feet in one and one-half miles; in the third, 1700 feet in less than three-quarters of a mile. The first was covered by our party in four and one-half hours, at a rate of 866 feet vertically and 3400 feet horizontally the hour. The second was covered in six and one-half hours, at the much slower rate of 477 feet vertically and 1260 feet horizontally the hour. The third was covered in two hours, at a more rapid rate than the second, though less than the first; at the rate, namely, of 850 feet vertically and 1766 horizontally the hour. From these figures it will be seen that the average grade becomes progressively heavier from camp to summit; but the going is the slowest between Anvil Rock and Gibraltar, indicating that it is also the most difficult. Cowlitz Cleaver and the Chute will be remembered in this connection, the latter offering the most arduous climb of the entire ascent. The Chute is, indeed, a fitting introduction to the rest station on Gibraltar aptly named Camp Misery. It is here that those who pass the 10,000 foot level usually drop out of line, if they drop out at all. It is hereabout that mountain sickness claims its victims. But for those who escape the nausea of it, the last part of the ascent is most surprisingly fatiguing. Here strong men may show signs of profound exhaustion under relatively trifling exertion. The symptoms are not uncommon. But why should they occur?

Mountain sickness may present itself under a variety of forms. Nausea is one, perhaps the most constant. Extreme weakness is another; and vertigo; and headache; and fever; and delirium; and, in rare cases, unconsciousness that may end in death. All of these symptoms may be encouraged by activity. But they may appear under conditions of complete repose. In the well known ascent of the balloon Zenith, April 15, 1875, the three balloonists became so weak after passing 25,000 feet that they could not adjust the bags of oxygen with which they were provided. At 26,250 feet, one of them, Tissandier, became speechless, sleepy, and soon after, unconscious. He alone survived to describe his sensations, among which, in this case, distress was entirely lacking.

Mountain sickness attacks different people at different al titudes. Below 10,000 feet cases are relatively rare. Above 10,000 feet they are more common, increasing in frequency with the altitude. They are due primarily to a diminished supply of oxygen, accompanying the rarification of the air. In the neighborhood of 10,000 feet of elevation, the amount of atmospheric oxygen becomes critically small for many people, who may there develop distressing physical symptoms. But this statement is not so simple as it may seem at first glance. If the story is to be more than an unhappy ending, it will be necessary to bring in a few more figures to develop the plot.

At sea level the weight of the atmosphere is just balanced by a column of mercury 760 millimeters high. This measures the barometric pressure. Approximately 21% of the atmosphere is oxygen, which thus accounts for 21% of the total barometric pressure. At sea level this amounts to 159 millimeters. At Camp Coursen (5700 feet), the barometric pressure is but 610 mm. The oxygen pressure is accordingly but 128 mm. At Anvil Rock (9600 feet), the barometric and oxygen pressures are respectively 520 mm. and 111 mm. At Gibraltar (12,700 feet), they are respectively 471 mm. and 99 mm. At the summit (14,400 feet), they are respectively 442 mm. and 93 mm. It thus appears that while the proportion of oxygen in the atmosphere is the same at all levels, its pressure, which represents its availability for breathing purposes, sinks to 80% of its sea level value at Camp Coursen, to 70% at Anvil Rock, to 62% at Gibraltar, and to less than 59% at the summit. This is only a little more than one-half of the supply of oxygen to which Portland residents are accustomed.

But these figures alone do not adequately convey the real significance of these reductions. To supplement them, it should be stated that the percentage of oxygen in the lungs (so-called alveolar oxygen) is about two-thirds that in the air inspired. At sea level, then, the alveolar oxygen will have a pressure of about 106 mm. of mercury; at Camp Coursen, 85 mm.; at Anvil Rock, 74 mm.; at Gibraltar, 66 mm.; at the summit 62 mm.

Now, the pressure of oxygen in the alveoli of the lungs is correlated significantly with the amount of oxygen in the blood. At sea level, where the alveolar oxygen pressure is 106 mm., the blood is almost saturated with oxygen. At Camp Coursen the oxygen pressure has fallen to 85 mm., but the amount of oxygen in the blood remains practically unchanged. At Anvil Rock the alveolar oxygen pressure has fallen to 74 mm.—but the blood is still 94% saturated. At Gibraltar (alv. pr. 66 mm.) it is 92% saturated. At the summit (alv. pr. 62 mm.) it is 90% saturated. All of these figures refer to a person at rest. With moderate exercise, the decline of blood oxygen is much more striking, reaching 90% at Camp Coursen, 82% at Gibraltar, and 80 at the summit.

It thus appears that the amount of oxygen in the blood is not seriously affected by the altitude below the 10,000 foot level. Above the latter it begins to fall off more and more rapidly. At Gibraltar, immediately after the very heavy grind up Cowlitz and the Chute, it falls to a concentration that is distinctly below normal and strikingly associated with the great fatigue of the ascent.

This fatigue is less than it might be, however, were it not for the fact that both heart and breathing mechanism respond to the lack of oxygen by increasing their own efficiency. As the atmospheric pressure lessens, the pulse quickens even in persons at rest. Exercise greatly accelerates the heart beat. This insures a more rapid transportation of oxygen to the tissues. Respiratory movements become more frequent and powerful, increasing the volume of air respired every minute and the amount of oxygen exposed to the blood in the lungs.

These movements of heart and respiratory muscles depend upon an increased acidity of the blood. It is well known that when carbon dioxide or lactic acid, both products of the decomposition of the tissues, reach in increased quantities, through the blood vessels, the nervous centers governing the activities of the heart and the respiratory muscles, these activities are accelerated.

So it happens that, though both of the acids are in a sense waste products, they are at the same time very useful to the body in adapting it to moderate changes in respiratory conditions.

Besides affecting the cardiac and respiratory movements, they control to some extent the carrying power of the blood for oxygen. The major portion of the oxygen in the blood is carried in loose chemical combination with the hemaglobin of the red blood corpuscles. An increase in carbon dioxide or lactic acid lessens the amount of oxygen that the blood can take up in the lungs, but it also performs the very important function of facilitating the delivery of oxygen by the blood to the tissues. Now if the reader will visualize a heart pumping faster than usual, and the blood consequently passing through the fine capillary network penetrating all parts of the body, at a greater speed than usual, then it will be clear that the added amount of blood coming every minute to any organ—the brain, for instance—can have an added usefulness for that organ only if some means is at hand for unloading with unusual rapidity the oxygen which it carries. That means is provided by the increased acidity of the blood.

On the other hand, diminished acidity makes it possible for the blood to take up more oxygen in the lungs, but less easy for it to part with that oxygen to the tissues. This latter fact has been found especially important. So it may be said that when the blood diminishes in acidity the tissues may obtain less oxygen, even though the amount of oxygen in the air remains constant.

The blood not only carries oxygen and carbon dioxide, and lactic acid on occasion, but it also carries phosphates and carbonates of soda, and other substances that tend to neutralize its acids and increase its alkalinity. They constitute the alkali reserve. It is well known that as the amount of carbon dioxide in the blood decreases, the alkali reserve normally decreases also. Now it happens that at higher altitudes, the carbon dioxide does materially decrease. It follows that the alkali reserve should decrease also. This is effected under normal conditions by the elimination of the excess through the kidneys. If the elimination is prompt and adequate, the normal balance of acid and alkali in the blood will be maintained. If, for any reason, it is delayed, the blood will become forthwith more alkaline. Since alkalinity tends to hinder the escape of oxygen from the blood to the tissues, the latter will suffer from lack of oxygen. Among other organs, the brain will be so affected. A deficiency of oxygen in the brain may lead to the nausea so characteristic of mountain sickness; to vertigo, delirium, loss of consciousness, according to circumstances.

Mountain sickness, then, appears to be sgnificantly associated with an oxygen deficiency. This deficiency is caused, first of all, by a diminished supply of atmospheric oxygen. But the fact that one person suffers at a given altitude as another does not—although the pressure of atmospheric oxygen is the same for both—means that in the former the elimination from the blood of those substances which make it alkaline is delayed, so that the tissues experience an added deficiency of oxygen for which the reduced oxygen pressure in the atmosphere does not account. In the case of the latter, not only do the kidneys function more satisfactorily in keeping the alkali reserve down to normal proportions, but it is probable that the concentration of acids in the blood is increased by delay in their elimination. It has already been pointed out not only that carbonic acid heightens the efficiency of the heart and breathing mechanism, but that it diminishes in concentration in the blood at higher altitudes.

Anything that prevents the loss of acids from the blood will tend to prevent the alkali reserve from increasing beyond normal proportions. Under certain conditions, moderate exercise helps in this direction; which means that moderate exercise may aid to some extent in warding off mountain sickness. On Pike's Peak (14,100 feet), for instance, it has been noticed that those who ascend on foot at a moderate pace adapt themselves more readily to the change of altitude than those who go by rail. And it has been shown further, that persons whose blood tends to give up oxygen readily to the tissues—owing perhaps to a retention and consequent augmentation of acid in the blood—acclimatize more rapidly than those whose blood gives up oxygen less easily.

Acclimatization is usually achieved in time by all save the most refractory cases. The human mechanism, admirably adapted to ordinary changes of oxygen pressure, finds itself often overtaxed by the almost violent changes incident to the ascent of such a mountain as Rainier. If one were to remain on the summit (as can be done quite comfortably on Pike's Peak) several days might elapse before the normal balance of acids and alkalis in the blood would be restored and the heart and kidneys and respiratory mechanism return to their normal activities. This adaptability is a very remarkable attribute of the living mechanism. But the adaptation takes time. In 1909 the Duke of the Abruzzi and several companions ascended, in the Himalayas, to 24,600 feet, where the atmospheric pressure was 312 mm. mercury, without suffering serious inconvenience. But this favorable result was doubtless due to the fact that the party had resided for the two months previous at an altitude of 17,000 feet and had become acclimated to the elevation. It is possible that Mt. Everest may be conquered in time, in spite of its 29,000 feet and a summit atmospheric pressure of less than 290 mm. mercury.

Individuals differ. Some may eat heartily without disaster, where others would speedily succumb; and drink freely and often where others cannot; just as they may climb at a pace which would soon cause others to fall by the way in distress. Certain facts, however, apply to all. One is that tea does not lessen fatigue, its essential merit being that it makes hot water more palatable. Another is that carbohydrate foods, among which sugar is important, are especially valuable on the climb. They burn completely in the blood to carbon dioxide and water, thus providing two great desiderata, energy and acids. And the energy and acids become rapidly available, within thirty or forty minutes from the time the carbohydrate is taken. Authorities have reccommended, accordingly, that such food be taken in small but frequent amounts during an ascent.

Whether lemon juice is a real aid to the mountain climber is a debatable question. It undoubtedly stimulates the flow of saliva, which relieves parched mouths. Like sugar, it burns in the body, liberating energy and weak acid. But it must be remembered that these effects are proportional to the amount taken, which is usually small. Fruits that contain acid citrates and the acid salts of other organic acids may be distinctly harmful, since derivatives of these acid salts, in the form of carbonates, give an alkaline reaction. To use them for mountain sickness, then, is to aggravate rather than allay the conditions that favor the disorder.

To summarize: In the ascent of such a mountain as Mount Rainier, one reaches a level—around 10,000 feet—where the oxygen of the atmosphere has diminished to such an extent as to initiate more or less serious disturbances of the acid-base equilibrium of the blood, which increase with the altitude. When the balance leans toward increased acidity, the effect is, within certain limits, beneficial—depending upon the increased delivery of oxygen to the tissues. When it leans toward alkalinity, the oxygen supply to the tissues diminishes to a point, it may be, where they cannot normally function. But, whatever the symptoms, lack of oxygen appears to be at the bottom of them all.

THE MAZAMAS

Hitting the trail in the morning,
Sniffing the mountain breeze,
Up on a wide, white snowfield,
Or down with the big fir trees;
We hike in our khaki and flannel,
And never a one but will say
The joy of the climb's in the climbing,
And the rest at the close of the day.

So here's to each perfect comrade
Who travels our upward trail,
With a voice to hearten the laggard
And a hand to him who would fail.
There's a peak to be won in the heat of the day,
A descent before the sun's decline;
But never a selfish thought can live
Above the timber line.—*Harold S. Babb.*

Reminiscences of Mt. Baker

DR. W. CLAUDE ADAMS

While the war was being fought on the battlefields of Europe we were waging a war in the United States against German propaganda. Sentiment was strong against the countless methods employed by the enemy to foist on us and others their nationalism. In the light of what we now know of their eagerness to excel and their arrogance in forcing themselves, their opinions and their goods on other people, we see through many things which happened years ago, the purpose and design of which was not apparent at the time. Now we pick up a book, for instance, "The Making of an American," by Jacob Riis, published in 1901, or "Imperial Germany" by Whitman, published as far back as 1897, and we read indictments against Germany in her attitude to Denmark and other countries which to us now only help to tell the tale of German kultur.

In looking over my Mazama records recently, I came across a little incident of the Mt. Baker climb in 1909, which was very interesting to me in regarding it from our wartime viewpoint. Believing that it will interest those who have never heard of it, and also refresh the minds of those who participated in the trip, I am delving into the past and bringing forth the story, the significance of which will not be lost, as the action of the offender in the story smacks of the German idea of predominance, which trait we have come so to despise.

The evening before the preliminary climb of Mt. Baker, one of the visitors in our camp, Dr. Bernard Hahn of Seattle, was seen surreptitiously gathering together material to make a flag to plant on the summit when they reached it on the morrow. Much to our surprise it proved to be a German flag. In spite of protestations, he brazenly took the flag up the mountain with him and planted it on the top, boasting about it when their party returned.

Resentment was rampant in camp and we vowed that we would tear the German flag down and place the stars and stripes on the top in its stead. Imagine our consternation to find that there was not a United States flag to be found in camp. Steps were taken at once to make one. Sadie Settlemeier, now Mrs. Chas. Whittlesey, was our Betsy Ross, and the girls commandeered all the red and blue bandana handkerchiefs and drew on the dunnage bags for a supply of white material. Soon we had manufactured a presentable flag, which was carried to the top by the late Clifford Lee, and, amid the cheers of the company who reached the summit, the German emblem was torn down and the Star Spangled Banner was given its rightful place.

Later, at one of the bonfires, a kangaroo court was held and was presided over by Judge Craven of Bellingham, and the prosecuting attorney was Richard W. Montague of Portland. One of the most notable cases tried was that of Dr. Hahn, who was charged with high treason, in that he had floated a foreign flag within the precincts of the camp.

The reason for the publication of the following notes on Mt. Baker at this time is that the trip of 1909 never has been written up for any number of "Mazama." For three or four years, including 1909, the magazine was not published. The club's treasury was depleted on account of having contributed money to help finance the Rusk expedition to Mt. McKinley, the object of which was to prove or disprove the authenticity of Dr. Frederick W. Cook's claim to have reached the summit.

On the preliminary climb, August 7, fourteen made the ascent, two of the party being women. And on August 11, the entire number of thirty-eight, led by John Lee, made the difficult climb and reached the summit elevation, 10,728 feet, without the use of a rope. The route followed by the Mazamas up the middle fork of the Nooksack river was the same as the famous Coleman party took in 1868 when it was led by Indian guides.

Pioneers in the Pacific Northwest were scaling the snow-capped peaks eight years before the first mountain-climbing club in America was organized. The Appalachian Club of Boston was organized in 1876. In our Pacific Northwest we boast of great snow-capped peaks ranging from 9000 to 16,000 feet in altitude and one of the most conspicuous of the mountains comprising this group is Kulshan, the Indian name for the mountain now bearing the name of the Englishman, Captain Baker. In 1868 Edward T. Coleman, an experienced alpine climber, became the first white man, of whom there is record, to scale the heights of Mt. Baker, which possesses so remarkably grand and rugged an outline.

Mt. Baker is more nearly active than any of the volcanic peaks in the northwest, as sulphurous fumes and steam are constantly rising out of the fumaroles and vents in the mountain. The steam vents are very numerous, while the two most active fumaroles are within the crater nearest the summit. Mr. Chas. F. Easton and later, Mr. M. W. Gorman and myself, descended into the crater, a black yawning chasm, and saw for ourselves the two fumaroles, one constantly emitting fumes and steam and the other intermittently. The sulphur fumes were very strong at times and the snow in and around the crater was yellow with sulphur crystals.

The moraine near the smaller of the two craters where we ate lunch was about 30 feet wide and 100 feet long. When the wind changed the fumes of sulphur came over so strong that some of our party became ill.

Chas. F. Easton, a Mazama living in Bellingham, has studied Mt. Baker in all its phases, having climbed it many times and from different sides. On one occasion he and three others were compelled to spend two days and nights on the mountain during a fierce blizzard and their lives were saved only by taking shelter in a cave which they dug in the ice and snow.

Mr. Easton has also conducted scientific investigations of great value to societies and clubs interested in the topography of our mountains. He has found evidence that Mt. Baker has materially changed shape since Coleman described its contour in 1868; the secondary peak (Sherman) having sunk lower by 500 feet than it was at that time. On the day of the San Francisco earthquake he was making a drawing of the summit of the mountain with the aid of a telescope. The day after the earthquake he found the outline of the second peak had changed so perceptibly that a new drawing had to be made, which proved that the tremor had had its effect on the mountain. On his next visit to the summit, he observed that the crater had been filled with rock and lava which had tumbled down from Sherman peak.

Coleman states in Harper's Monthly, Vol. 39, 1869, that the two peaks were of the same height at the time of his visit, but now there is a difference between them of more than 500 feet. Baker at one time stood 1000 feet higher than at the present time.

The pinnacle of loose lava and rocks which the Mazamas found impossible to scale in 1906 has also crumbled away, so that, were it not for the rolling rocks, the ascent might now be made from that side.

Mr. Easton and Mr. W. P. Hardesty set stakes and kept data during our stay at Baker and estimated that at least two of the glaciers were moving at the rate of more than 50 feet a year, their rapid movement being due to the fact that the temperature of the underlying rock formation is greater than that of other mountains on the Pacific Coast.

One sight on the mountain I shall never forget and the impression of it is still as vivid in my mind as on the day in which I saw it. On the mountain one evening fell a sunset of surpassing beauty, the like of which few of us had ever seen, so wonderful and yet so weird. We were seated at our evening meal in the woods of Camp Gorman when gradually we became conscious of a roseate glow all about us, enveloping and tinting the landscape with an ineffable softness of light and color. Everyone seemed mystified by the peculiar effect of the light and spoke of it wonderingly.

About this time two of our party came down the mountain and called us to come and see the wonderful coloring. Everyone hastened up to the lookout point, and there we beheld a sight of supernatural charm. The lofty summit was bathed in a soft mellow glow and it was with a feeling of awe that we were silent in the presence of this phenomenon, not unlike the Aurora Borealis.

Mr. Coleman, first white man known to have made the ascent of Mt. Baker, expressed his feelings when the mountain crest was gained: "We felt at heaven's gate, in the immediate presence of the Almighty. My companions, to whom, for the first time, such wonderful scenery was unfolded, were deeply impressed. The remembrance of the dangers they had escaped, the spectacle of the overwhelming desolation around—these combined evidences of the Creator's power filled their hearts with extreme emotion. With one accord we sang the doxology. No profane thought could be cherished, no idle jest uttered, on this, one of the high altars of the earth."

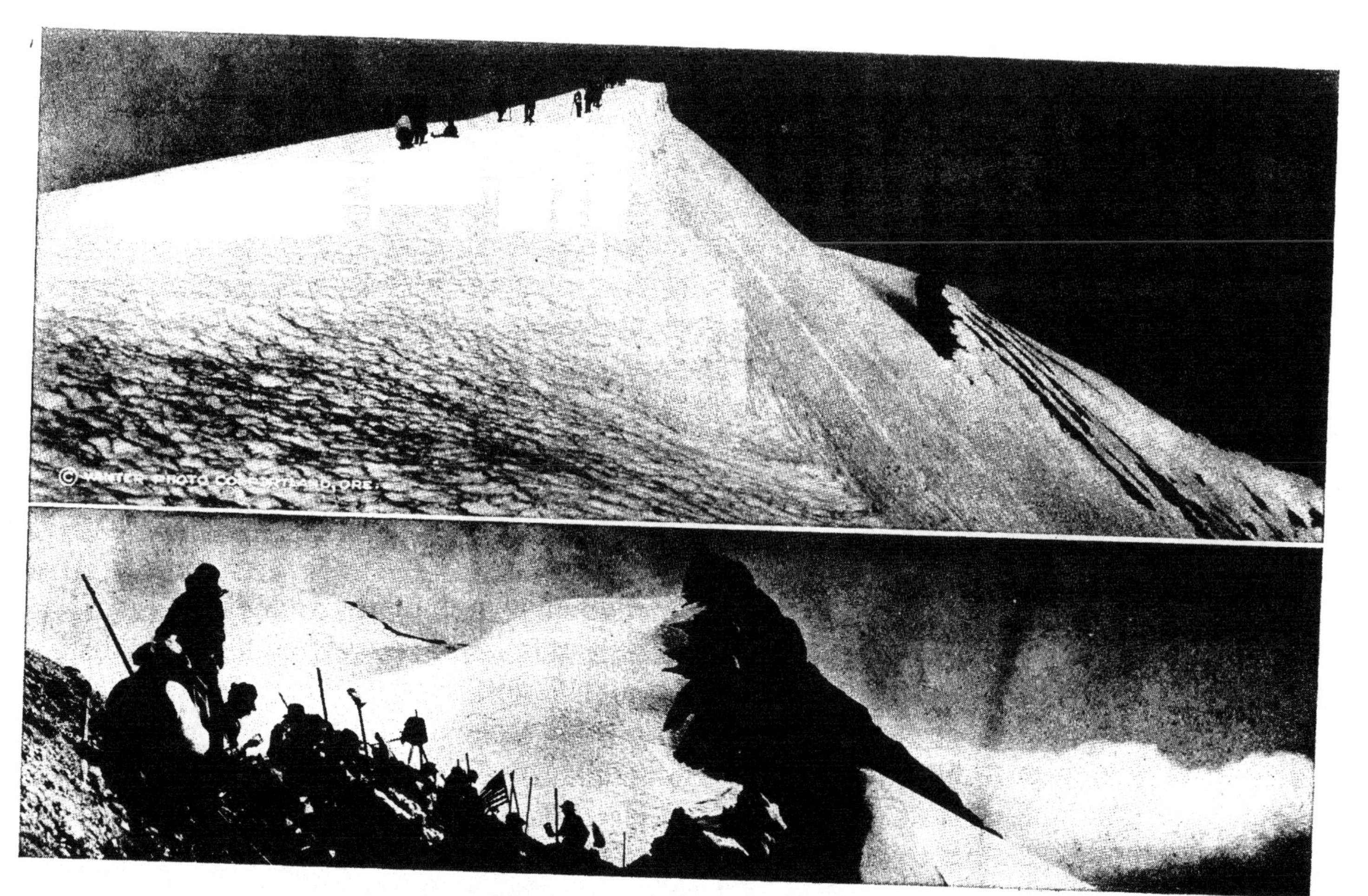

UPPER—SUMMIT OF MT. BAKER.
LOWER—PARTY EATING LUNCH AT THE CRATER.

1—"Sun" Mountain, South Side.
2—Showing character of rock climbing on "Sun" Mountain.
3—Upper St. Mary's Lake, opposite "Sun" Mountain.

Mountain Climbing In Glacier Park

By FRANCIS BARBOUR WYNN
August, 1919

Fortunate for the traveler and nature lover is the fact that America's national playgrounds present very striking differ ences. The Grand Canyon of the Colorado breathes the spirit of awesomeness and weird color effects; Yosemite charms by the unsurpassed beauty of its waterfalls, precipitate cliffs and verdure; at Sequoa are the giant trees, immortals of their race, which have stood the ravages of time and forest fires but came near yielding to the ruthless hand of commercialism; at Crater Lake you look down upon a body of water filling the crater of an enormous extinct volcano; in Rainier Park is seen the most massive and majestic of our mountain peaks with its twenty wonderful glaciers and bewildering wealth of flowers; in Yellowstone the uncanny things hold attention—roaring and spurting geysers, bubbling mudpots and wild animals grown tame; in Colorado the matchless canyons and gushing mountain streams; and Glacier Park is characterized by its Indians and Indiana lore, superb mountan scenery—rugged peaks, wonderful lakes, forests, glaciers and flowers, strikingly like the Tyrolese Alps.

To the nature-lover this park offers a variety of attraction not presented in any of the other parks. Whilst train or automobile may bring one close to the main points of scenic interest, the motor car can never desecrate the trails leading to the high mountain passes, hdden lakes and floral rewards worth many times the effort to reach them. It is *par excellence* the hiker's paradise. The tenderfoot will do the trails by horse, but the real lover of the outdoors will go afoot. The trails are now sufficiently marked so that no one need lose his way in going from camp to camp—the distance ranging from five to twenty miles.

But it is chiefly as a rendezvous for mountain climbers that I wsh to consider Glacier Park. Many have followed its trails and visited its innumerable and beautiful lakes, but the mountains of the region are almost unexplored. Until this year not a single record box had been placed upon any summit. The ignorance of the guides and those in the camps and hotels about about the mountains is appalling. I was not able to find a single person who had ever climbed any of the beautiful, rugged peaks. Six years ago I made the ascent of Mount Jackson, and this year I was resolved to climb Going to the Sun Mountain, or "Sun" Mountain as it is commonly called. At the Chalets there was a sort of tradition that the mountain had been clmbed twenty years ago; and five years previously; but I could not even learn the names nor the route taken in making the ascent. On two previous visits to the region I had been upon the west, south and east sides of the peak, and was convinced that these approaches were impossible. It was therefore decided by our little party (Sidney F. Daily, Daniel Boone Clapp and Frank B. Wynn) that we would make the attempt from the north side. We estimated the total distance from "Sun" chalets and return would be thirty mles. To conserve our strength for the climb we took horses to the timberline on the north—nine miles. This approach presented a cirque of many narrow ledges and sliding shale. In the morning we had noted goat feedng high up on these ledges. We made satisfactory progress until one o'clock, when the lead brought us up against sheer walls. We made several attempts up chimneys, each of which brought us to a precipitate jump off of three thousand feet down upon Sexton Glacier. This was our situation at three o'clock when it was decided that we would have to abandon the climb for that day. The other two gentlemen went on down leaving me to reconnoiter for the subsequent attempt. I followed a narrow ledge for a half-mile, around to where we had seen the goats in the morning. Here another chimney was found which I believed negotiable. The other two men were too far away to hail, and the hour was too late to overtake them and bring them to this point of attempt. So I determined, contrary to mountain rules, to try it alone. The chimney was steep, and, at several points, difficult to negotiate. Many times I wished for a rope and somebody to give me a boost. Aside from a disfigured camera, and some tremulousness incident to the physical strain, I reached the summit without mishap. There was, of course, great joy in the achievement but sadness over the fact that I had not my colleagues with me.

The top of "Sun" mountain is a well weathered plateau, in the center of which is a huge pile of disintegrating rocks quite similar to the pinnacle found upon the summit of Shasta. This little peak is plainly visible from Gunsight pass on the west, 12 miles away. There was no evidence anywhere of any former climber. I built a cairn, placed the Mazama Record Box within it and chained the box to a large upright stone. Then placing my mountain stick by the cairn and my hat upon it, I took a photograph as the only evidence in proof of the ascent. Besides signing and dating the record, the book contained the following original

MOUNTAIN-TOP PRAYER

Dear Lord, I thank Thee for this view
Of paradise.
The fearsome trail was hard to do,
But worth the price.
The arching canopy of art
In Heaven wrought,
Encompasses the very heart
Of beauteous thought.
In Nature's lap of forest green,
Rests tranquilly
The shimmering lake; the glinting stream
Leaps jously.
The serried ranks of snow-clad peaks
Attention stand,
Like faithful, white-robed sheiks
Await command.
Thy handiwork! How wondrous and
How beautiful!
My soul enraptured bids my hand
Be dutiful!
The trudging up yon toilsome trail
How well repaid!
'Twill help me in lfe's sore travail,
Hath courage made!
For strength of limb and will to do
And try again,
I thank Thee, Lord, and pledge anew
My faith. Amen!

And now a most interesting postscript to this narrative of the ascent of "Sun" mountain, which shows that the placing of the Mazama Record Box upon the top of this superb peak has already borne good fruit. After my return to Indiana I received authentic proofs of the climb in a letter from H. R. W. Horn, of Defiance, Ohio. I quote in part from his letter as follows:

"It is with much pleasure and pride that I can inform you that
Dr. H. H. Goddard, Columbus(Ohio
Howard S. Riddle, Columbus, Ohio
R. E. Wilson, Defiance, Ohio

and myself were successful in climbing Going-to-the-Sun mountain on August 14, 1919. On the peak we found a cairn containing a tin box with an Official Record Book of the Mazama Mountain Club. I am enclosing several pictures taken on the summit which will prove to you our successful climb of the mountain. We saw your description of the route followed on the Hotel Register at Sun Camp and our climb was made according to your directions. I have become much enthused over mountan climbing and can say most of my trips in the future will be spent in this grand sport."

The wonder of an ancient awe
Takes hold upon him when he sees
In the cold autumn dusk arise
Orion and Pleiades;

Or when along the southern rim
Of the mysterious summer night
He marks, above the sleeping world,
Antares with his scarlet light.
—*Bliss Carmean.*

Larch Mountain Ascent

The following account of the Larch Mountain ascent on Sunday, October 12, 1919, was contributed by Muriel Kennedy, an eastern visitor enjoying her first outing with the Mazamas.—Editor.

That glorious sun beamed and sent thrills and life into the very soul of that little crowd who watched on the mountain the birth of that new day, Sunday, October 12. A most gorgeous rainbow guarded the shadow of Larch on the solid fog in the West.

Another climb to the top of the lookout showed the scene entirely changed by sunlight. Another snow-cap, Mt. Jefferson, was added to the list, quite like a pyramid. The summit of St. Helens resembled a saddle heaped with tons of snow. Adams had been obscured by ungrateful clouds, but Rainier proudly supported a black cloud like a hat. In the foreground miles and miles of beautiful forests bordered acres and acres of devastated land, which had been recently swept by fires.

They told me I should see the range by moonlight, that it was very beautiful. The lookout was six flights almost vertical, with about twenty steps to the flight. I took two steps up and, shall I say cried? I wanted to scream. Mountain stiffness is a feeling all its own. For a minute, the climb seemed impossible, but later I proceeded stepping up first with the left foot and dragging the other. Tears with many steps. Four snow-capped mountains were very beautiful, namely, Hood, St. Helens, Rainier and Adams. Hood, which was the nearest, being only twenty miles away, looked wonderful. Up until that time, the mountains had never appeared so close to me like many say, and none did then except Hood. I felt as though I could almost touch it. At the farthest it seemed only two blocks away. I shall never forget it. The other peaks were pretty also. Toward the Columbia and Portland a fleecy bank of clouds prevailed. But Larch is not one of the highest; it is only 4,045 feet in height.

We rested again at the fire, then the crowd began to move over to the cobble-pointed pinnacle which is the highest spot of the mountain. This highest point is the sunrise observatory. It is about a quarter of a mile from the grove or camping place. When it was suggested that we go over there, I thought, "Sunrise be hanged. I can't." But on a trip I'm never a piker, so we went, and I shall never regret it. The sunrise alone was enough to pay one for the journey. The wall opposite the trail was almost perpendicular for hundreds and hundreds of feet. The whole sky was magnificent. Snowy-white dapples of clouds hung low. The eastern horizon grew rose-colored. One black-blue cloud in the North resembled a whale.

While watching, the eastern scene turned to blood-red, bordered at the top by the blackest cloud I have ever seen. The black slowly faded away and the deep red lifted upward. A dark yellow streak appeared. Toward the left, Rainier, St. Helens and Adams, snow-clad, blushed a soft crimson. Toward the west, pale pink and blue, beautfully blended, adorned the sky; and the once white dapples artistically turned to pink. A few minutes later the red melted into a bank of gold, changing the pink to amber. The range of mountains against the gold showed black. As the mellow golden varied to a dazzling yellow, the crowd was speechless, for "Old Sol" was about to join us. Between two peaks which seemed to be especially designed for the occasion marked the spot of intense interest. Screams and shouts and cheers and hat twirling hailed the bursting of the first rays from the brilliant ball of fire that gently lifted and floated upward from between the mountains.

The stars are forth, the moon above the tops
Of the snow-shining mountains.—Beautiful!
I linger yet with Nature, for the night
Hath been to me a more familiar face
Than that of man: and in her starry shade
Of dim and solitary loveliness,
I learned the language of another world.

—*Byron.*

What the Indians Tell

F. H. SAYLOR

The Birth of Gold, as Told by the Piutes, the Southern Utes and Columbia River Indians; The Creation, a Phaeton's Fall, a Deluge and a Friendly Fish.

THE BIRTH OF GOLD, PIUTE INDIAN VERSION

Nearly everything in the terrestrial sphere knew a deification by the Indians. While the sun and the moon generally take the most prominent places in their myths, the planets, various stars and the terrifying comets were not forgotten, but formed a considerable portion of its composition. The two former have many legends concerning them; some where they are mentioned direct, and again, spoken of under totemic designations. Of the planets, Venus held first place, and the last is best remembered in two ways: As a monster serpent coming to wreck and destroy, or as a woman, young, beautiful and arrayed most gorgeously—one whose long flowing hair, let loose to kiss the breezes, looked like threads of waving gold. The Piutes say that this class of deity brought about the placement of the precious metals and gems in the earth, such not having been stored therein at the time of its creation. At this epoch the moon did not form a part of the known luminaries of the heavens, but subsequently came as a wanderer from space beyond the glitter of the stars, which at this time seem to have been led by the planet Venus, as the morning and evening star. The Piutes are not alone in this belief, as the tribes about Lake Tahoe, the Southern Utes, and a portion of the Columbia River families relate similar legendary accounts. Such ideas are not altogether confined to these localities, nor to this continent, as the myths of the Peruvians parallel them, and in sections of the old world kindred legends were told.

According to the story of the Piutes, the Sun-god was possessed of a magnificent robe, the glow and sparkle of which caused the rays of sunshine. This he wore only during the time he traveled through space in daytime, leaving it at night in the custody of a brave whom he trusted. A siren, like a Queen of Sheba, living in a far-off land, hearing of the dress of the ruler of the day, came to view, and in viewing not only coveted but determined to secure it for herself. Unable to captivate the Sun-god, she concluded to try her wiles upon the custodian of the robe while its owner slept. This watcher, beguiled by her winsome ways, allowed her to enwrap herself within the garment's folds, but, not content with self-admiration and unmindful of the praise of the enamored brave, she resolved that those who guard the blue while night shades the hours should be witness of her grandeur. Leaving the bewitched guard to bewail her actions and mourn his rashness, she set forth to exhibit the splendor of her raiment. As the folds of the glittering robe trailed behind her like a billowed sea and her luxuriant hair spread and rose and fell as a field of golden grain swayed by zephyred force, onlooking stars in rapture stood in contemplation of the beauteous scene. The commotion awakening the Sun-god, he became enraged; not so much for being disturbed in his slumbers, but because of the happening which brought it about. Rivalry up to that hour had been unknown to him and its presence roused him to a state of fury. Especially was he wrathy because the admiration another received was through plumage stolen from himself. Avowing that presumption, as well as unfaithfulness, should receive dire punishment, he called for his "thurbesay," which was none other than the rainbow; and, tipping an arrow, with lightning's flash he shot the destructive messenger athwart the adventurous maiden's path, producing blindness.

Deprived of sight, she fell at last to reach the earth. To regain her lost estate, she there wandered over rugged and higher elevations in the hope of being able to grasp something which would bring her back to the upper ether. As she climbed from peak to peak, and from hill to hill, the gold, silver and gems of many a sparking hue fell from tresses of hair, or were torn from the resplendent robe by jagged rock and marked her path as she moved from place to place.

Repentant tears came from sightless eyes to fall upon the earth, and in their sinking below the surface, ornaments were caught, and enfolded by them sank therewith. Tear-drops were transformed into quartz, becoming the wrap of the brilliants which brought the goddess low, sundering relations pleasant, and changing happy hours to those of sorrow; thus, hidden in a house of darkness only to be exposed again to light, centuries following, as the root of evil to mankind.

Each gleaming particle as it fell;
So runs the legend old—
Sank deep into the mountain's breast
To deck, or vein its heart with gold.

Her piteous plight moved the stars to intercede for her pardon, and, consenting to condone, the Sun-god permitted them to lift her on high again, but conditioned as he did so that no more should she wander save around the earth; that from her face alone should brightness come thereafter, and that to be a borrowed lustre. There, ever sighing for past radiance pictured in her memory, she seeks to penetrate the veil of darkness obscuring vision and locate the earth and scope thereon where the gorgeous apparel once worn was lost. But unable to discern correct direction, it is seldom that her face is presented to its full. When full facing, one can see the stains thereon caused by tear-wet hands not free from the cling of dross of earth, yet none tell her of its condition, for all know that she could not remove them if she would. Thus a presumptuous wandering comet was a-rest in roaming and became what mankind now calls the moon.

The disobedience of the unfaithful brave received a punishment almost equal in severity, as he was confined upon a lonely rock, there to remain forever. In the pole star he is seen by the Indian, never moving; ever watching; regretting a trust betrayed from age to age. Always maintaining the position decreed as his fitting fate.

Version of the Southern Ute Indians

The Southern Utes say that before the tall pines had grown to heights no higher than blades of grass, there lived a people who though few in numbers were not without individuals among them possessing the powers by which divinity moulds, and makes and obliterates. The controlling spirit or chief among them was Ku-ku-lu-yah, or "Bird that flies far." to him, Ah-le-u-to, or "Daughter of the Clouds," was first in thought and foremost in heart, she being his only child. Willingly she bowed to paternal wishes in all ways save those of love, but in affairs wherein the latter played a part she would not listen to dictation. She had given her heart into the keeping where a father would not have it dwell. Next to the chief stood How-al-ak-wah,

"The Winding Water," who was a man of magic, both feared and revered because of his skill in mixing healing or destructive potions, and exorcising the "Masachee Tomanowis" or "Spirit of Evil," whether it lurked in fountain, in hill, in dale, or beset mankind with its unwonted presence as a disease.

The arrow which pierced the bosom of a father's pride had attached to its feathered end a garland thong, connecting it with a mate sent flying from Cupid's bow to make captive the man of medicine; two shots, as one, bringing the stronger to bend the knee, while rosy blushes and drooping, love-lit eyes spoke the secret of the other. Over the wooing of one the old chieftain scowled, and was sore troubled that his daughter was anxious that the lover whom he hated should win and become her husband. How to prevent an obnoxious grafting upon his family tree of a limb he would much prefer see cast aside long perplexed the old chief, but at last he conjured up what he believed would be an expedient whereby the charm might be brought about. He would give a feast and thereat offer his daughter to the brave most worthy, such to be the one who would perform certain exploits which he would name. Well he knew that How-al-ak-wah, by right of rank, would have precedence over any others who saw fit to blindly dare conditions not to be defined until after acceptance of them had been made. And he was equally sure that his contemplated victim would worship at uncertainty's shrine. Runners soon gave notice to the people of the chief's intent, and all hastened to attend the royal banquet. Eating, dancing and smoking brought good cheer to the assembled guests, each of the warriors present being ready to applaud the terms of trial before their purport had been revealed. This was what the crafty father had calculated upon, and he purposed using it to the ending, as he believed, of the hated magician, as he reasoned that the latter's show of bravery would give way to one of cowardice when he learned what was required of him, thus removing him from the list of eligible candidates for the hand of his daughter. New terms not so difficult could then be dictated, and the warrior successful in their accomplishment would become the husband of his daughter.

Taking Ah-le-u-to by the hand, the old chief stepped within the circling fires, when silence prevailed. Calm and stately he remained for a moment, and then began slowly but distinctly to speak, saying: "The time comes when all must close their eyes and go away to the spirit land, no more to be seen again.

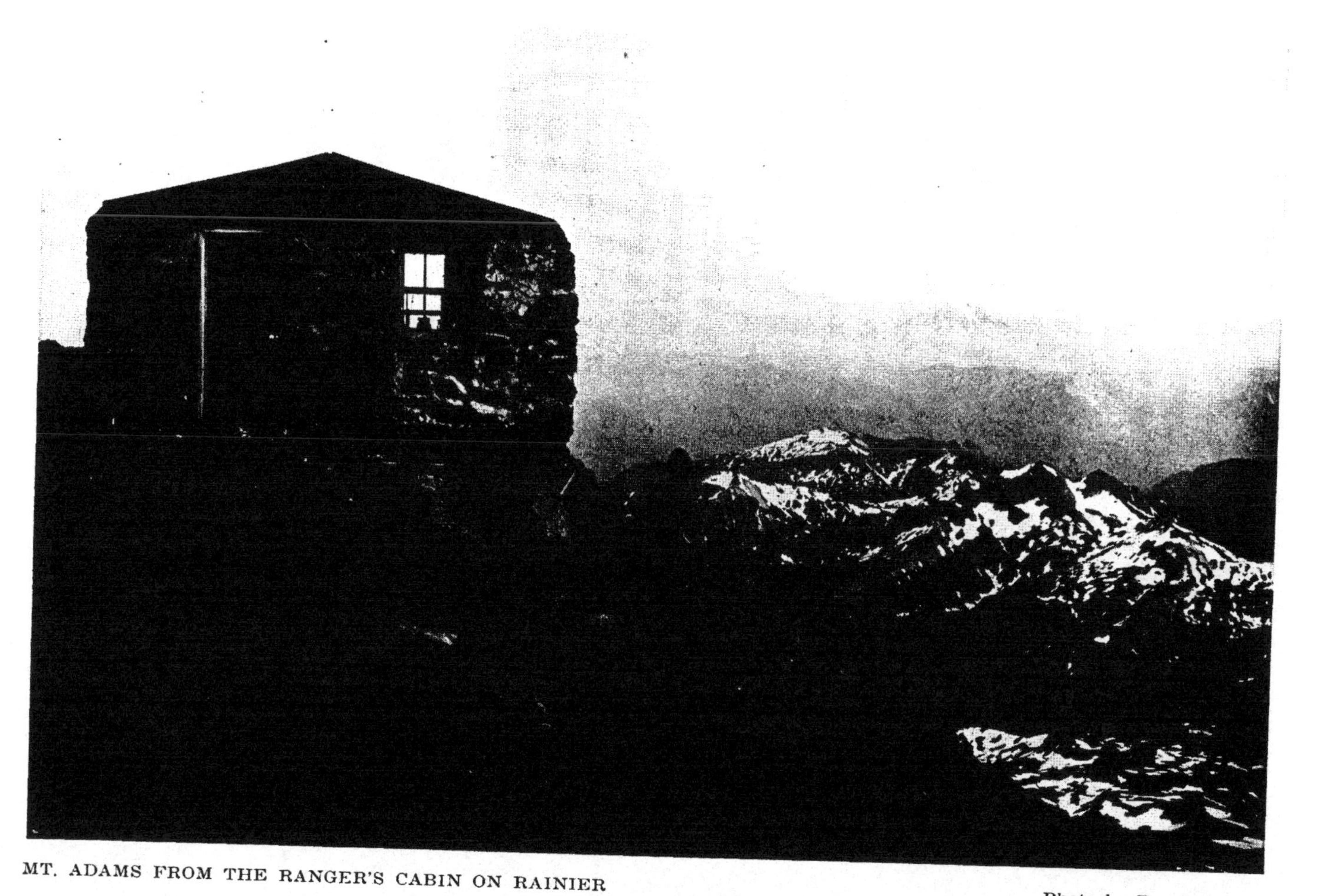

MT. ADAMS FROM THE RANGER'S CABIN ON RAINIER

Photo by Boychuk.

On he way to Tatoosh Range.

The moons which I have lived have whitened my hair like the robe of snow on the summit of yonder peak, everlasting in its changeless change. Many children came to my lodge as sun succeeded sun, but all of them have gone the path of mystery except this flower, Ah-le-u-to, to brighten and give happiness to a father's heart. Through her must my successor come. Who among you dares to war with fate and arise to command when I lay it down? He who does, and knows not failure, the remainder of you shall obey. Let him who would win a chieftaincy when no more I rule, and wed a woman young and beautiful, stand forth!"

A hundred fearless braves arose to spring across the fires, but quicker still than they was How-al-ak-wah, who was before them. Welling up and resounding far and wide then rose a mighty cheering of the champion. As it floated away upon the winds to give place to quiet, those standing round the circle removed their adorning necklaces made from the claws of jungle beasts, grewsome symbols of ambition well accomplished, and laid them at the feet of the man of magic who had dared the worst. Not unmindful of the homage shown him, still giving it no recognition, like a rock untouched by a blast of raging storm the magician stood with folded arms before the exultant chieftain, ready to attempt any difficulties he might propose, and as he presumed it would be, so it was.

Thus the hating father spoke to him as each one listened with bated breath: "Mark this test, How-al-ak-wah, of thy magic, and if thou wouldst win Ah-le-u-to for thy bride, and with her the right to rule when a father's hand no longer grasps the tomahawk of command, perform it! Thou shalt brew a potion which shall be as bright as the wigwam of the Sun-god; one which shall be a blessing, or bring a curse upon all mankind; worthless, yet more powerful than anything now known; stranger than the strangest substance that the oldest brave remembers. Brew this before the dawn shall come again and all you wish for shall be granted; chieftaincy shall be yours, and my daughter shall go with you and bless futurity upon the earth. Fail, and from the tribe you shall be banished to go your way alone until time with you shall be no more. I have spoken."

While this harangue was being delivered the magician maintained perfect composure and bearing bold; when finished, his eyes sought those of the almost fainting maiden; then, with a smile, as if to encourage, he dropped his arms and slowly, though proudly, went forth from the glare-lit encampment. Those left behind watched his retiring figure with awe-stricken faces as it disappeared in the gloom cloaking the mountain up whose sides he took his way, and when no longer within their view, endeavored to picture for him success in the fitful glowing of the dying embers of the fires about them. Soon the light of kindled flame was seen upon the crest of higher bluffs above, showing that the magician had begun his task, he appearing to be stirring with a spear a something within a cauldron. Again the flames sank to nothingness, when the mournful chant of a death song was heard coming from its vicinity. So absorbed were the watchers that they did not notice the stealing away of one of their number into the shrouding darkness. From lowlier pitch the chanting began to assume a louder tone, and one indicative of victory. As it rang out loud and clear the fire's expiring rays renewed their brightness and intensity of burning until the face of fair Luna looked like a painted one when compared with the natural bloom of youth and loveliness, as its light exposed to view and betrayed the fact that Ah-le-u-to had stolen away to the magician's side.

While the man of medicine stirred the contents of the cauldron his loved one commingled her voice with his in song. Slow and solemn was the measure, then with resonance round and full with happier strain, again in wailing cry of farewell to hope, then in glee it caught up the breezes. All at once a burst of thunder seemed to shake the earth's foundations and from the cauldron a moulten mass was lifted skyward, glittering with more brightness than thrice ten thousand times the glow of the many stars that twinkle in the blue above. As a pillar of fire it gleamed, transforming the azure robe of night into noontime's sightful hour; then, sundering its bonds, it fell as lightning flashes from the clouds to seek and strike a lower level, and in its fall scattering far and wide. An explosion had overturned the kettle in which the magician was brewing the potion asked for, when its contents arose and fell like a golden river plunging down the slopes, racing past boulders, hiding among the crevices of the rocks and sinking into the softer ground, producing a strange, hitherto unknown thrilling within the hearts of the watching ones at the mountain's foot. The peace their bosoms knew before gave place to greed and malice; lips yet to utter other than pleasant words parted to give vent to sounds of rage and howls more blood-curdling than those of maddened beasts. Brother fought against brother, frenzied fathers smote their children, heedless of all save How-al-ak-wah's magic potion, each endeavoring to secure as much of it as possible.

Darkness suddenly fell upon the scene when the few remaining survivors in this, the world's first mad scramble for wealth, were caught within the wrap of a roaring, raging stormcloud arising, and carried to a location far away. Once its inky folds became awry, disclosing a gorgeous view in the westward horizon; one showing the gathering together of the stars to greet a bride and groom, then closing as if the portals of the happy hunting grounds had been passed by master magician and lovely maid. When morning came a little old man appeared among the transported people and to them said: "For unnumbered snows shall the contents of How-al-ak-wah's wondrous mixture be hidden from your eyes. Seek to find it, and a paler face than yours shall step between, and in its coming a punishment you shall know for your endeavor, for it shall strike you to the earth as you brought low your brothers. When it finds, you may accept of and use, as it will give you. Beware!" Since then the red race disturbs not the earth in search of gold, and it would prevent if possible the paleface miner's doing so, for it is believed that in its uncovering harm in some manner will afflict them.

Version of the Columbia River Indians

The perpetual snow mantling, beauty of outline and the glaciers, together with the incomparable scenic combination of lake, river, wooded hills and expanse of verdure skirting its base, has endeared Mt. Hood to all those residing in sight of its high uplifted head, as well as winning the lasting admiration of travelers who occasionally behold it. Like all of the sentinels of the Cascades, it has been the basis of numerous traditions current among the aboriginal inhabitants whose hunting grounds were adjacent to the Columbia river, possibly having more legends connected therewith than any one other mountain of the range. It also stands unique not only among its fellows, but among the mountains of the world, from the fact that at times it casts a double shadow. To the white man this freak of nature presents a beautiful picture only, being to him but an added charm thrown around the grand old eminence to lend increasing pleasure while in contemplation of its grandeur. To the superstitious mind of the Indian, however, the phenomenon displays a reflex of an incident which they believed happened when mankind was in its years of youth and from it he reads the future of his race. To him it has a meaning, and hopefully he awaits the coming of the hour when promises spoken in the long dead past shall be fulfilled.

The Indian narrator of the legend will assert that the incident connected with the first appearance of this double shadow brought about a remarkable change in the stature of his forefathers, telling one that antedating such occurrence the people then living grew to heights as tall as the noble firs which kiss the clouds with their higher reaching boughs. The tyee, or chief among them, was of more gigantic build than his subject people, his head towering above them so much that his warriors could walk under his outstretched arm without disturbing his eagle's plumes with which they adorned their hair. He was not only majestic in appearance but most pure of soul. He was a kind and impartial judge and always solicitous for the welfare of those over whom he ruled.

Passing days brought a son and heir to his wigwam, and as this boy's life increased he became more like the source from whence he sprung. He was of commanding presence, his bravery unquestioned and his character without stain. A young man well and rightly equipped to grasp the tomahawk of authority and rule a people. Already had he grown nearly to the prime of manhood without caring to select from the many handsome maidens among the people one he would make his bride. Yet that period of his existence was destined to come, for few there be who escape from the impulse for companionship with one of the opposite sex, which when grown to fulness marks the measure of their joy or brings acutest pain. To this rule the son was no exception, for one day there came from the eastern sky the vision of a lovely maid who seemed to step from splendor's seat out of the low-hung morning star. It was a comet goddess, and one so blessed with brilliancy that the sunlight paled before the radiance of her smile. Seeing her, the young warrior stood transfixed, his heart aflame with thrills. Attracted, she halted. Then from his lips fell fervent plea that no more, unloved, should she roam alone in space, but dwell henceforth upon the earth as his royal mate and queen. Wounded by Cupid's dart herself, it took but little urging of a suit so strange and born in haste to win a heart already won. With mutual consent came outstretched arms to clasp heart to heart, when lo! between them rose a shape of aspect dread, veiling the ardent swain from a would-be bride. 'Twas the daughter of the Thunderer, Ma-sah-chee Tamanowis, goddess of jealousy, who, in rage and hate, had interfered with love's young dream.

Not content with obscuring sight that glow the eye and heart, the rancorous spite of the intruder must be further vented. Seizing the golden tresses of the comet maid, she struck them from her head. As the angered fiend threw them here and there, or stamped upon them, they were ground into the rocks, or carried away by the winds into open crevices, there to lie, not forever, but until the miner of a race to come should find, extract and make or mar the happiness of a people. While gold may be the means of bringing pleasure to its possessor, it still contains the contaminating touch of her who murdered joy and love, and through this, those who covet it for selfish motives are afflicted with a band of misery around the heart that darkens and blanks it from the happier glow.

Awe-stricken the youthful chieftain stood, then burst his heart with flood of grief. The father, bewailing his untimely end, wove a mantle pure and white about his form, renewing the robe each passing year as a symbol of his grief. Thus Mt. Hood was formed, and a grander tomb before or since no soul has ever known. Thus was brought about the birth of gold found scattered far and wide away from lover's last resting place.

"And ever as the summer comes the mystic queen,
Forbidden ever to return as comet to the sky,
Steals silently from out the east, at rising of the sun,
To look upon her lover's mantled form
And meditate, alone, that sweet, sad morn
When first they met; and still the hag, hell born,
Pursues and draws obscuring veil o'er each; to realms unknown
They thus return. The tale is true, for every mortal eye,
When blessed with sight, may yet behold that very scene."

Following the entombment of the departed brave, the Ma-sah-chee Tamanowis, her jealousy still unappeased, gathered great stones and hurled them toward the place of sepulchre to break its covering and expose to view the form of him lost to earth. In their fall these missiles struck, thus killing some and dwarfing those that remained to their present size. Before further damage could be done by her the Sun-god stopped her devilish work, but too late to undo what she had already done. The stricken people were told not to grieve, for in the future the dead would break away from bondage and live again, and all would resume their pristine forms.

As Mt. Hood seems to rumble, or emit a cloud of smoke, the Indian thinks he hears the quickening of a soul whose re-habiliment will be to them a renewal of departed excellence; their entering under the leadership of a reanimated warrior, who claims a bride in spite of jealousy, into everlasting happy hunting grounds. Disappointed often, yet they stoically wait and hope.

To prove the story handed down for unnumbered years, they point to the double shadow cast by Mt. Hood at times, claiming that the brighter one is the lovely comet maid in spirit form coming to greet her lord and lover when he again awakens, and that the darker one is Ma-sah-chee Tamanowis, the fiend, ever present, if possible, to intrude upon and blast the bloom and blossom of a happy hour.

The Creation, a Phaeton's Fall, a Deluge and a Friendly Fish

In order that the reader may better appreciate this legend, it might be well to state the conditions under which it came to the writer; and it might be also said that it was the first interview he ever had with an Indian regarding his legendary past.

The date was in the sixties; location, a country village in the Willamette Valley, Oregon. The story-teller was Us-tow, chief of the Wah-pa-to Indians, who was not only chief of his tribe but its last survivor. He was called "Dave" by the whites. When the cup which inebriates was denied him he was always law-abiding—otherwise, not. He was trusted by his race when sober, nearly always being called upon to represent them when matters concerning them were brought to the attention and adjudication of the Indian Agent at the Grand Ronde Agency. The writer became well acquainted with him through a donation of apples and other edibles and the barter of some old clothes in exchange for bows, arrows and moccasins.

In that epoch of the writer's life about all he became conversant with concerning an outside world of an ancient age was learned at his mother's knee while she read "Peter Parley" or told Bible stories, and it was no more than natural that he should feel his importance upon finding that Dave was ignorant along such lines. In the exchange of confidences, the writer, among other stories, related the biblical account of creation and that of the deluge. During their recital Dave sat like the proverbial boulder, unable to move or give forth sound. Tales, however, being ended, the dumb found his voice, and gave the version of time remote, coming down to him from an age when his fathers were "little children," a relation that not only astonished a listener, but caused him to gravely question Dave's veracity.

According to his story there existed in the beginning an era of great darkness, all that there was during such period being the Sahale Tyee, or Great Spirit above, and below, a vast breadth of waters, calm and lone in their boundaries. As ages multiplied, the Sahale Tyee grew tired of immobility, silence and black night, when he descended from his place above to the face of the tide beneath. Striking it with his bow, it began to swirl and toss, when the earth pushed up and out from its turmoil and formed an island of great proportions. Unable to see these sudden and strange changes without a light, an eye Divine looked about for something which could be made subservient to his purposes. Finding some seaweed, he waved it back and forth until it became dry, and then rolled it into a couple of ball-shaped masses which were ignited by their being blown upon. These luminaries created, they were hung in the heavens, where they are now seen as the sun and moon. Having a sufficiency of light, the earth was supplied with vegetation and peopled with fish, creeping things and fowls; and last of all was created a race generally termed by the Indians as an animal people, a race possessing the qualities of demi-gods with human instincts and intellects, yet known by animal names, and in many instances accredited as animals reasoning and acting like human beings, their characteristics being similar to the gods of ancient Egypt and Greece.

That the sun and moon should have such a warder, a couple of effigies were made out of clay, the first being formed while the aura of day shed its light upon the earth, and in drying partook of its nature, fiery at times, yet could be the most beneficent. The latter, left to dry and vivify under the glow of the pale moon, was featured much lighter; her children, which each dwelt in a star, inheriting her disposition rather than that of the sun's warder, their father. At one time one of these came to the earth and sought to play with the children of the animal people. This youth was known by the name of Sea Otter. At first there was no objection to his becoming a party in the games played, but finding that he was more expert than the animal children and took all the honors, a jealousy arose among the latter, when the victor was taunted with an uncertainty as to his parentage. Aggrieved at his treatment, Sea Otter returned to the sky and went to the lodge of his father. While his parent was asleep during the pass of the night, the son amused himself by handling the sun as he would a ball, finally taking it into his head that he could carry it through the heavens as well as his father. The idea once conceived, he could not rest until consent was given for him to do so. The trail was by a spider's-web bridge; an arch like the bend of a bow when strung was not very wide. Before starting, Sea Otter was cautioned as to what he should do and not do while on the trip. He must ascend slowly, lest he tire and be unable to securely hold his charge; careful when highest lest a false step bring disaster; hasten not thereafter, for the course was steep and the sure foot loses certainty of safety in speed.

Sea Otter got very tired ere noonday arrived, and by the time he reached the place where he must begin a descent kept changing his load from one shoulder to the other to relieve his weary arms. The glare of the sun, in shifting it before his eyes, brought about an uncertainty of sight, thereby preventing his clearly viewing the finely woven path, when he missed his footing and lost his hold upon the burden carried. As the fiery ball struck the earth it burst into innumerable fragments and started an earth-wide conflagration. Sea Otter also fell from the bridge and in his fall was the first to die. As in the fate of Phaeton,

"The strong winds bearing him beyond the breast of earth,
Where, plunging headlong, with robe aflame,
Like the shooting star, which marks the heavens
With its brightness as it falls, his career
Found ending in an awaiting, engulfing sea."

The Sahale Tyee, noting the conflagration, sent a mighty wave to extinguish it. Those surviving the first of the destructive elements, endeavored to seek safety from the latter, among them being a woman, she being aided by a monster fish. It observed her plight, and remembering her former kindness to it by frequently providing it with food, swam to her rescue, telling her to get upon its back and it would save her. This she did, the succoring fish all the while keeping its back high above the crest of the deluge until the overwhelming tide had served its purpose and was again cradled in ocean's deep. Danger past, the fish swam to the shore of a large river, which it entered, when the woman, the sole survivor of the catastrophe, stepped again upon a drowned earth to begin thereon a dispensation new.

This part of the legend reminds one of the parting of Orion and the Dolphin:

> "Farewell, thou faithful, friendly fish! Would that I could reward thee; but thou canst not wend with me, nor I with thee. Companionship we may not have. May Galatea, queen of the deep, accord thee her favor, and thou, proud of the burden, draw her chariot over the smooth mirror of the sea."

Again it recalls the legend of the Hindoo king, Satravrata, who, with a few others, took refuge from a deluge covering the earth in an ark, the god Vishnu, in the form of a fish, taking care that the ark sailed in safer waters by conducting it around by a cable tied to its horn.

At the time the legend was told by Dave, the small boy listening was of that age when he thinks that his little brother or sister, following him upon the stage of life, is a present from the doctor. The boy could not understand how such a present, especially a double one, could arrive, when the man of medicine had met death in the flood. Upon telling his mother that he thought Dave had "storied" to him, she reminded him of the story of Jonah and the whale, and explained that possibly another great fish might have swallowed the doctor, keeping him safe for a time and then spewing him again out upon dry land.

Associated Mountaineering Clubs of North America

The membership in the Bureau has shown steady increase and now numbers 31 clubs and societies with over 60,000 individual members, as follows:

American Alpine Club, Philadelphia and New York.
American Forestry Association, Washington.
American Game Protective Association, New York.
American Museum of Natural History, New York.
Adirondack Camp & Trail Club, Lake Placid Club, N. Y.
Appalachian Mountain Club, Boston and New York.
Boone and Crockett Club, New York.
British Columbia Mountaineering Club, Vancouver.
Colorado Mountain Club, Denver.
Dominion Parks Branch, Dept. of the Interior, Ottawa.
Field and Forest Club, Boston.
Forest Service, U. S. Dept. of Agriculture, Washington.
Fresh Air Club, New York.
Geographic Society of Chicago.
Geographical Society of Philadelphia.
Green Mountain Club, Rutland, Vermont.
Hawaiian Trail and Mountain Club, Honolulu.
Klahhane Club, Port Angeles, Wash.
Mazamas, Portland, Oregon.
Mountaineers, Seattle and Tacoma.
National Association of Audubon Societies, New York.
National Parks Association, Washington.
Nat'l Park Service, U. S. Dept. of the Interior, Washington.
New York Zoological Society, New York.
Prairie Club, Chicago.
Rocky Mountain Climbers Club, Boulder, Colorado.
Sagebrush and Pine Club, Yakima, Wash.
Sierra Club, San Francisco and Los Angeles.
Tramp and Trail Club, New York.
Wild Flower Preservation Society of America, New York.

The common bond uniting all is the desire for the preservation of our finest scenery from commercial ruination. We are working in co-operation with the National Park Service for the creation, development and protection of our National Parks and Monuments. In our annual Bulletin attention is called to what various departments of the Government are doing for the mountaineer and traveler, and mention is made of the claims of scenic regions to become national parks or monuments. When these projects are considered by the Government, we present the views of our members, and give publicity to the plans of the Government.

We have encouraged and assisted our clubs in forming and increasing reference and circulating collections of books for the use of their members. We are calling public attention to many important but little known scenic regions by illustrated magazine articles, and by illustrated lectures before leading clubs and societies.

LeRoy Jeffers, Secretary
Librarian American Alpine Club,
476 Fifth Ave., New York.

Report of Local Walks Committee

The Local Walks Committee for the Mazama year 1918-1919 has nothing very special to report. In the performance of its functions the committee has been content to follow pretty closely the lines of policy inaugurated by its predecessors.

The chairman introduced one innovation in appointing a vice-chairman to assist him. This innovation, we think, might well become an established practice, as the duties of the chairman are time-consuming and exacting. W. P. Hardesty, who for so many years and so ably had officiated as chairman of the committee, very kindly consented to act in the capacity of vice-chairman and did so act until his private business called him away from the city in April. For the remainder of the year Eugene H. Dowling was the vice-chairman. The chairman appreciates much the valuable assistance rendered by both of these gentlemen, and likewise that furnished by the other members of the committee.

The July trip to Mt. Hood, which was put on by the committee this year as usual, was in every way a success. There were climbing parties from both the north and south sides, and 120 people attained the summit. Special interest attached to this trip this year as it was held on the twenty-fifth anniversary of the organization of the club. Four of the 198 immortals who stood on the summit of Mt. Hood on July 19, 1894, and there formally organized the Mazamas, were with us again this year. These sturdy veterans were C. H. Sholes, Rev. Earl M. Wilbur, Charles M. Meredith and Willis W. Ross. Mr. Sholes had been the chairman of the executive committee in the preliminary organization of the club and later served as its president for five terms. Mr. Wilbur had presided at the meeting on the summit when the club was formally organized. All four of these gentlemen made interesting reminiscent talks.

The following is a complete list of local walks taken during the year:

Date		Time	Place Visited	Leader	Attendance
1918					
Oct.	13	½ day	Fulton-Oswego	Jacques Letz	40
	20	1 day	Vancouver-Hidden Station	(Mrs. Ira Harper) (Augustus High)	38
	27	1 day	Bull Run-Walker's Prairie	W. P. Hardesty	20
Nov.	3	½ day	Mt. Scott-Clackamas	Eugene Dowling	35
	10	1 day	Gresham-Troutdale	Crissie Young	23
	17	½ day	East St. Johns	J. I. Teesdale	43
	24	½ day	Palatine Hill—Taylor's Ferry Road	Harold Babb	49
Dec.	1	½ day	Oswego Lake	Minna Backus	58
	8	½ day	Milwaukee-Oak Grove	Rhoda Ross	29
	14-15	2 days	Bull Run-Aschoff Hotel	Evelyn Hardinghaus	53
	22	1 day	Dundee-Mistletoe Trip	Crissie Young	68
	29	½ day	Mt. Tabor-Woodstock	Elma L. Fish	44
1919					
Jan.	5	½ day	Beaverton-Sylvan	Pasho Ivaneff	50
	12	½ day	Council Crest-Riverview	J. Homer Clark	31
	19	½ day	Tualatin-Fulton	Colista M. Dowling	47
	26	½ day	Oswego-Oregon City	Jean Richardson	42
Feb.	2	½ day	Castle-Eagle Point	A. B. Williams	68
	9	1 day	Clackamas-Sycamore	Marion Schneider	29
	13	½ day	Moonlight Walk	Fred Johnson	10
	16	1 day	Willamette-Pete's Mountain	Harold S. Babb	15
	23	½ day	Skyline Ridge-King's Heights	Ralph Tucker	28
Mar	2	1 day	Skyline Blvd.	L. Adele Bornt and C. M. Pendleton	19
	9	½ day	Mt. Sylvania-Oswego Lake	James Ormandy	76
	13	½ day	Moonlight Walk	Anne C. Grassl	46
	15-16	2 days	Larch Mountain	(R. W. Ayer) (L. E. Anderson)	120
	23	1 day	Cooper Mountain	John A. Lee	61
	30	1 day	Troutdale-Gresham	(H. L. Plumb) (Ed Berglund)	39
April	6	½ day	Columbia Blvd.-Rocky Butte	Lola Creighton	43
	13	1 day	Cottrell-Sandy River	P. G. Payton	37
	16	½ day	Moonlight Walk-Terwilliger Blvd.	Colista M. Dowling	12
	20	½ day	Linnton Road-Willamette Heights	Cinita Nunan	28
	27	1 day	Bethany-Holcomb Lake	J. I. Teesdale	52
May	4	1 day	Vancouver-Livingston Hill	Eric Bjorklund	29
	10-11	2 days	Bull Run-Badger Creek-Aschoff's	Mary Gene Smith	82
	14	½ day	Moonlight Walk	James Ormandy	12
	18	1 day	Canemah-Linn's Mill	Jacques Letz	51
	25	1 day	Hillsboro-North Plains-Logie Trail	George Meredith	46
June	30-31-1	3 days	Upper Clackamas River	John A. Lee	24
	1	½ day	Willamette Hgts.-St. Johns	Cecil Pendleton	19
	8	1 day	Rooster Rock-Bull Run.	W. W. Ross	63
	14-15	2 days	Greeleaf Peak	(Agnes Lawson) (Crissie C. Young)	50
	22	1 day	Chehalem Mountain	L. W. Waldorf	44
	29	1 day	Burlington Skyline Blvd.	A. S. Peterson.	65

July	4-5-6	3 days	Whatum Lake-Chinidere and Indian Mountains	Harold S. Babb	36
	4-5-6	3 days	Dowling Farm	Eugene H. Dowling	30
	19	½ day	Moonlight Walk-Blasted Butte	A. Boyd Williams	60
	12-13	1½ days	Larch Mt.-Wahkeena Falls	(R. W. Ayer) (E. F. Peterson)	112
	19-20	2 days	Mt. Hood—South Side	Committee	98
	18-19-20	2½ days	Mt. Hood—North Side	Committee	29
	26-27	2 days	Estacada-Clear Creek	Chas. E. Warner	17
Aug.	3	1 day	Oswego Lake-Oregon City	J. Homer Clark	19
	10	1 day	Gresham Butte	Dr. Wm. Amos	17
	17	1½ days	Blue Lake-Columbia River	E. H. Dowling	15
	24	1 day	Sauvie's Island	J. I. Teesdale	5
	31	1 day	Mountain View-Cedar Mills	Eugene Dowling	23
Sept.	6-7-8	3 days	Neakahnie-Mt. Short Sand Beach	Committee	75
	14	1 day	Gladstone-Clackamas River-Barton	Harold S. Babb	21
	21	1 day	Latourelle Falls-Pepper Mt.	Chas. E. Warner	46
	28	1 day	Forest Grove-David's Hill	Mary Knapp Lee	22
Oct.	4-5	2 days	Table Mountain	Edw. C. Sammons	28

Local Walks Committee Financial Report

October 13, 1918, to October 5, 1919, inclusive.

RECEIPTS

Amount collected on local walks		$151.70
Profit from Mt. Hood trip		157.47
Total		$309.17

EXPENDITURES

Printing and mailing schedules	$115.30	
Commissary and other supplies	53.90	
Total		169.20
Credit Balance		$139.07

JOHN A. LEE, Chairman.

Report of Certified Public Accountant Who Examined the Financial Affairs of the Mazamas

INCOME AND PROFIT & LOSS ACCOUNT

For the Period From October 7, 1918, to October 6, 1919.

INCOME:		
Members' Dues	$1,245.00	
Life Membership	50.00	
		$1,295.00
Miscellaneous:		
Interest on Liberty Bonds	20.00	
Key Sales	9.25	
Picture Sales	13.00	
		42.25
NET INCOME FROM COMMITTEE TRANSACTIONS:		
Income:		
Annual Outing, Mt. Rainier	$2,680.28	
Mt. Hood Outing	157.47	
Local Walks	98.90	
	2,936.65	
Less—		
Loss on Magazine Publication	289.58	
		2,647.07
Gross Income		$3,984.32
EXPENSES:		
Club Room Rent	$ 445.00	
Telephone Rent and Tolls	75.15	
Printing and Stationery—General	241.65	
Entertainment	20.85	
Lecture Expense	34.88	
Associated Club Dues	15.00	
Insurance	7.38	
Floral Contributions—Deceased Members	18.00	
Furniture Repairing and Renovating	67.15	
Sundries	50.54	
		975.60
Net Income		$3,008.72

Balance Sheet

As at October 6, 1919.

ASSETS

Cash at Bank—General Fund	$4,000.98
United States Liberty Bonds	600.00
Club Room Furniture and Camp Equipment	900.00
	$5,500.98

LIABILITIES

Surplus	$5,500.98
	$5,500.98

Portland, Oregon, October 29, 1919.

THE MAZAMA COUNCIL,
Portland, Oregon.

Dear Sirs:

In accordance with your instructions I have audited the accounts of the Mazamas for the fiscal year ended October 6, 1919, and present herewith my report. After meeting all expenses, the operations of the Club for the period under review resulted in net profits of $3,008.72, which are set forth in the accompanying Income and Profit & Loss Account. The Balance Sheet, given on this page, reflects the financial condition of the Club as at October 6, 1919.

The cash funds have been verified by a certificate from the bank. The United States Liberty Bonds are filed in a safety deposit vault.

The accounts of the Treasurer and the various Committees were examined and found to be in order.

Yours truly,

ROBERT F. RISELING,
Certified Public Accountant.

Address of the Retiring President

FELLOW MEMBERS:

As retiring president and member of what will be known in Mazama history as the "Ladies' Executive Council," I wish to make a very brief report of the activities of the Mazamas during the past year. Greatly to the surprise of the early critics, the closing year has been one of the most successful and prosperous ever enjoyed by the club and the credit for this belongs to a great extent to the six lady members of the council, even if they were ably assisted by The Three Wise Men. The club membership grew during the year from 411 to 465. The cash in the treasury increased from $1092 to $4000. The small quarters in the Northwestern Bank Building were changed to our present beautiful suite of rooms in the Chamber of Commerce Building, where we have ample room for council meetings, official club meetings and social gatherings of all kinds. Our banquet and dance in honor of our returning service men and women was attended by a large number and was a delightful affair. Our local walks are becoming increasingly popular and enjoyable.

Our Rainier summer outing was one of the most successful ever given by the Mazamas, both as regards attendance and general satisfaction. A noteworthy feature was the participation in the outing of a very large number of eminent men from all parts of the country, whose presence gave us one of the most brilliant series of camp-fire sessions ever experienced by the Mazamas.

In closing, I thank the ladies and gentlemen of the retiring Council for their unfailing kindness and courtesy to me and I wish the incoming president a year as full of pleasure in his duties as I have enjoyed.

EDGAR E. COURSEN.

MT. RAINIER FROM CAMP COURSEN.

Photo by Boychuk.

MR. THEODORE ROOSEVELT
MR. H. L. PITTOCK
MISS HILDA PLEBECK
MR. JOHN D. MEREDITH

In Memoriam

MR. THEODORE ROOSEVELT

Born October 27, 1858; died January 6, 1919. Was for several years an honorary member of the Mazamas.

MR. H. L. PITTOCK

Born March 1, 1835; died January 28, 1919. Was a charter member of the Mazamas and was chosen President of the club in 1897.

MISS HILDA PLEBECK

Died September 12, 1919. Was elected a member of the Mazamas August 19, 1919.

MR. JOHN D. MEREDITH

Born November 17, 1888; died August 15, 1919. Was elected a member of the Mazamas August 8, 1916. He served in France with Base Hospital No. 46, and had just returned and joined the members at the Annual Outing on Rainier. He lost his life while descending Little Tahoma.

Book Reviews

Edited by MINNIE R. HEATH.

"THE APPLEWOMAN OF THE KLICKITAT" The apple-growing country of Washington is presented as a fresh, new sort of frontier in which an eastern woman develops a quarter-section of government land into an orchard.

Her untiring interest in the success of the venture led her into many experiences new, not only to her, but to women in general. Intertwined with the story of apple-culture are many delightfully portrayed incidents of the lives of her neighbors and friends, the genuineness of her presentation being attested by the comment of a reader familiar with the environs, to the effect that "I'm glad that someone has ben able to tell just what really happened in that section, for it isn't often that the inhabitants, both native and otherwise, are so correctly estimated."

It is a book full of interest for both East and West.

ANNA VAN RENSSELAER MORRIS. "The Applewoman of the Klickitat." 1918. Duffield & Company, 211 West 33rd Street, New York.

"IN THE WILDS OF SOUTH AMERICA" Countless dangers encountered and endless hardships cheerfully endured in the quest for knowledge of a country widely known and little understood—thus may be summarized the adventures related in the more than four hundred pages of this book, most of which teem with attention-compelling narrative of the author's impressions gathered during six years of scientific investigation of these great areas, and of the animal and vegetable life maintained there.

Who has not been thrilled by the stories of the Incas and the Spanish Conquistadores? What boy or girl is not familiar with the "Conquest of Peru"? Of course everyone knows that South America has boa constrictors, jaguars, condors, rubber trees, mountains twenty-two thousand feet high and the greatest river in the world. Mr. Miller, the explorer-scientist, in this book dedicated to "my wife," seems to be constantly endeavoring to picture to one at home the facts observed and the sentiments inspired by his voyage through and over this vast wonderland.

The following extracts from the preface form a very fair key to the book: "Six years of almost continuous exploration in South America explorations into the tropical jungles of the Amazon, Paraguay, Orinoco and other of South America's master rivers, and to the frigid heights of the snow-crowned Andes"—"To start at the sudden, long-drawn hss of a boa or the lightning-like thrust of the terrible bush-master, the largest of poisonous snakes"—"ascents of the stupendous mountain ranges where condors soar majestically above the ruins of Incan greatness."

Among his trips is included the one with our great nature-lover, Theodore Roosevelt, in 1913 and 1914.

A delightful bit of description, and somewhat characteristic of the book, has to do with the town of Cali, Colombia, where children were seen bathing and ducks swimming in a gutter stream from which a housewife dipped a pitcher of water for domestic use. And yet "Embroidery and music are the chief diversions" (of women) and "it was remarkable to notice how many pianos there were, when we consider that each instrument has to be brought over the Andes slung on poles and carried by mules."

With all the dangers and hardships, the descriptions are so interesting and so intimate that one is led to feel as the author felt when in concentration camp preparatory to departure for the war zone of Europe—"almost daily my thoughts go back to the great wonderland that lies south of us, and which I have learned to love. Speed the day when I may again eagerly scan the horizon for a first faint tinge of its palm-fringed shore-line."

C. E. WARNER.

LEO E. MILLER, of the American Museum of Natural History. "IN THE WILDS OF SOUTH AMERICA." 1918. Chas. Scribner's Sons, New York.

"THE LAND OF TOMORROW"

There are endless opportunities in Alaska for the man with courage enough to seize them, is the opinion of the author of this fascinating little book. The land for which we paid two cents an acre has practically untouched resources. The gold, copper and coal mines, the immense fisheries, and the reindeer meat industry are capable of development far beyond their present output.

The author was formerly United States Commissioner at St. Michael's, Alaska, and his travels for pleasure and in the course of his work enabled him to see a great deal of the country. Consequently he is able to give the reader a summary of the people, social life, and customs of our northern territory which is truly amazing.

He does not neglect the Alaskan scenic beauty, which has made people call it the "Eighth Wonder of the World." The description of Mt. McKinley and of the smaller peaks are of especial interest to mountaineers. Mt. Katmai and the wonderful "Valley of Ten Thousand Smokes" are also briefly described.

For one who desires a picture of Alaska as it is today, this book will meet his need.

CLARENCE A. HOGAN.

"THE GRIZZLY, OUR GREATEST WILD ANIMAL"

This gives the experience of the author's acquaintance with the grizzly during his many years of life among the Rocky Mountains. The book contains fascinating bear stories and explains that the grizzly's true character is defensive and not aggressive. During the greater part of his life Mr. Mills has lived in the grizzly bear country and camped for months without a gun. He has trailed them and studied their habits, observed their sagacity, and has found them animals of wonderful endurance, masters of strategy, sensing danger from afar, and ever ready for something new in their environment. "He is an expert in eluding his pursuer, he rivals the fox in concealing his trail, in confounding the trailer and escaping with his life."

Lovers of animal life will enjoy reading this book, and after doing so will admire the grizzly and will be ready to agree with the naturalists that it would be a glorious thing if everybody appreciated his real character.

C. N. MORGAN.

ENOS A. MILLER. "The Grizzly, Our Greatest Wild Animal." Houghton Mifflin Co. Illustrated. $2.00.

"ADVENTURES IN ALASKA" This is a series of eight thrilling short stories, each complete in itself, representing phases of Alaskan life all the way from Fort Wrangell to Behring Sea. Its author is Dr. Young, "Sour Dough Preacher," "Mushing Parson," "Alaska Sky Pilot." The fact that he confesses pride in these names bestowed upon him by the people he served shows something of the character of the man—has that rare combination, the experience of forty years of Alaska pioneer life coupled with the descriptive power to make you see and feel what he has seen and felt during his long experience as a frontier missionary.

His descriptive style is typical of the Northland and his church association appears in these narratives only as a part of the framework in which are shown beautiful word-pictures of strong rugged men and women in a grand and beautiful but severe country.

In his foreword he expresses the hope that these stories of Alaska "will afford healthy-minded young people a true idea of some phases of human and animal life there." This hope is certainly realized, for the tales are true in detail beyond doubt, and so written as to hold us enthralled while we read of "Bunch Grass Bill, the Nome saloon-keeper, "Louie Paul and the Hootz" (brown bear), "Old Snook" and other characters and episodes which serve to complete the volume.

An item of added interest to Mazamas lies in the fact that Dr. Young was an intimate friend of John Muir and owned the dog "Stickeen," subject of the little book of that name, by Muir.

CHARLES E. WARNER.

S. HALL YOUNG. "Adventures in Alaska." 1919. Fleming H. Revell Company, New York.

"THE BOOK OF NATIONAL PARKS" This is a work which will readily appeal to the wide-awake American who loves the out-of-doors. The author describes our national parks, not as meaningless scenery, but as a thrilling story of creation. He explains how much more vital and personal an interest we will take in our parks when we really know and understand them. He then gives the geologic facts concerning the parks in a most instructive and interesting manner. How few Americans realize that our parks excel in scenic quality the combined scenery in all the rest of the world together. To quote the author: "They are the gallery of masterpieces and the museums of the ages."

NELLIE C. CROUT.

ROBERT STERLING YARD. "The Book of National Parks." 1919. Illustrated. Chas. Scribner's Sons. $3.00.

"NEW RIVERS OF THE NORTH" An exceedingly interesting account of the travels and adventures of the author and Auville Eager on a trip to the head waters of the Frazer, the Peace and the Hay rivers in Northwestern Canada. To the lover of the great outdoors this book will have a strong appeal and makes one wish that he had been one of the party, sharing alike the hardships as well as the joys and pleasures of the trails, the waters and the mountains. May this adventurous, pioneering spirit of our New World never cease.

F. M. REDMAN.

HULBERT FOOTNER. "New Rivers of the North." $2.00. George H. Doran Company, New York.

"CALIFORNIA DESERT TRAILS" The desert, with its seemingly endless sand and skies, is made vitally interesting because of the companionable way in which the author takes the reader with him through the various phases of desert scenery and life, by day and by night. The author disclaims any intention of making the book one of scientific research, yet in describing his impressions of the desert country, one finds much material of instructive value.

The author states, "But I confess that the fascination of the untamed desert has proved to be of too subtle a quality for words of mine to render."

Appendix A is a digest of "Hints on Desert Traveling" and teems with valuable suggestions.

Appendix B concerns "Noticeable Plants of the Desert," which are classified and briefly described.

The book gives one a feeling of enjoyment in the author's extended trip across the Colorado Desert, which lies mainly in the state of California and contains such characteristics as palm oases, canyons, cacti, shrubs, flowers and bustling towns at places where irrigation has reclaimed the desert.

M. R. H.

J. SMEATON CHASE. "California Trails." Houghton Mifflin Co. 1919. $3.00.

Teach me your mood, O patient stars;
Who climb each night the ancient sky,
Leaving on space no shade, no scars,
No trace of age, no fear to die.—*Emerson.*

Mazama Membership List, November 1, 1919

ACTON, HARRY W., 519 West 121st St., New York, N. Y.

ACTON, MRS. HARRY W., 519 West 121st St., New York, N. W.

ADAMS, DR. W. CLAUDE S., 1010 East 28th St., N., Portland, Ore.

AITCHISON, CLYDE B., Interstate Commerce Commission, Washington, D. C.

AKIN, DR. OTIS F., 919 Corbett Building, Portland, Ore.

ALLARD, NAN F., Foot of Miles St., Portland, Ore.

ALLEN, ENID C., 917 Andrus Building, Minneapolis, Minn.

ALMY, LOUISA, Box 426, Dillon, Montana.

AMOS, DR. WM. F., 1016 Selling Building, Portland, Ore.

ANDERSON, LEROY E., 206 Mercantile Place, Los Angeles, Calif.

ANDERSON, WM. H., 4464 Fremont Ave., Seattle, Washington.

ANDRAE, GERTRUDE ELOISE, 206 East 71st St., Portland, Ore.

APPLEGATE, ELMER I., Klamath Falls, Ore.

ASCHOFF, ADOLF, Marmot, Ore.

ASCHOFF, OTTO, Linnton, Ore.

ATKINSON, R. H., American Chain Co., 603 Beck Building, Portland, Ore.

ATLAS, CHAS. E.

AVERILL, MARTHA M., 1144 Hawthorne Ave., Portland, Ore.

AYER, ROY W., 689 Everett St., Portland, Ore.

AYER, LEROY Jr., P. O. Box 88, Crawfordsville, Ore.

BABB, HAROLD S., 583 Miller Ave., Portland, Ore.

BACKUS, LOUISE, 122 East 16th St., Portland, Ore.

BACKUS, MINNA, 122 East 16th St., Portland, Ore.

BAGLEY, FRANK S.

BAILEY, A. A., JR., 644 East Ash St., Portland, Ore.

BAILEY, VERNON, 1834 Kalorama Ave., Washington, D. C.

BALLOU, O. B., 80 Broadway, Portland, Ore.

BALMANNO, JACK H., 611 East 56th St., N., Portland, Ore.

BALOGH, W. A., 125 Sixth St., Portland, Ore.

BANFIELD, ALICE, 570 East Ash St., Portland, Ore.

BARCK, DR. C., 205-207 Humboldt Building, St. Louis, Mo.

BARNES, M. H., May Apartments, Apt. No. 44, Portland, Ore.

BARNES, MRS. M. H., May Apartments, Apt. No. 44, Portland, Ore.

BARRINGER, ALICE, 415 Tenth St., Portland, Ore.

BARRINGAR, MAUDE, 1207 Dearborn St., Caldwell, Idaho.

BATES, MYRTLE, 448 East 7th St., Portland, Ore.

BEAN, MRS. IDORA M., 113 D St., La Verne, Calif.

BEATTIE, BYRON J., 830 Rodney Ave., Portland, Ore.

BELL, HALLIE, Carlton Hotel, No. 515, Portland, Ore.

BENEDICT, LEE, 185 East 87th St., N., Portland, Ore.

BENEDICT, MAE, 185 East 87th St., Portland, Ore.

BENTALL, MAURICE, General Delivery, Hathaway, Montana.

BENZ, CHAS. A., P. O. Box 1433, Missoula, Montana.

BERG, MRS. G. ALBERT, Minerva, Iowa.

BIDWELL, EDMUND, 701 Corbett Building, Portland, Ore.

BIGGS, ROSCOE G., 284 East 6th St. N., Portland, Ore.

BISSELL, GEO. W., 223 W. Emerson St., Portland, Ore.

BLACKINTON, PAULINE, 169 Sixteenth St., Portland, Ore.

BLAKNEY, C. E., R. F. D. No. 2, Box 151, Milwaukie, Ore.

BLUE, WALTER, 1306 East 32nd St., N., Portland, Ore.

BLUMENAUER, FLORENCE, 1133 Rodney Ave., Portland, Ore.

BODWAY, W. P.

BORNT, LULU ADELE, 641 East 13th St., Portland, Ore.

BOWERS, NATHAN A., 501 Rialto Building, San Francisco, Calif.

BOWIE, ANN, 361 Eleventh St., Portland, Ore.

BOWIE, MARGARET, 361 Eleventh St., Portland, Ore.

BOYCE, EDWARD, 207 St. Clair St., Portland, Ore.

BOYCHUK, WALTER, 174 Meade St., Portland, Ore.

BRENNAN, THERESA, 380 Montgomery St., Portland, Ore.

BREWSTER, WM. L., 1022 Gasco Building, Portland, Ore.

BROCKMAN, GUS., 329 Burnside St., Portland, Ore.

BRONAUGH, JERRY ENGLAND, Gasco Building, Portland, Ore.

BRONAUGH, GEORGE, 350 North 32nd St., Portland, Ore.
BROWN, ALBERT S., 676 Riverside Drive, New York, N. Y.
BROWN, G. T., 500 East Morrison St., Portland, Ore.
BRUNELL, EVA, 18 Abbott St., Worcester, Mass.
BUCK, C. J., 549 East 39th St., N., Portland, Ore.
BULLIVANT, ANNA, 269 Thirteenth St., Portland, Ore.
BUNNAGE, R. H., 696 Sherrett St., Portland, Ore.
BURGLUND, E. E., 201 Union Ave., N., Portland, Ore.
BUSH, FRANK H., 1224 E. 31st St. N., Portland, Ore.
BUSH, J. C., 683½ E. Morrison St., Portland, Ore.
BJORKLUND, ERIC, 711 East Flanders St., Portland, Ore.
BARNES, E. L., 658 Schuyler St., Portland, Ore.
BUERNIE, CLARA MACGREGOR, Box 136, Portland, Ore.

CALHOUN, MRS. HARRIET S., 38 Delaware Ave., Detroit, Mich.
CALDWELL, CHARLOTTE, 309 San Rafael St., Portland, Ore.
CAMPBELL, GRACE, 600 E. Fiftieth St. N., Portland, Ore.
CAMPBELL, DAVID, care Mrs. Mary Campbell, Monmouth, Ore.
CAMPBELL, P. L., 1170 Thirteenth Ave., East, Eugene, Ore.
CARY, N. LEROY, U. S. Forest Service, Portland, Ore.
CARL, MRS. BEULAH MILLER, 629 East Ash St., Portland, Ore.
CARROLL, RANDOLPH S., 250 N. 24th St., Portland, Ore.
CASE, GEORGENE M., 3700 California St., San Francisco, Calif.
CATCHING, EVA, Carlton Hotel, Portland, Ore.
CECIL, K. P., Portland, Ore.
CHAMBERLAIN, RUTH, 685 Elliott Ave., Portland, Ore.
CHAMBERS, MARY H., 729 Eleventh Ave., E., Eugene, Ore.
CHASE, J. WESTON, Brix Lumber Co., Pittock Block, Portland, Ore.
CHENOWETH, MAY, 104 East 24th St., N., Portland, Ore.
CHRISTIANSON, WM. D., 134 Coburn St., Brantford, Ont.
CLARK, WM. D.
CHURCH, WALTER E., 1170 Thirteenth Ave., E., Eugene, Ore.
CHURCHILL, ARTHUR M., 1229 Northwestern Bank Building, Portland, Ore.
CLARK, J. HOMER, 92 Front St., Portland, Ore.
COLBORN, MRS. AVIS EDWARDS, Clovis, New Mexico.
COLLINS, W. G., 510 32nd Ave., South, Seattle, Washington.
**COLVILLE, PROF. F. V., Dept. of Agriculture, Washington, D. C.
CONNELL, DR. E. DEWITT, 628 Salmon St., Portland, Ore
CONWAY, D. J., 4705 Sixtieth St., S. E., Portland, Ore.
CONWAY, T. RAYMOND, 4705 Sixtieth St., S. E., Portland, Ore.
COOK, ARTHUR, 243 W. Park St., Portland, Ore.
COOK, F. R., 430 East 40th St. North, Portland, Ore.
CORNING, H. I., 255 Cherry St., Portland, Ore.
COURSEN, EDGAR E., 658 Lovejoy St., Portland, Ore.
COWPERTHWAITE, JULIA, Station E, Portland, Ore.
COWIE, LILLIAN G., 37 Wellesley Court, Portland, Ore.
CREIGHTON, LOLA I., 920 East Everett St., Portland, Ore.
CROUT, NELLE C., 1326 Tillamook St., Portland, Ore.
CURRIER, GEORGE H., Leona, Ore.
**CURTIS, EDWARD S., 614 Second Ave., Seattle, Wash.
CUSHMAN, CLYDE H., 458 E. 21st St. N., Portland, Ore.
CUTTING, RUTH M., 615 Elliott Ave., Portland, Ore.
COOK, VERA E., 1798 Woolsey St., Portland, Ore.

**DAVIDSON, PROF. GEORGE, 530 California St., San Francisco, Calif.
DAVIDSON, R. J., 1391½ Sandy Blvd. Portland, Ore.
DA , BESSIE, 690 Olive St., Eugene, Ore.
**DILLER, PROF. JOS., U. S. Geological Survey, Washington, D. C.
DILLINGER, MRS. C. E., 547 East 39th St., Portland, Ore.
DOWLING, EUGENE H., 742 Belmont St., Portland, Ore.
DOWLING, MRS. COLISTA M., 742 Belmont St., Portland, Ore.
DUDLEY, ALEXANDER P., 1240 East 30th St. N., Portland, Ore.
DUDLEY, MRS. ALEXANDER P., 1240 East 30th St., Portland, Ore.
DUFFY, MARGARET C., 1724 North Steel St., Tacoma, Wash.
DYER, R. L., 1323 Terry Ave., Seattle, Wash.

EMMRICH, ARTHUR J., 690 East 67th St., N., Portland, Ore.
ENGLISH, NEDSON, 267 Hazel Fern St., Portland, Ore.
ERREN, H. W., 285 Ross St., Portland, Ore.

ESTES, MARGARET P., 692 East 43rd St., N., Portland, Ore.

EVANS, WM. W., 744 Montgomery Drive, Portland, Ore.

FAGSTAD, THOR, Cathlamet, Wash.

FALLMAN, NINA A., 151 Park St., Portland, Ore.

FARRELL, THOS. G., 328 East 25th St., Portland, Ore.

FARRELLY, JANE, 1072 East 29th St. N., Portland, Ore.

FELLOWS, LESTER O., 4309 74th St. S. E., Portland, Ore.

FETY, TOMINE, 247 Grant St., Portland, Ore.

FINLEY, MRS. IRENE, 651 East Madison St., Portland, Ore.

FINLEY, WM. L., 651 East Madison St., Portland, Ore.

FISH, ELMA, 259 East 46th St., Portland, Ore.

FLEMING, MARGARET A., 214 Post Office Building, Portland, Ore.

FLESHER, J. N., Carson, Wash.

FORD, G. L., 104 Fourth St., Portland, Ore.

FORMAN, W. P., 128 North 18th St., Portland, Ore.

FORSYTH, JAMES R., 1028 Williams Ave., Portland, Ore.

FOSTER, FORREST L., 354 East 49th St., S. E., Portland, Ore.

FOSTER, HERBERT J., 1537 Curtiss Ave., Portland, Ore.

FOSTER, W. C., 224 Glenn Ave., Portland, Ore.

FRANKLIN, F. G., Willamette University, Salem, Ore.

FRANING, ELEANOR, 549 N. Broad St., Galesburg, Ill.

FRIES, SAMUEL M., 691 Flanders St., Portland, Ore.

FULLER, MARGARET E., 115 East 69th St., Portland, Ore.

GARDNER, BERNICE J., Apt. 33, Knickerbocker Apts., 410 Harrison St., Portland, Ore.

GARRETT, GEO., 646 Cypress St., Portland, Ore.

GASCH, MARTHA M., 9 East 15th St. N., Portland, Ore.

GEORGE, LUCIE M., 345 Clay St., Portland, Ore.

GILBERT, HAROLD S., 384 Yamhill St., Portland, Ore.

GILBERTSON, MARTHA, 656 Flanders St., Portland, Ore.

GILE, ELEANOR, 622 Kearney St., Portland, Ore.

GILMOUR, W. A., Title & Trust Bldg,. Portland, Ore.

GIRSBERGER, MABEL R., Modoc Lumber Co., Chiloquin, Ore.

GLISAN, RODNEY L., 612 Spalding Bldg., Portland, Ore.

GOLDAPP, MARTHA OLGA, 455 East 12th St., Portland, Ore.

GOLDSTEIN, MAX, 575 Third St., Portland, Ore.

*GORMAN, M. W., Forestry Building, Portland, Ore.

GRAF, S. H., 2260 Monroe St., Corvallis, Ore.

GRASSL, MRS. CHAS. W., 547 East 39th St., Portland, Ore.

GRAVES, HENRY S., U. S. Forest Service, Washington, D. C.

**GREELEY, GEN'L A. W., General Delivery, Center Conway, N. H.

GRENFELL, MRS. W. H., 1628 Belmont St., Portland, Ore.

GRIFFIN, MARGARET A., 1605-6 Pioneer Building, Robert St., St. Paul, Minn.

GRIFFITH, B. W., 417 Boyd St., Los Angeles, Calif.,

HAFFENDEN, A. H. S., 4236 49th Ave. S. E., Portland, Ore.

HALLINGBY, OLGA, 767 East Flanders St., Portland, Ore.

HANSEN, BESSIE L., 577 Kerby St., Portland, Ore.

HANSEN, RUTH E., 577 Kerby St., Portland, Ore.

HANSEN, ROSWELL J., Box 366, Vancouver, Wash.

HARBISON, RUTH L., 556 Fifth St., Hillsboro, Ore.

HARDESTY, WM. P., 617 Chamber of Commerce Building, Portland, Ore.

HARDINGHAUS, EVELYN, Weaver Hotel, Portland, Oregon.

HARNOIS, PEARL E., 1278 Williams Ave., Portland, Ore.

HARPER, IRA H., 2801 H Street, Vancouver, Wash.

HARPER, MRS. IRA H., 2801 H Street, Vancouver, Wash.

HARRIS, CHARLOTTE M., 1195 East 29th St., N., Portland, Ore.

HARTNESS, GEORGE, 671 Clackamas St., Portland, Ore.

HARZA, L. F., 505 Harvester Bldg., Chicago, Ill.

HATCH, LAURA, 36 Bedford Terrace, Northampton, Mass.

HAWKINS, E. R., 17 Union Station, Portland, Ore.

HAZARD, JOSEPH T., 4050 First Ave. N. E., Seattle, Wash.

HEATH, MINNIE R., 665 Everett St., Portland, Ore.

HEDENE, PAUL F., 720 East 22nd St., N., Portland, Ore.

HEINZE, AMY A., 261 Fourteenth St., Portland, Ore.

HELFRICH, CHARLES S.

HEMPY, M. RAYMOND, M. A. A. C., Portland, Ore.

Henderson, G. P., 1087 Belmont St., Portland, Ore.

HENDRICKSON, J. HUNT, Spalding Building, Portland, Ore.

HENRY, E. G., Newberg, Ore.

HENTHORNE, MARY C., 1834 East Morrison St., Portland, Ore.

HERMANN, HELEN M., 965 Kerby St., Portland, Ore.

HEYER, A. L., JR., 744 Hastings St. West, Vancouver, B. C.

HIGH, AUGUSTUS, 300 West 13th St., Vancouver, Wash.

HILTON, FRANK H., 504 Fenton Building, Portland, Ore.

HIMES, GEORGE H., Auditorium, Portland, Ore.

HINE, A. R., 955 East Taylor St., Portland, Ore.

HITCH, ROBERT E., Box 652 Juneau, Alaska.

HODGSON, CASPAR W., Rockland Ave., Park Hill, Yonkers, N. Y.

HOGAN, CLARENCE A., 591 Borthwick St., Portland, Ore.

HOLDEN, JAMES E., 1652 Alameda Drive, Portland, Ore.

HOLLISTER, HELEN, 550 East Main St., Portland, Ore.

HOLMAN, F. C., 558 Lincoln Ave., Palo Alto, Calif.

HORN, C. L., Wheeldon Annex, Portland, Ore.

HOWARD, HAZEL, 682 East 42nd St., North, Portland, Ore.

HOWARD, ERNEST E., 1012 Baltimore Ave., Kansas City, Mo.

HOWLAND, LUTHER H., 1207 East Flanders St., Portland, Ore.

IVANAKEFF, PASHO, 246 Clackamas St., Portland, Ore.

IVEY, RALPH S., R. F. D., Milwaukie, Ore.

JACOBS, MARY B., 315 Eleventh St., Portland, Ore.

JAEGER, J. P., 131 Sixth St., Portland, Ore.

JANE, GWENDOLEN, 1540 Hawthorne Ave., Portland, Ore.

JEPPESEN, ALICE, 891 Albina Ave., Portland, Ore.

JOHNSON, FRED J., Box 233, Salmon, Idaho.

JOHNSON, H. G., Brockton, Montana.

JOHNSTON, AMY, 545 East 23rd St., North, Portland, Ore.

JONES, F. I., 507 Davis St., Portland, Ore.

JOYCE, ALICE V., 591 Marshall St., Portland, Ore.

KACH, F. G.

KERN, EMMA B., 335 Fourteenth St., Portland, Ore.

KERR, DR. D. T., 556 Morgan Building, Portland, Ore.

KETCHUM, VERNE L., U. S. Shipping Board, Securities Builiding, Seattle, Wash.

KLEPPER, MILTON REED, Multnomah Hotel, Portland, Ore.

KOEMMECKE, MARIE, 1278 Williams Ave., Portland, Ore.

KOOL, JAN, 1309 Yeon Building, Portland, Ore.

KREBS, H. M., 285 Ross St., Portland, Ore.

KREINER, ROSE, 374 Third St., Portland, Ore.

KRESS, CHARLOTTE, Campbell-Hill Hotel, Portland, Ore.

KRUSE, JOHANNA, R. F. D., Route A, Portland, Ore.

KUENEKE, ALMA R., 869 Clinton St., Portland, Ore.

KUNKEL, HARRIET, 405 Larch St., Portland, Ore.

KUNKEL, KATHERINE, 857 Garfield Ave., Portland, Ore.

LA MADE, ERIC, 455 West Park St., Portland, Ore.

LADD, HENRY A., care Ladd & Tilton Bank, Portland, Ore.

LADD, W. M., care Ladd & Tilton Bank, Portland, Ore.

LANDIS, MARTHA, 2019 East Main St., Portland, Ore.

LANE, JOHN L., 2057 87th Ave., Oakland, Calif.

LANE, MRS, JOHN L., 2057 87th Ave., Oakland, Calif.

LAWFFER, G. A., 104 Fourth St. Portland, Ore.

LAWSON, AGNES G., 767 Montgomery Drive, Portland, Ore.

LEADBETTER, F. W., 795 Park Ave., Portland, Ore.

LEE, JOHN A., 505-6 Concord Building, Portland, Ore.

LEE, MARY KNAPP, 656 Flanders St., Portland, Ore.

LERDALL, ELMER

LETZ, JACQUES, State Bank of Portland, Ore.

LEWIS, CLYDE E., 407 Fourth St., Portland, Ore.

LIBBY, HARRY C., 422 East Stanton St., Portland, Ore.

LIND, ARTHUR, care U. S. National Bank, Portland, Ore.

LOUCKS, ETHEL MAE, 466 East 8th St., North, Portland, Ore.

LEE, FAIRMAN B., 1217 Sixth Ave., Seattle, Wash.

LUETTERS, F. P., 133 Vine Street, Roselle, N. J.

LUND, WALTER, 191 Grand Ave. N., Portland, Ore.

LUTHER, DR. C. V., 401 Selling Building, Portland, Ore.

LYON, GEORGIA E., 297 Broadway, Chicopee Falls, Mass.

McARTHUR, LEWIS A., 561 Hawthorne Terrace, Portland, Ore.

McBRIDE, AGNES, P. O. Box 383, Oswego, Ore.

McCLELLAND, ELIZABETH, 267 Shawnee Path, Akron, Ohio.

McCOLLOM, DR. J. W., 553-557 Morgan Building, Portland, Ore.

McCORKLE, J. F., 506 Washington St., Portland, Ore.

McCOY, SALLIE E., 211 Lumbermans Building, Portland, Ore.

McCREADY, SUE O., Box 147, Vancouver, Wash.

McCULLOCH, CHARLES E., 1410 Yeon Building, Portland, Ore.

McDONALD, MRS. LAURA H., 354 East 49th St., S., Portland, Ore.

McISAAC, R. J., Parkdale, Ore.

McKAUGHAN, HENRIETTA, 375 Sixteenth St., Portland, Ore.

McLAUGHLIN, SADIE, 648 Kline St., Portland, Ore.

McMASTER, RUTH E., 660 East Oak St., Portland, Ore.

McNEIL, FLORENCE, 607 Orange St., Portland, Ore.

McNEIL, FRED H., care The Journal, Portland, Ore.

MacDOUGALL, CHARLOTTE, Alexander Hotel, Spokane, Wash.

MACKENZIE, WM. R., 1002 Wilcox Building, Portland, Ore.

MAHONEY, MRS. HELENA C., 1238 Commonwealth Ave., Boston, Mass

MAHONEY, PAUL, 1238 Commonwealth Ave., Boston, Mass.

MARBLE, W. B., 3147 Indiana Ave., Chicago, Ill.

MARCOTTE, HENRY, D. D., 218 E. 56th St., Kansas City, Mo.

MARCY, EDITH, 309 First National Bank Building, The Dalles, Ore.

MARSH, J. W., Underwood, Wash.

MARSHALL, BERTHA, 1445 B St., San Diego, Calif.

MATTSON, DON F., 412 Oregon Building, Portland, Ore.

MEARS, HENRY S., 494 Northrup St., Portland, Ore.

MEARS, S. M., 721 Flanders St., Portland, Ore.

MEREDITH, MRS. C. M., 735 Hillsboro Ave., Portland, Ore.

MEREDITH, DAISY LORENA, 263 Miles St., Portland, Ore.

MEREDITH, GEORGE, 11th and Burnside Sts., Portland, Ore.

MEREDITH, HELEN E., 735 Hillsboro Ave., Portland, Ore.

**MERRIAM, DR. C. HART, 1919 Sixteenth St. N. W., Washington, D. C.

MERTEN, CHARLES J., 307 Davis St., Portland, Ore.

MILES, S., Room 1411, 80 Maiden Lane, New York, N. Y.

MILLER, JESSE, 726 E. 20th St., Portland, Ore.

MILLS, ENOS A., Long's Peak, Estes Park, Colorado.

MONROE, HARRIETT E., 1431 East Salmon St., Portland, Ore.

MONTAGUE, JACK R., 1310 Yeon Building, Portland, Ore.

MONTAGUE, RICHARD W., 1310 Yeon Building, Portland, Ore.

MOORE, DUNCAN, 1303 Chamber of Commerce Building, Chicago, Ill.

MORGAN, MRS. CHRISTINE N., Box 144, Palms, Calif.

MORKILL, ALAN BROOKS, 1971 Oak Bay Ave., Victoria, B. C.

MUELLHAUPT, OSCAR W. T., 407-409 U. S. National Bank Building, Portland, Ore.

MURPHY, JOHN, 973 East Stark St., Portland, Ore.

McELROY, FLORENCE, 954 Gladstone Ave., Portland, Ore.

NALLEY, JOHN F., 129 Rhode Island Ave., N. E., Washington, D. C.

NEELS, CARL, 495 Jefferson St., Portland, Ore.

NELSON, BUELL C., 128 North Eighteenth St., Portland, Ore.

NELSON, L. A., West Coast Lumbermens Ass'n, 1207 Yeon Building, Portland, Ore.

NEWELL, BEN W., Ladd & Tilton Bank, Portland, Ore.

NEWLYN, MRS. HAROLD V., 689 Northrup St., Portland, Ore.

NEWTON, JOSEPHINE, 1350 Pine St., Philadelphia, Pa.

NIECHANS, MARGARET, 353 Harrison St., Portland, Ore.

NICKELL, ANNA, 410 Stanley Apts., Seattle, Wash.

NILSSON, MARTHA E., 320 East 11th St., N., Portland, Ore.

NISSEN, IRENE, 969 East 23rd St., N., Portland, Ore.

NORDEEN, EDITH, 361 Graham St., Portland, Ore.

NORMAN, OSCAR M., 698 East 62nd St., N., Portland, Ore.

NOTTINGHAM, JESSIE RAY, 271 East 16th St., N., Portland, Ore.

NUNAN, CINITA, 489 W. Park St., Portland, Ore.

O'BRYAN, HARVEY, 602 McKay Building, Portland, Ore.

*O'NEILL, MARK, Worcester Block, Portland, Ore.

OGLESBY, ETTA M., 818 Lombard St., Portland, Ore.

ORMANDY, HARRY M., 501 Weidler St., Portland, Ore.

ORMANDY, JAMES A., 501 Weidler St., Portland, Ore.

OLSON, RUTH, 919 Borthwick St., Portland, Ore.

OTIS, EMILY, 525 Yeon Bldg., Portland, Ore.

PAETH, WILLIAM J., U. S. Forest Service, Portland, Ore.

PARKER, ALFRED F., 374 East 51st St., Portland, Ore.

PARKER, JAMIESON, 374 East 51st St., Portland, Ore.

PARKER, ROSE F., Butterfield Bros., Portland, Ore.
PARKER, MRS. W., Box 34, Route 1, Milwaukie, Ore.
PARSONS, MRS. M. R., Mosswood Road, University Hill, Berkeley, California.
PATTULLO, A. S., 500 Concord Building, Portland, Ore.
PAUER, JOHN, 1625 26th St., Sacramento, Calif.
PAYTON, PERLEE G., 3916 64th St., S. E., Portland, Ore.
PEARCE, MRS. LLEWELLYN C., 1137 E. Yamhill St., Portland, Ore.
PENDLETON, CECIL M., 285½ First St., Portland, Ore.
PENLAND, JOHN R., Box 345, Albany, Ore.
PENWELL, ESTHER, 95 East 74th St., Portland, Ore.
PETERSON, AUGUST, Y. M. C. A., Portland, Ore.
PETERSON, ARTHUR S., 780 Williams Ave., Portland, Ore.
PETERSON, E. F., 780 Williams Ave., Portland, Ore.
PETERSON, H. C., M. A. A. C., Portland, Ore.
PETERSON, LAURA H., 395½ Clifton St., Portland, Ore.
PIERCE, MARIE M., 1406 W. 39th St., Portland, Ore.
PILKINGTON, THOMAS J., Sebastopol, California.
PHILLIPS, MABEL F., R. F. D. 43, Box 18, Salem, Ore.
PLATT, ARTHUR D., 211 East 55th St., Portland, Ore.
PLUMB, H. L., care Forest Service, Portland, Ore.
PLUMMER, AGNES, 3rd and Madison Sts., Portland, Ore.
PRENTYS, R. P., King-Davis Apts., Portland, Ore.
PREVOST, FLORENCE, Highland Court Apts., Portland, Ore.
PUGH, LAURA E., 4811 34th Ave., Portland, Ore.

RAUCH, G. L., 902 Yeon Building, Portland, Ore.
REDDEN, CECIL V., 314 W. 8th St., Vancouver, Wash.
REDMAN, FRANK M., 1014 Northwestern Bank Building, Portland, Oregon.
REED, MRS. ROSE COURSEN, 308 Eilers Building, Portland, Ore.
**REID, PROF HARRY FIELDING, Johns Hopkins University, Baltimore, M. D.
RENFRO, JOE H., 41 Jessup W. St., Portland, Ore.
RENFRO, MRS. BESSIE M., 41 Jessup W. St., Portland, Ore.
RENSTROM, HENRIK, Beach Rd. 14, Squantum, Mass.
RHODES, EDITH G., 935 E. 26th St., N., Portland, Ore.
RICE, EDWIN L., 1191 E. Yamhill St., Portland, Ore.
RICHARDSON, EDWARD L., 10 S. LaSalle St., Chicago, Ill.
RICHARDSON, JEAN, 131 East 19th St., Portland, Ore.
RICHMOND, STANLEY C.
RIDDELL, GEO. X., 689 Everett St., Portland, Ore.
RILEY, FRANK BRANCH, Chamber of Commerce Building, Portland, Ore.
RISELING, ROBERT F., 1427 N. W. Bank Building, Portland, Ore.
ROBERTS, ELLA PRISCILLA, 109 East 48th St., Portland, Ore.
ROBINSON, DR. EARL C., 660 Morgan Building, Portland, Ore.
ROEMER, LOWELL, 4405 East 89th St., S. E., Portland, Ore.
ROSENKRANS, F. A., 335 East 21st St., Portland, Ore.
ROSS, RHODA, 1516 East Oak St., Portland, Ore.
ROSS, WILLIS W., 272 Stark St., Portland, Ore.
RYAN, MILDRED L., Portland, Ore.

SAKRISON, C. H., 356 Fargo St., Portland, Ore.
SAMMONS, E. C., 69 East 18th St., Portland, Ore.
SCARPF, GRETCHEN, 429 Northeast 46th St., Portland, Ore.
SCHNEIDER, KATHERINE, 260 Hamilton Ave., Portland, Ore.
SCHNEIDER, MARION, 260 Hamilton Ave., Portland, Ore.
SCHROEDER, LAURA G., 514 Flint St., Portland, Ore.
SCOTT, ISABELLA J., 593 East 8th St. N., Portland, Ore.
SEARCY, ROBERT D., San Francisco., Calif.
SELF, NORA, Camas, Wash.
SEVERIN, WILLIAM C. E., Box 34, Route ,1 Milwaukie, Ore.
SEYMOUR, DARWIN CY, First National Bank, Portland, Ore.
SHELTON, ALFRED C., 1390 Emerald St., Eugene, Ore.
SHEPARD, F. E., 490 East 33rd St., Portland, Ore.
SHERMAN, LENA, 1123 N. E. 22nd St., Portland, Ore.
SHERMAN, MINET E., 774 Everett St., Portland, Ore.
SHIPLEY, J. W., Underwood, Wash.
*SHOLES, CHAS. H., Box 243, Portland, Ore.
SHOLES, MRS. C. H., Box 243, Portland, Ore.
SIEBERTS, CONRAD J., 683 E. Stark St., Portland, Ore.
SIEBERTS, MRS. CONRAD J., 683 E. Stark St., Portland, Ore.
SILL, J. G., 511 Merchants Trust Building, Portland, Ore.
SILVER, ELSIE M., 100 Sixth St., Portland, Ore.

SMEDLEY, GEORGIAN E., 262 E. 16th St., Portland, Ore.

SMITH, ADRIAN E., 127 East 39th St., Portland, Ore.

SMITH, MARY GENE, Campbell-Hill Hotel, Portland, Ore.

SMITH, W. E., 589 East 12th St. N., Portland, Ore.

SMITH, KAN, 2908 Fifteenth Ave., S., Seattle, Wash.

SMITH, LEOTTA, 842 East Stark St., Portland, Ore.

SMITH, PROF. WARREN D., 941 E. 19th St., Eugene, Ore.

SNEAD, J. L. S., 572 E. Broadway, Portland, Ore.

SNOKE, ESTHER, 380 Tenth St., Portland, Ore.

SPAETH, DR. J. DUNCAN, Princeton, N. J.

STARKWEATHER, H. G., 602 Broadway Building, Portland, Ore.

STARR, NELLIE S., 6926 45th Ave., S. E., Portland, Ore.

STEVENTON, JOSEPHINE, 720 Oberlin, St., Portland, Ore.

STONE, DR. W. E., Purdue Univesity, Lafayette, Ind.

STONE, MRS. W. E., 146 North Grant St., Lafayette, Ind.

STRINGER, A. R., JR., 179 Bancroft Ave., Portland, Ore.

STROOP, D. VINCENT, U. S. Forest Service, Portland, Ore.

STUDER, GEORGE A., 608 Schuyler St., Portland, Ore.

SULLIVAN, F. F., 305 Madison Park Apts., Portland, Ore.

TAYLOR, VERA E., 814 Spalding Building, Portland, Ore.

TENNESON, ALICE M., High School, Yakima, Wash.

THATCHER, GUY W., 302 Sacramento St., Portland, Ore.

THAXTER, B. A., 391 East 24th St., Portland, Ore.

THOMAS, E. H., Parkdale, Ore.

THOMAS, EMMA M., 770 21st St., Oakland, Calif.

THORINGTON, DR. J. M., 2031 Chestnut St., Philadelphia, Pa.

THORNE, H. J., 755 East 26th St., N., Portland, Ore.

TOMPKINS, MARGARET, 285 Couch St., Portland, Ore.

TREICHEL, CHESTER H., 624 Idaho Building, Boise, Idaho.

TUCKER, RALPH J., 389½ Sixteenth St., Portland, Ore.

UPSHAW, F. B., 401 Concord Building, Portland, Ore.

VAN BEBBER, L., 503 Fenton Building, Portland, Ore.

VAN ZANDT, DEAN, 1408 Fourteenth Ave., Seattle, Wash.

VENSTRAND, EVA E., 493 East 9th St., N., Portland, Ore.

VIAL, LOUISE ONA, 241 Prospect St, Berkeley, Calif.

WALDORF, LOUIS W., 724 East 59th St. N., Portland, Ore.

WALTER, WILLIAM S., 53 North 21st St., Portland, Ore.

WARD, JOHN S., 170 E. 121st St., New York, N. Y.

WARNER, CHARLES E., The Portland News, Portland, Ore.

WEBB, ONEITA, 514 Jefferson St., Portland, Ore.

WEER, J. H., West Coast Grocery Co., P. O. Box 1563, Tacoma, Wash.

WELCH, JENNIE, Welches, Ore.

WENNER, B. F., Bradley Road, West Dover, Ohio.

WHITE, WILLIAM, Suite 1000, 1211 Chestnut St., Philadelphia, Pa.

WILBURN, VESTA, care Richmond Paper Co., Seattle, Wash.

WILDER, GEORGE W., 226 Fourteenth St., Portland, Ore.

WILLARD, CLARA, 619 High St., Bellingham, Wash.

WILLIAMS, A. BOYD, King-Davis Apts., Portland, Ore.

WILLIAMS, MRS. A. BOYD, King-Davis Apts., Portland, Ore.

WILLIAMS, GEO. M., 713 F St., Centralia, Wash.

WILLIAMS, GERTRUDE, 314 W. 8th St., Vancouver, Wash.

*WILLIAMS, JOHN H., 2671 Filbert St., San Francisco, Calif.

WILSON, CHARLES W., Bellevue, Idaho.

WILSON, MAUDILEEN, 197 N. E. 66th St., Portland, Ore.

WILSON, RONALD M., U. S. Geological Survey, Washington, D. C.

WING, MARY, 1124 Macadam Rd., Portland, Ore.

WINN, ETHEL, 415 Yamhill St., Portland, Ore.

WISE, DR. T. P., 568 Elizabeth St., Portland, Ore.

WOLBERS, HARRY L., 577 Kerby St., Portland, Ore.

WOODWORTH, C. B., 214 Spalding Building, Portland, Ore.

WYNN, DR. FRANK B., 421 Hume Mansur Building, Indianapolis, Ind.

WALSH, GRAYCE, Portland, Ore.

WALSH, ROBERT P., St. Louis, Mo.

YORAN, W. C., 912 Lawrence St., Eugene, Ore.

YOUNG, CRISSIE C., 520 Elizabeth St., Portland, Ore.

YOUNGKRANTZ, EDITH M., U. S. Forest Service, Walla Walla, Wash.

ZANDERS, RUTH, M. A. A. C., Portland, Ore.

ZEIDLHACK, FELIX S., 349 Harrison St., Portland, Ore.

---

*Life Members
**Honorary Members

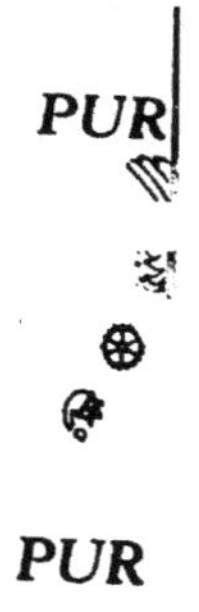

PURDUE
LIBRARY
SCIENCE
TECHNOLOGY
AGRICULTURE
PURDUE

PURDUE
LIBRARY
LIBRARY
SCIENCE
TECHNOLOGY
AGRICULTURE
PURDUE

PURDUE
LIBRARY
LIBRARY
SCIENCE
TECHNOLOGY
AGRICULTURE
PURDUE

PURDUE
LIBRARY
SCIENCE
TECHNOLOGY
AGRICULTURE
PURDUE

PURDUE
LIBRARY
LIBRARY
SCIENCE
TECHNOLOGY
AGRICULTURE
PURDUE

PURDUE
LIBRARY
LIBRARY
SCIENCE
TECHNOLOGY
AGRICULTURE
PURDUE

CPSIA information can be obtained
at www.ICGtesting.com
Printed in the USA
FSOW03n0709041016
25719FS